LOEB CLASSICAL LIBRARY
FOUNDED BY JAMES LOEB 1911

EDITED BY
JEFFREY HENDERSON

SEXTUS EMPIRICUS
IV

LCL 382

SEXTUS EMPIRICUS

AGAINST THE PROFESSORS

WITH AN ENGLISH TRANSLATION BY

R. G. BURY

HARVARD UNIVERSITY PRESS
CAMBRIDGE, MASSACHUSETTS
LONDON, ENGLAND

First published 1949

LOEB CLASSICAL LIBRARY® is a registered trademark
of the President and Fellows of Harvard College

ISBN 978-0-674-99420-1

Printed on acid-free paper and bound by
The Maple-Vail Book Manufacturing Group

CONTENTS

PREFATORY NOTE

This volume contains the first six books of *Adversus Mathematicos*, *viz. Adv. Grammaticos*, *Adv. Rhetores*, *Adv. Geometras*, *Adv. Arithmeticos*, *Adv. Astrologos*, *Adv. Musicos*. As in the previous volumes the text is based on that of Bekker, the chief deviations being indicated in the footnotes. The remaining five books of *Adversus Mathematicos* are the works found in Vols. II and III of this edition. Books vii and viii are in Vol. II and the rest in Vol. III.

The Translator is indebted to the Editor, Dr. Rouse, for kindly help in explaining the passages (in *Adv. Gram.* 79 ff.) which deal with προσῳδία. He is very grateful also to Prof. R. Hackforth for many valuable suggestions.

AGAINST THE PROFESSORS

ΠΡΟΣ ΜΑΘΗΜΑΤΙΚΟΥΣ

Α

ΠΡΟΣ ΜΑΘΗΜΑΤΙΚΟΥΣ—ΠΡΟΣ ΓΡΑΜΜΑΤΙΚΟΥΣ

1 Τὴν πρὸς τοὺς ἀπὸ τῶν μαθημάτων ἀντίρρησιν κοινότερον μὲν διατεθεῖσθαι δοκοῦσιν οἵ τε περὶ τὸν Ἐπίκουρον καὶ οἱ ἀπὸ τοῦ Πύρρωνος, οὐκ ἀπὸ τῆς αὐτῆς δὲ διαθέσεως, ἀλλ' οἱ μὲν περὶ τὸν Ἐπίκουρον ὡς τῶν μαθημάτων μηδὲν συνεργούντων πρὸς σοφίας τελείωσιν, ἢ ὥς τινες εἰκάζουσι, τοῦτο προκάλυμμα τῆς ἑαυτῶν ἀπαιδευσίας εἶναι νομίζοντες (ἐν πολλοῖς γὰρ ἀμαθὴς Ἐπίκουρος ἐλέγχεται, οὐδὲ ἐν ταῖς κοιναῖς ὁμιλίαις καθαρεύων),
2 τάχα δὲ καὶ διὰ τὴν πρὸς τοὺς περὶ Πλάτωνα καὶ Ἀριστοτέλη καὶ τοὺς ὁμοίους δυσμένειαν πολυμαθεῖς γεγονότας. οὐκ ἀπέοικε δὲ καὶ διὰ τὴν πρὸς Ναυσιφάνην τὸν Πύρρωνος ἀκουστὴν ἔχθραν· πολλοὺς γὰρ τῶν νέων συνεῖχε καὶ τῶν μαθημάτων σπουδαίως ἐπεμελεῖτο, μάλιστα δὲ ῥητορικῆς.
3 γενόμενος οὖν τούτου μαθητὴς ὁ Ἐπίκουρος ὑπὲρ τοῦ δοκεῖν αὐτοδίδακτος εἶναι καὶ αὐτοφυὴς φιλόσοφος ἠρνεῖτο ἐκ παντὸς τρόπου, τήν τε περὶ

[a] §§ 1-8 are Introductory.

AGAINST THE PROFESSORS

BOOK I

§§ 1-40—AGAINST THE PROFESSORS; §§ 41-END—AGAINST THE GRAMMARIANS

[a] THE case against the *Mathematici* (or Professors of 1
Arts and Sciences) has been set forth in a general way, it would seem, both by Epicurus and by the School of Pyrrho, although the standpoints they adopt are different. Epicurus took the ground that the subjects taught are of no help in perfecting wisdom; and he did this, as some conjecture, because he saw in it a way of covering up his own lack of culture (for in many matters Epicurus stands convicted of ignorance, and even in ordinary converse his speech was not
correct). Another reason may have been his hostility 2
towards Plato and Aristotle and their like who were men of wide learning. It is not unlikely, too, that he was moved by his enmity against Nausiphanes, the disciple of Pyrrho, who kept his hold on many of the young men and devoted himself earnestly to the Arts
and Sciences, especially Rhetoric. Epicurus, then, 3
though he had been one of this man's disciples, did his best to deny the fact in order that he might be thought to be a self-taught and original philosopher, and tried hard to blot out the reputation of Nausi-

αὐτοῦ φήμην ἐξαλείφειν ἔσπευδε, πολύς τε ἐγίνετο τῶν μαθημάτων κατήγορος, ἐν οἷς ἐκεῖνος ἐσεμνύ-
4 νετο. φησὶ γοῦν ἐν τῇ πρὸς τοὺς ἐν Μυτιλήνῃ φιλοσόφους ἐπιστολῇ "οἶμαι δὲ ἔγωγε τοὺς βαρυστόνους καὶ μαθητήν με δόξειν τοῦ πλεύμονος εἶναι, μετὰ μειρακίων τινῶν κραιπαλώντων ἀκούσαντα," νῦν πλεύμονα καλῶν τὸν Ναυσιφάνην ὡς ἀναίσθητον. καὶ πάλιν προβὰς πολλά τε κατειπὼν τἀνδρὸς ὑπεμφαίνει τὴν ἐν τοῖς μαθήμασιν αὐτοῦ προκοπὴν λέγων "καὶ γὰρ πονηρὸς ἄνθρωπος ἦν καὶ ἐπιτετηδευκὼς τοιαῦτα ἐξ ὧν οὐ δυνατὸν εἰς
5 σοφίαν ἐλθεῖν," αἰνισσόμενος τὰ μαθήματα. πλὴν ὁ μὲν Ἐπίκουρος, ὡς ἄν τις εἰκοβολῶν εἴποι, ἀπὸ τοιούτων τινῶν ἀφορμῶν πολεμεῖν τοῖς μαθήμασιν ἠξίου, οἱ δὲ ἀπὸ Πύρρωνος οὔτε διὰ τὸ μηδὲν συνεργεῖν αὐτὰ πρὸς σοφίαν, δογματικὸς γὰρ ὁ λόγος, οὔτε διὰ τὴν προσοῦσαν αὐτοῖς ἀπαιδευσίαν· σὺν γὰρ τῷ πεπαιδεῦσθαι καὶ πολυπειρότεροί[1] παρὰ τοὺς ἄλλους ὑπάρχειν φιλοσόφους ἔτι καὶ ἀδιαφόρως ἔχουσι πρὸς τὴν παρὰ τοῖς πολλοῖς
6 δόξαν. καὶ μὴν οὐδὲ δυσμενείας χάριν τῆς πρός τινας (μακρὰν γὰρ αὐτῶν τῆς πρᾳότητός ἐστιν ἡ τοιαύτη κακία), ἀλλὰ τοιοῦτόν τι ἐπὶ τῶν μαθημάτων παθόντες ὁποῖον ἐφ' ὅλης ἔπαθον τῆς φιλοσοφίας. καθὰ γὰρ ἐπὶ ταύτην ἦλθον πόθῳ τοῦ τυχεῖν τῆς ἀληθείας, ἰσοσθενεῖ δὲ μάχης[2] ἀνωμαλίᾳ τῶν πραγμάτων ὑπαντήσαντες ἐπέσχον, οὕτω καὶ ἐπὶ τῶν μαθημάτων ὁρμήσαντες ἐπὶ τὴν ἀνάληψιν αὐτῶν, ζητοῦντες καὶ τὸ ἐνταῦθα μαθεῖν ἀληθές,

[1] πολυπειρότεροί cj. Bekk. : πολυπειροτέρους mss., Bekk.
[2] μάχης: fortasse μάχῃ καὶ.

phanes, and became a violent opponent of the Arts and Sciences wherein Nausiphanes prided himself. Thus, in his *Letter to the Philosophers*[a] *in Mytilenê*, 4
Epicurus says, " I quite suppose that 'the bellowers' will fancy that I am even a disciple of 'the Stockfish,' having sat under him in the company of some crapulous striplings " ; where he calls Nausiphanes a "Stockfish," as being without sense. And again, after proceeding further and abusing the man at length, he hints at his proficiency in Arts and Sciences when he says—" In fact he was a sorry fellow and exercised himself on matters which cannot possibly lead to wisdom," alluding thereby to Arts and Sciences. Such, 5
in fact,—as we may conjecture—were the sort of motives which decided Epicurus to make war on the Arts and Sciences. The School of Pyrrho, on the other hand, were not moved either by the view that these subjects are of no help to gaining wisdom (for that is a "dogmatic" assertion) or by any lack of culture attaching to themselves ; for in addition to their culture and their superiority to all other philosophers in breadth of experience they are also indifferent to the opinion of the multitude. Nor is the reason to be 6
found in ill-will towards any (for that sort of vice is wholly alien to their gentle character), but in the fact that in respect of the Arts and Sciences they have met with the same experience as they did in respect of philosophy as a whole. For just as they approached philosophy with the desire of attaining truth,[b] but, when faced by the equipollent conflict and discord of things, suspended judgement,—so also in the case of the Arts and Sciences, when they had set about mastering them with a view to learning here also the

[a] *Cf.* Diog. Laert. x. 8.   [b] *Cf. P.H.* i. 12.

τὰς δὲ ἴσας εὑρόντες ἀπορίας, οὐκ ἀπεκρύψαντο.
7 διόπερ καὶ ἡμεῖς τὴν αὐτὴν τούτοις ἀγωγὴν μεταδιώκοντες πειρασόμεθα χωρὶς φιλονεικίας τὰ πραγματικῶς λεγόμενα πρὸς αὐτοὺς ἐπιλεξάμενοι θεῖναι.

Τὸ μὲν οὖν διδάσκειν ἀπὸ τίνος ἐγκύκλια προσηγόρευται μαθήματα καὶ πόσα τὸν ἀριθμόν ἐστι, περιττὸν ἡγοῦμαι, πρὸς ἱκανὴν ἤδη τὴν περὶ τούτων ἔχοντας κατήχησιν γινομένης ἡμῖν τῆς
8 διδασκαλίας. ὃ δέ ἐστιν ἀναγκαῖον ἐπὶ τοῦ παρόντος, ὑποδεικτέον ὅτι τῶν λεγομένων πρὸς τὰ μαθήματα τὰ μὲν καθολικῶς λέγεται πρὸς πάντα τὰ μαθήματα τὰ δ' ἰδίως[1] πρὸς ἕκαστα, καὶ καθολικώτερον μὲν τὸ περὶ τοῦ μηδὲν εἶναι μάθημα, ἰδιαίτερον δὲ πρὸς μὲν γραμματικούς, εἰ τύχοι, περὶ τῶν τῆς λέξεως στοιχείων, πρὸς δὲ γεωμέτρας περὶ τοῦ μὴ δεῖν ἐξ ὑποθέσεως λαμβάνειν τὰς ἀρχάς, πρὸς δὲ μουσικοὺς περὶ τοῦ μηδὲν εἶναι φωνὴν μηδὲ χρόνον. ἴδωμεν δὲ τάξει πρῶτον τὴν καθολικωτέραν ἀντίρρησιν.

Α'.—ΕΙ ΕΣΤΙ ΜΑΘΗΜΑ

9 Τὴν μὲν οὖν γενομένην παρὰ τοῖς φιλοσόφοις περὶ μαθήσεως διαφωνίαν πολλὴν καὶ ποικίλην οὖσαν οὐ τοῦ παρόντος ἐστὶ καιροῦ ἐπικρίνειν· ἀπόχρη δὲ παραστῆσαι ὡς εἴπερ ἔστι τι μάθημα, καὶ τοῦτο ἀνυστὸν ἀνθρώπῳ, τέσσαρα δεῖ προομο-

[1] τὰ δ' ἰδίως scripsi : τὰ δὲ ὡς MSS., Bekk.

[a] *i.e.* the Subjects of general, pre-professional, education,

truth, they found difficulties no less serious, which they did not conceal. Accordingly, we too will 7
pursue the same method as they and endeavour, in no spirit of controversy, to select and set forth the substantial arguments against the Professors of culture.

Now I deem it superfluous to explain why the "cyclical studies" [a] are so called, and how many of them there are, when our exposition is addressed to those who are already sufficiently instructed in these matters. But it is necessary for our present purpose 8
to indicate this point,—that of the arguments brought against the Arts and Sciences some are of a general character directed against all these subjects, while others are of a special character, being against the subjects taken separately; thus, the argument that no Art or Science exists is of a more "general" character, whereas more "special" is the argument against, say, the Grammarians concerning the elements of speech, and that against the Geometers denying the right to assume the principles, and that against the Musicians which questions the existence of tone and time. First, then, in point of order let us consider the more general refutation.

Chapter I.—Does a Subject of Learning exist?

It is no part of our present task to pronounce upon 9
the long and varied dispute regarding learning which has been carried on by the philosophers. It is sufficient to lay down that if any subject of learning exists, and if it is attainable by man, four things must

including astronomy, geometry, music, grammar, and rhetoric.

λογήσασθαι, τὸ διδασκόμενον πρᾶγμα, τὸν διδάσκοντα, τὸν μανθάνοντα, τὸν τρόπον τῆς μαθήσεως. οὔτε δὲ τὸ διδασκόμενον ἔστιν οὔτε ὁ διδάσκων οὔτε ὁ μανθάνων οὔτε ὁ τρόπος τῆς μαθήσεως, καθάπερ ὑποδείξομεν· οὐκ ἄρα ἔστι τι μάθημα.

Β'.—ΠΕΡΙ ΤΟΥ ΔΙΔΑΣΚΟΜΕΝΟΥ

10 Καὶ δὴ περὶ τοῦ πρώτου λέγοντες πρῶτόν φαμεν ὡς εἴπερ διδάσκεταί τι, ἤτοι τὸ ὂν τῷ εἶναι διδάσκεται ἢ τὸ μὴ ὂν τῷ μὴ εἶναι. οὔτε δὲ τὸ ὂν τῷ εἶναι διδάσκεται οὔτε τὸ μὴ ὂν τῷ μὴ εἶναι, καθάπερ παραστήσομεν· οὐκ ἄρα διδάσκεταί τι. καὶ δὴ τὸ μὲν μὴ ὂν τῷ μὴ εἶναι οὐκ ἂν διδάσκοιτο· εἰ γὰρ διδάσκεται, διδακτόν ἐστι, διδακτὸν δὲ καθ-
11 εστὼς τῶν ὄντων γενήσεται, καὶ διὰ τοῦτο ἔσται μὴ ὄν τε καὶ ὄν. οὐχὶ δέ γε δυνατόν ἐστι τὸ αὐτὸ καὶ ὂν [τε] καὶ μὴ ὂν ὑπάρχειν· οὐκ ἄρα τὸ μὴ ὂν τῷ μὴ εἶναι διδάσκεται. τῷ τε μὴ ὄντι οὐδὲν συμβέβηκεν, ᾧ δὲ μηδὲν συμβέβηκεν, οὐδὲ τὸ διδάσκεσθαι συμβήσεται· ἓν γάρ τι ἦν τῶν συμβεβηκότων καὶ τὸ διδάσκεσθαι. τοίνυν οὐδὲ ταύτῃ
12 διδακτόν ἐστι τὸ μὴ ὄν. καὶ μὴν τὸ διδασκόμενον φαντασίαν κινοῦν εἰς μάθησιν ἡμῖν ἔρχεται, τὸ δὲ μὴ ὂν ἀδυνατοῦν φαντασίαν κινεῖν οὐδὲ διδακτόν ἐστιν. ἔτι δ' οὐδ' ὡς ἀληθὲς τὸ μὴ ὂν διδακτόν ἐστιν· οὔτε γὰρ τῶν μὴ ὄντων ἐστὶ τἀληθές, οὔτε τι ἀληθὲς ὡς μὴ ὂν διδακτόν ἐστιν. εἰ δὲ μηδὲν ἀληθὲς ὡς μὴ ὂν διδακτόν ἐστι (τῶν γὰρ ὄντων ἐστὶ τἀληθές), ἀδίδακτον ἄρα τὸ μὴ ὄν.

---

a *Cf. P.H.* iii. 252.

b *Cf. P.H.* iii. 256 ff.

first be agreed upon[a]—the subject taught, the teacher, the learner, the method of learning. But, as we shall show, neither does the subject exist nor the teacher nor the learner nor the method of learning; therefore no subject of learning exists.

Chapter II.—Concerning the Subject taught

Now in dealing with the first point first we assert 10
that if anything is taught either the existent *qua* ex-
istent is taught or the non-existent *qua* non-exist-
ent.[b] But, as we shall establish, neither is the existent
qua existent taught nor the non-existent *qua* non-
existent; therefore nothing is taught. Now the
non-existent *qua* non-existent will not be taught;
for if it is taught it is teachable, and being teachable
it will become an existent, and because of this it 11
will be both non-existent and existent. But it is not
possible for the same thing to be both existent and
non-existent; therefore the non-existent *qua* non-
existent is not taught.—Also, the non-existent has no
property, and what has no property will not have the
property of being taught; for being taught is one
of the properties. Hence, on this ground, too, the
non-existent is not capable of being taught.—More- 12
over, that which is taught comes to be learnt by us
through exciting a presentation, but the non-existent
being unable to excite a presentation is likewise in-
capable of being taught.—Further, the non-existent
cannot be taught as being true; for neither is the
true a non-existent, nor is anything true capable of
being taught as a non-existent. But if nothing true
can be taught as non-existent (since the true is an
existent), then the non-existent is incapable of being

13 εἰ δὲ μηδὲν ἀληθὲς διδάσκεται, πᾶν τὸ διδασκό-
μενον ψεῦδός ἐστιν· ὅπερ ἀλογώτατον ὑπάρχει.
οὐ τοίνυν τὸ μὴ ὂν διδάσκεται. ἤτοι γὰρ τὸ δι-
δασκόμενον ψεῦδός ἐστιν ἢ ἀληθές. ἀλλὰ ψεῦδος
μὲν ἀλογώτατον, τὸ δὲ ἀληθὲς ὂν ὑπῆρχεν. οὐκ
14 ἄρα τὸ μὴ ὂν διδακτόν. καὶ μὴν οὐδὲ τὸ ὂν τῷ
εἶναι διδακτόν ἐστιν, ἐπειδήπερ τῶν ὄντων πᾶσι
φαινομένων ἐπ' ἴσης πάντα ἔσται ἀδίδακτα. ᾧ
ἀκολουθήσει τὸ μηδὲν εἶναι διδακτόν· δεῖ γὰρ
ὑποκεῖσθαί τι ἀδίδακτον, ἵνα ἐκ τοῦ γινωσκομένου
γένηται ἡ τούτου μάθησις. τοίνυν οὐδὲ τὸ ὂν τῷ
εἶναι διδάσκεται.

15 Ὁ δὲ ὅμοιος τῆς ἀπορίας γενήσεται τρόπος καὶ
πρὸς τοὺς ἐροῦντας τὸ οὔτι ἢ τὶ διδάσκεσθαι. εἰ
γὰρ τὸ οὔτι διδάσκοιτο, ἔσται ᾗ διδάσκεται τί,
καὶ διὰ τοῦτο ⟨τὸ⟩[1] αὐτὸ τἀναντία οὔτι καὶ τὶ
ἔσται, ὅπερ ἦν τῶν ἀδυνάτων. τῷ τε οὔτινι οὐδὲν
συμβέβηκεν, διὸ οὐδὲ τὸ διδάσκεσθαι· καὶ γὰρ
16 τοῦτο τῶν συμβεβηκότων ἐστίν. οὐ τοίνυν τὸ οὔτι
διδάσκεται. κατὰ δὲ τὴν αὐτὴν ἀναλογίαν καὶ τὸ
τὶ τῶν ἀδιδάκτων γενήσεται· εἰ γὰρ διὰ τοῦτο
διδακτὸν ἔσται ὅτι ⟨τὶ⟩[2] ἔστιν, οὐδὲν ἀδίδακτον
17 ἔσται, ᾧ ἕπεται τὸ μηδὲν εἶναι διδακτόν. καὶ μὴν
εἰ διδάσκεταί τι, ἤτοι διὰ τῶν οὐτινῶν διδαχθή-
σεται ἢ διὰ τῶν τινῶν. ἀλλὰ διὰ μὲν τῶν οὐτινῶν
οὐχ οἷόν τε διδαχθῆναι· ἀνυπόστατα γάρ ἐστι τῇ
διανοίᾳ ταῦτα κατὰ τοὺς ἀπὸ τῆς στοᾶς. λείπεται

[1] ⟨τὸ⟩ add. cj. Bekk.

[2] ⟨τὶ⟩ add. Heintz.

taught. And if nothing true is taught, everything 13
that is taught is false, which is most irrational. The non-existent, therefore, is not taught. For what is taught is either false or true. But to say it is false is most irrational, and what is true is existent. Therefore
the non-existent is not capable of being taught.—Nor, 14
indeed, is the existent *qua* existent capable of being taught, for, since existents are equally evident to all men, they will all be incapable of being taught. Wherefrom it will follow that nothing is capable of being taught; for something untaught must be assumed in order that the learning of it may be derived from what is known. Neither, then, is the existent *qua* existent taught.

The same method of scepticism will be used also 15
against those who will say that either "nothing" is taught or "something." For if "nothing" should be taught, it will be "something" inasmuch as it is taught, and because of this the same thing will be two opposite things, nothing and something, which is impossible.—Besides, "nothing" has no property, and therefore it has not that of being taught; for
this is, in fact, one of the properties. So, then, 16
"nothing" is not taught.—And, by parity of reasoning, "something" also is one of the things which cannot be taught. For if it is to be capable of being taught because it is "something," nothing will be untaught; wherefrom it follows that nothing is
capable of being taught.—Moreover, if something is 17
taught it will be taught by means either of the "nothings" or of the "somethings"; but it is not possible for it to be taught by means of the "nothings," for these have no reality for the mind according to the Stoics. It remains, then, that learning

οὖν διὰ τῶν τινῶν γίνεσθαι τὴν μάθησιν. ὃ πάλιν
18 ἄπορόν ἐστιν· ὥσπερ γὰρ αὐτὸ τὸ διδασκόμενον
κατὰ τοῦτο διδάσκεται καθὸ τί ἐστιν, οὕτως ἐπεὶ
καὶ τὰ ἐξ ὧν ἡ μάθησις τινά ἐστι, γενήσεται
διδακτά. καὶ ταύτῃ μηδενὸς ὄντος ἀδιδάκτου[1]
ἀναιρεῖται ἡ μάθησις.

Γ'.—ΠΕΡΙ ΣΩΜΑΤΟΣ

19 Ἄλλως τε, ἐπεὶ τῶν τινῶν τὰ μέν ἐστι σώμα-
τα τὰ δὲ ἀσώματα, δεήσει τὰ διδασκόμενα τινὰ
ὄντα ἤτοι σώματα εἶναι ἢ ἀσώματα· οὔτε δὲ
σώματα δύναται ὑπάρχειν οὔτε ἀσώματα, ὡς
παραστήσομεν· οὐκ ἄρα ἔστι τινὰ διδασκόμενα.
20 τὸ μὲν οὖν σῶμα, καὶ μάλιστα κατὰ τοὺς στωικούς,
οὐκ ἂν εἴη τῶν διδακτῶν· δεῖ γὰρ τὰ διδασκόμενα
λεκτὰ τυγχάνειν, τὰ δὲ σώματα οὐκ ἔστι λεκτά,
διόπερ οὐ διδάσκεται. εἴπερ δὲ τὰ σώματα μήτε
αἰσθητά ἐστι μήτε νοητά, δῆλον ὡς οὐδὲ διδακτὰ
γενήσεται. αἰσθητὰ μὲν οὖν οὐκ ἔστιν, ὡς ἐκ
21 τῆς ἐννοίας αὐτῶν συμφανές. εἰ γὰρ σύνοδός ἐστι
κατὰ ἀθροισμὸν μεγέθους καὶ σχήματος καὶ
ἀντιτυπίας τὸ σῶμα, ὡς φησὶν Ἐπίκουρος, ἢ τὸ
τριχῇ διαστατόν, τουτέστι τὸ ἐκ μήκους καὶ
πλάτους καὶ βάθους, καθάπερ οἱ μαθηματικοὶ
λέγουσιν, ἢ τὸ τριχῇ διαστατὸν μετὰ ἀντιτυπίας,
ὡς πάλιν Ἐπίκουρος, ἵνα τούτῳ διορίζῃ τοῦ κενοῦ,[2]
22 ἢ ὄγκος ἀντίτυπος, ὡς ἄλλοι,—ὅπως δ' ἂν ἔχῃ,
ἐπεὶ κατὰ σύνοδον πολλῶν ἰδιωμάτων νοεῖται, ἡ
δὲ πλειόνων ἐπισύνθεσις οὐχ ἁπλῆς τινὸς καὶ

[1] ἀδιδάκτου Heintz : διδακτοῦ MSS., Bekk.
[2] τοῦ κενοῦ Heintz : τὸ κενόν MSS., Bekk.

takes place by means of the "somethings," which too is dubious. For just as the thing taught is itself 18
taught *qua* "something," so also, since the things from which the learning is derived are "somethings," they too will be taught. And thus, since nothing is untaught, learning is abolished.

Chapter III.—Concerning the Corporeal

Furthermore, since some of the "somethings" are 19
corporeal, others incorporeal, the things taught, being "somethings," will have to be either corporeal or incorporeal[a]; but, as we shall establish, they cannot be either corporeal or incorporeal; therefore there are no "somethings" which are taught. Now the 20
corporeal—according to the Stoics in special—will not be capable of being taught; for things taught must be "meanings,"[b] but corporeals are not "meanings" and, consequently, are not taught. And if corporeals are neither sensibles nor intelligibles it is plain that they will not be capable of being taught. Now they are not sensibles, as is manifest from our conception of them. For if the corporeal is, as Epi- 21
curus says, a combination of magnitude, shape and solidity massed together,—or, as the Mathematicians assert, that which is three-dimensional, that is to say compounded of length, breadth and depth,—or that which has three-dimensions *plus* solidity (another definition of Epicurus: in order to distinguish it thus from void),—or, as others say, a solid mass,—be this 22
as it may, inasmuch as the corporeal is conceived as a combination of a number of separate factors, and the combining of many is not the operation of a simple

[a] *Cf. P.H.* iii. 255.
[b] The Stoic *lekta*; see *P.H.* ii. 81 n.

ἀλόγου αἰσθήσεώς ἐστιν ἔργον ἀλλὰ λογικῆς διανοίας, [· εἰ δὲ λογικῆς διανοίας,][1] οὐκ ἔσται τῶν
23 αἰσθητῶν τὸ σῶμα. κἂν αἰσθητὸν δὲ [πάλιν] αὐτὸ ὑποθώμεθα, πάλιν ἐστὶν ἀδίδακτον. τὸ γὰρ αἰσθητὸν πάλιν, ᾗ αἰσθητόν ἐστιν, οὐ διδάσκεται· οὐδεὶς γὰρ λευκὸν ὁρᾶν μανθάνει, οὐδὲ γλυκέος γεύεσθαι, οὐδὲ θερμοῦ ἅπτεσθαι, οὐκ εὐώδους ὀσφραίνεσθαι, ἀλλ' ἔστι ταῦτα τῶν ἀδιδάκτων
24 καὶ φυσικῶς ἡμῖν προσόντων. λείπεται οὖν νοητόν τε λέγειν τὸ σῶμα καὶ ταύτῃ διδακτόν. ὅπως δ' ἂν ἀληθὲς εἴη, σκοπῶμεν. εἰ γὰρ μήτε μῆκός ἐστι κατ' ἰδίαν τὸ σῶμα μήτε πλάτος ἢ βάθος, τὸ δὲ ἐξ ἁπάντων νοούμενον, ἀνάγκη πάντων ἀσωμάτων ὄντων καὶ τὸ ἐξ αὐτῶν συστὰν ἀσώματον νοεῖν καὶ οὐ σῶμα, διὰ δὲ τοῦτο
25 καὶ ἀδίδακτον. πρὸς τῷ τὸν νοοῦντα τὸ ἐκ τούτων συνεστὼς σῶμα πρότερον ὀφείλειν αὐτὰ ταῦτα νοεῖν, ἵνα κἀκεῖνο δυνατὸς[2] ᾖ νοεῖν. ⟨ἀλλ' οὐ δύναται ταῦτα νοεῖν·⟩[3] ἢ γὰρ περιπτωτικῶς αὐτὰ νοήσει ἢ κατὰ μετάβασιν ἀπὸ περιπτώσεως. οὔτε δὲ περιπτωτικῶς· ἀσώματα γάρ ἐστι, καὶ τῶν ἀσωμάτων οὐκ ἀντιλαμβανόμεθα περιπτωτικῶς, ἀεὶ κατὰ θίξιν γινομένης τῆς περὶ τὴν αἴσθησιν ἀντιλήψεως. καὶ μὴν οὐδὲ κατὰ μετάβασιν ἀπὸ περιπτώσεως, τῷ μηδὲν ἔχειν αἰσθητὸν ἀφ' οὗ μετιών τις ποιήσεται τούτων ἐπίνοιαν. τοίνυν οὐδὲ τὰ ἐξ ὧν τὸ σῶμα νοεῖν δυνάμενοι πάντως οὐδὲ διδάσκειν τοῦτο ἰσχύσομεν.

26 Ἀλλὰ περὶ μὲν τῆς τοῦ σώματος νοήσεώς τε

[1] [· εἰ . . . διανοίας,] del. Heintz.
[2] δυνατὸς cj. Bekk. : δυνατὸν MSS., Bekk.
[3] ⟨ἀλλ' οὐ δύναται ταῦτα νοεῖν·⟩ add. cj. Bekk.

and irrational sense but of a rational intellect, the corporeal will not be one of the sensibles. And even 23
if we assume that it is sensible, yet even so it is still incapable of being taught. For the sensible also, *qua* sensible, is not taught ; for no one learns to see what is white, or to taste the sweet, or feel the hot, or smell the odorous, but these sensations are untaught and belong to us naturally.—It remains, then, to say that 24
the corporeal is intelligible and thus capable of being taught. But let us consider what is the truth of the matter. If the corporeal is neither length by itself nor breadth nor depth, but that which is conceived as compounded of them all, then, since these are all incorporeal, one must necessarily conceive the compound formed from them as being incorporeal and not corporeal, and therefore incapable of being taught. Besides, the man who conceives the body 25
compounded of these dimensions must first conceive the dimensions themselves, in order that he may also be able to conceive the body. ⟨But these he cannot conceive ;⟩ for he will conceive them either by way of sense-experience or by way of transference from sense-experience. Not, however, by way of experience ; for they are incorporeals, and we do not apprehend incorporeals by experience, since sensuous apprehension always takes place by way of contact. Nor yet by way of transference from experience, since one possesses no sense-object by transference from which one might form the conception of the dimensions. Thus, since we cannot even conceive the elements which go to form the corporeal, we certainly shall not be able to teach it.

But we have treated more exactly of the conception 26

καὶ ὑποστάσεως ἐν τοῖς σκεπτικοῖς ὑπεμνήσαμεν ἀκριβέστερον· νυνὶ δὲ ἀποστάντες τούτων τῶν ἐλέγχων ἐκεῖνο λέγωμεν ὅτι τῶν σωμάτων κατὰ τὸ ἀνωτάτω διττή τίς ἐστι διαφορά· τὰ μὲν γὰρ αὐτῶν αἰσθητὰ καθέστηκε τὰ δὲ νοητά. καὶ εἰ τὸ διδασκόμενόν ἐστι σῶμα, πάντως ἤτοι νοητόν
27 *ἐστιν ἢ αἰσθητόν. ἀλλ' οὔτε αἰσθητὸν εἶναι δύναται διὰ τὸ πᾶσιν ἐπ' ἴσης ὀφείλειν φαίνεσθαι καὶ πρόδηλον ὑπάρχειν, οὔτε νοητὸν διὰ τὸ ἀδηλεῖσθαι καὶ δι' αὐτὸ τοῦτο ἀνεπικρίτως διαφωνεῖσθαι παρὰ πᾶσι τοῖς φιλοσόφοις, τῶν μὲν ἄτομον τοῦτο λεγόντων ὑπάρχειν τῶν δὲ τμητόν, καὶ τῶν τμητὸν φαμένων εἶναι ἐνίων μὲν εἰς ἄπειρον τέμνεσθαι τοῦτο ἀξιούντων, ἐνίων δὲ εἰς ἐλάχιστον καὶ ἀμερὲς καταλήγειν. οὐκ ἄρα διδακτόν ἐστι τὸ σῶμα.*

28 *Καὶ μὴν οὐδὲ τὸ ἀσώματον. πᾶν γὰρ καὶ ὁποῖον ἄν τις ἀσώματον λέγῃ διδάσκεσθαι, ἐάν τε τὴν Πλατωνικὴν ἰδέαν, ἐάν τε τὸ παρὰ τοῖς στωικοῖς λεκτόν, ἐάν τε τόπον ἢ κενὸν ἢ χρόνον ἢ ἄλλο τι τῶν τοιούτων, ἵνα μηδὲν προπετὲς περὶ τῆς ὑποστάσεως αὐτῶν λέγωμεν, μηδ' ἑτέρας σκέψεις ἐφ'*[1] *ἑτέραις διεξοδεύωμεν παριστάντες τὸ ἀνυπόστατον ἑκάστου, [ὃ] προδήλως μὲν ἐπιζητεῖται καὶ*

ἔς τ' ἂν ὕδωρ τε νάῃ καὶ δένδρεα μακρὰ τεθήλῃ

ζητήσεται παρὰ τοῖς δογματικοῖς, τῶν μὲν εἶναι ταῦτα διαβεβαιουμένων τῶν δὲ μὴ εἶναι, τῶν δὲ ἐπεχόντων· τὸ δὲ τὰ ἔτι ἐπίδικα καὶ ἐν μετεώροις

[1] ἐφ' scripsi: ἐν MSS., Bekk.

and reality of the corporeal in our *Skeptika*[a]; so now let us leave aside these criticisms and proceed to state that the most generic distinction amongst corporeals is twofold; some of them are sensibles, others intelligibles. And if the thing taught is corporeal it must certainly be either an intelligible or a sensible. But it cannot be a sensible, since then 27 it ought to be equally apparent to all men and be pre-evident; nor can it be an intelligible owing to the fact that it is non-evident, and for this very reason there is an unsettled controversy about it amongst all the philosophers, some of them saying that it is indivisible, others that it is divisible; while of those who assert its divisibility some claim that it is divided to infinity, others that the division stops at what is minimal and atomic. Therefore the corporeal is not capable of being taught.

Nor, indeed, is the incorporeal. For in every case, 28 whatever be the incorporeal which one declares to be taught,—whether it be the Platonic Idea, or the "lekton" of the Stoics, or place or void or time or anything else of the kind,—without making any rash assertion about the reality of these things, or pursuing divers inquiries one after another in order to prove the non-reality of each of them, we may say that it is manifest that the problem of their reality is a matter of discussion amongst the Dogmatists, and will be so

> Long as the waters flow on and the tall trees cease not to burgeon,[b]

since some of them stoutly assert that these things exist, others that they do not exist, while still others suspend judgement; thus it is absurd to say that

[a] *Cf. Adv. Phys.* i. 359 ff.; *P.H.* iii. 38 ff.
[b] *Cf. P.H.* ii. 37 n.

ἀμφισβητήσεσι κείμενα ὡς σύμφωνα καὶ ὁμόλογα διδάσκεσθαι λέγειν τῶν ἀτόπων ἐστίν.

29 Εἰ οὖν τῶν ὄντων τὰ μέν ἐστι σώματα τὰ δὲ ἀσώματα, δέδεικται δὲ οὐδέτερα τούτων διδασκόμενα, οὐδὲν διδάσκεται.

Ἐπιχειρητέον δὲ καὶ οὕτως. εἰ διδάσκεταί τι, ἤτοι ἀληθές ἐστιν ἢ ψεῦδος. οὔτε δὲ ψεῦδος διδακτόν ἐστιν, ὡς αὐτόθεν ὁμόλογον, οὔτε ἀληθές· τὸ γὰρ ἀληθὲς ἄπορον, ὡς ἐν τοῖς σκεπτικοῖς ὑπομνήμασι δέδεικται, καὶ τῶν ἀπόρων οὐκ ἔστι
30 μάθησις. οὐκ ἄρα ἔστι τι τὸ διδασκόμενον. καθόλου τε, εἰ διδάσκεταί τι, ἤτοι τεχνικόν ἐστιν ἢ ἄτεχνον. καὶ ἄτεχνον μὲν ὂν οὐκ ἔστι διδακτόν, τεχνικὸν δὲ εἴπερ καθέστηκεν, αὐτόθεν μὲν φαινόμενον οὔτε τεχνικόν ἐστιν οὔτε διδακτόν, ἄδηλον δὲ καθεστὼς διὰ τὸ ἀδηλεῖσθαι πάλιν ἐστὶν ἀδίδακτον.

Ὧι συναναιρεῖται καὶ ὁ διδάσκων διὰ τὸ μὴ ἔχειν ὃ διδάξει [ᾗ ἄδηλόν ἐστιν], ὅ τε μανθάνων διὰ τὸ μὴ ἔχειν ὃ μάθῃ. ὅμως δ' οὖν καὶ περὶ ἑκατέρου τούτων κατ' ἰδίαν ἐπελθόντες ἀπορήσομεν.

Δʹ.—ΠΕΡΙ ΤΟΥ ΔΙΔΑΣΚΟΝΤΟΣ ΚΑΙ ΜΑΝΘΑΝΟΝΤΟΣ

31 Εἰ γὰρ ἔστι τις τούτων, ἤτοι ὁ ἄτεχνος τὸν ὁμοίως ἄτεχνον διδάξει ἢ ὁ τεχνίτης τὸν ὁμοίως τεχνίτην ἢ ὁ ἄτεχνος τὸν τεχνίτην ἢ ἀνάπαλιν. οὔτε δὲ ὁ ἄτεχνος τὸν ἄτεχνον δύναται διδάσκειν, ὡς οὐδὲ ὁ τυφλὸς τὸν τυφλὸν ὁδηγεῖν, οὔτε ὁ

things which are undecided and subjects of unsettled controversy are taught, as though they were unanimously agreed upon and undisputed.

If, then, of existents some are corporeal, others 29
incorporeal, and it has been shown that neither of these is taught, nothing is taught.

We may also argue thus: If a thing is taught it is either true or false.[a] But neither is the false taught (as the very notion compels us to admit), nor yet the true; for the true is dubious, as has been shown in our *Treatise on Scepticism*, and of things dubious there is no learning. Therefore there is nothing 30
which is taught.—Also, in general, if a thing is taught it is either a matter of art or without art. And if it is without art it cannot be taught; while if it is a matter of art, either it is self-evident and consequently no matter of art nor capable of being taught, or else it is non-evident and thus, too, incapable of being taught because of its being non-evident.

This involves also the abolition of the teacher because he will have nothing to teach, and of the learner because he will have nothing to learn. Notwithstanding, we shall proceed to state the difficulties about each of these separately.

Chapter IV.—Concerning the Teacher and the Learner

For if anyone of these exists either the non-expert 31
will teach him who is similarly non-expert, or the expert him who is similarly expert, or the non-expert the expert, or *vice versa*.[b] But the non-expert cannot teach the non-expert (just as the blind cannot lead

[a] *Cf. P.H.* iii. 253.
[b] *Cf. P.H.* iii. 259 ff.

τεχνίτης τὸν ὁμοίως[1] τεχνίτην· οὐδέτερος γὰρ αὐτῶν ἐδεῖτο μαθήσεως, καὶ οὐ μᾶλλον οὗτος ἐκείνου ἢ ἐκεῖνος τούτου χρείαν ἔχει πρὸς τὸ μανθάνειν, τοῖς ἴσοις περιουσιαζόμενοι. οὔτε ὁ
32 ἄτεχνος τὸν τεχνίτην· ὅμοιον γὰρ ὡς εἴ τις λέγοι τὸν βλέποντα ὑπὸ τοῦ πεπηρωμένου ὁδηγεῖσθαι. καὶ γὰρ ὁ ἄτεχνος πρὸς τὰ τεχνικὰ τῶν θεωρημάτων πεπηρωμένος οὐκ ἂν δύναιτό τινα διδάσκειν ἃ μηδὲ τὴν ἀρχὴν οἶδεν, καὶ ὁ τεχνίτης διαβλέπων ἐν τοῖς τεχνικοῖς θεωρήμασι καὶ γνῶσιν αὐτῶν
33 ἐσχηκὼς οὐ δεήσεται τοῦ διδάξοντος. λείπεται οὖν τὸν τεχνίτην τοῦ ἀτέχνου διδάσκαλον εἶναι λέγειν. ὃ τῶν προτέρων ἐστὶν ἀτοπώτερον· ὅ τε γὰρ τεχνίτης συνηπόρηται ἡμῖν τοῖς τῆς τέχνης θεωρήμασιν ἐν τῷ σκεπτικῷ τόπῳ, ὅ τε ἄτεχνος οὔτε ὅτε ἐστὶν ἄτεχνος δύναται γενέσθαι τεχνίτης, οὔτε ὅτε ἐστὶ τεχνίτης ἔτι γίνεται τεχνίτης ἀλλ'
34 ἔστιν. ἄτεχνος μὲν γὰρ ὢν ὅμοιός ἐστι τῷ ἐκ γενετῆς τυφλῷ ἢ κωφῷ, καὶ ὃν τρόπον οὗτος οὐδέποτε ἢ εἰς χρωμάτων ἢ εἰς φωνῶν ἔννοιαν ἐλθεῖν πέφυκεν, οὕτως οὐδὲ ὁ ἄτεχνος, ἐφ' ὅσον ἔστιν ἄτεχνος, τετυφλωμένος καὶ κεκωφωμένος πρὸς τὰ τεχνικὰ θεωρήματα οὔτε ἰδεῖν οὔτε ἀκοῦσαί τι τούτων οἷός τε ἐστίν· τεχνίτης δὲ γενόμενος οὐκέτι διδάσκεται ἀλλὰ δεδίδακται.

35 Μετακτέον δὲ τὰς ἀπορίας ἐκ τῶν περὶ μεταβολῆς καὶ πάθους γενέσεώς τε καὶ φθορᾶς προεγκεχειρημένων ἡμῖν ἐν ταῖς πρὸς τοὺς φυσικοὺς ἀντιρρήσεσι. τὰ δὲ νῦν συγχωρήσαντες τοῖς ἀπὸ τῶν μαθημάτων εἶναί τι τὸ διδασκόμενον πρᾶγμα

[1] *τὸν ὁμοίως* scripsi: *ὁμοίως τὸν* MSS., Bekk.

the blind), nor the expert the similarly expert; for neither of these requires to learn, and this one has no more need of learning than that one, nor that one than this one, both being equally full of knowledge. Nor can the non-expert teach the expert; for that 32 would be like saying that the man who has eyesight is led along by the man who has none. For in fact the non-expert is blind as regards the technical principles, and thus will be unable to teach things of which he knows nothing at all; while the expert, having clear vision of the technical principles and having gained knowledge of them, will not require a teacher.—It remains, then, to say that the expert 33 teaches the non-expert. But this is even more absurd than the former suppositions; for along with the principles of his art the existence of the expert himself has been shown in our Sceptic text-book to be open to doubt, and the non-expert neither can become an expert while he is non-expert, nor when he is an expert does he any longer become an expert, but is one. For if he is non-expert he is like a man blind 34 or deaf from birth, and just as such a man cannot ever form a notion of colours or of sounds, so the non-expert, in so far as he is non-expert, being blind and deaf to the technical principles is incapable of seeing or hearing any of them; and if he has become an expert he is no longer being taught but has been taught.

We may also borrow criticisms from those we have 35 already brought forward in our arguments against the Physicists concerning change and affection and becoming and perishing.[a] For the present let us concede to the Professors that there exists a subject

[a] *Cf. P.H.* iii. 102 ff., *Adv. Phys.* ii. 39 ff.

καὶ εἶναί τινα τὸν ὑφηγούμενον, ὡσαύτως δὲ καὶ τὸν μανθάνοντα, τὸ μετὰ τοῦτο ἀπαιτῶμεν τὸν
36 τρόπον τῆς μαθήσεως. ἢ γὰρ ἐναργείᾳ γίνεται ἢ λόγῳ τὰ τῆς διδασκαλίας. ἀλλὰ τούτων ἡ μὲν ἐνάργεια τῶν δεικτῶν ἐστί, τὸ δὲ δεικτὸν φαινόμενον, τὸ δὲ φαινόμενον, ᾗ φαίνεται, κοινῶς πᾶσι ληπτόν, τὸ δὲ κοινῶς πᾶσι ληπτὸν ἀδίδακτον· οὐκ
37 ἄρα τὸ ἐναργείᾳ δεικτὸν διδακτόν. ὁ δὲ λόγος ἤτοι σημαίνει τι ἢ οὐ σημαίνει. καὶ μηδὲν μὲν σημαίνων οὐδὲ διδάσκαλός τινός ἐστι, σημαίνων δὲ ἤτοι φύσει σημαίνει τι ἢ θέσει. καὶ φύσει μὲν οὐ σημαίνει διὰ τὸ μὴ πάντας πάντων ἀκούειν, Ἕλληνας βαρβάρων καὶ βαρβάρους Ἑλλήνων ἢ
38 Ἕλληνας Ἑλλήνων ἢ βαρβάρους βαρβάρων· θέσει δὲ εἴπερ σημαίνει, δῆλον ὡς οἱ μὲν προκατειληφότες τὰ καθ' ὧν αἱ λέξεις κεῖνται καὶ ἀντιλήψονται τούτων, οὐ τὸ ἀγνοούμενον ἐξ αὐτῶν διδασκόμενοι, τὸ δ' ὅπερ ᾔδεισαν ἀνανεούμενοι, οἱ δὲ χρῄζοντες τῆς τῶν ἀγνοουμένων μαθήσεως οὐκέτι.

Εἰ οὖν οὔτε τὸ διδασκόμενον ἔστιν οὔτε ὁ διδάσκων οὔτε ὁ μανθάνων οὔτε ὁ τρόπος τῆς μαθήσεως, δῆλον ὡς οὐδὲ μάθημα, οὐδὲ ὁ μαθήματος προ-
39 εστώς. ἀλλ' ἐπεὶ οὐ καθολικὴν μόνον πρὸς πάντας τοὺς μαθηματικοὺς ὑπεσχόμεθα ποιήσασθαι τὴν ἀντίρρησιν ἀλλὰ καὶ εἰδικωτέραν πρὸς ἕκαστον, ὑποτιθέμενοι τὸ εἶναί τι μάθημα καὶ δυνατὴν ὑπάρχειν τὴν μάθησιν, σκοπῶμεν εἰ καὶ τὸ ἑκάστου

---

[a] *Cf. P.H.* ii. 214.

taught and that the instructor exists and likewise the learner, and let us next investigate the method of learning. Now teaching takes place either by means 36
of sense-evidence or by means of speech. But of these sense-evidence is concerned with ostensible things, and the ostensible is apparent, and the apparent, in so far as it appears, is perceptible by all alike, and what is perceptible by all alike is incapable of being taught; therefore what is shown by sense-evidence is not capable of being taught.—And speech 37
either signifies or does not signify something. Now if it signifies nothing, neither does it teach anything; while if it signifies, it signifies a thing either by nature or by convention. But it does not signify by nature since all men do not understand the speech of all,—Greeks that of barbarians and barbarians that of Greeks, or Greeks that of Greeks or barbarians that of barbarians.[a] And if it signifies by convention, it is 38
plain that those who have apprehended beforehand the objects to which the terms are conventionally applied will also understand those terms, not that they are taught by them what they did not know, but rather as reviving what they did know; but those who lack learning about the things not known will fail to do so.

If, then, the subject taught does not exist, nor the teacher, nor the learner, nor the method of learning, it is clear that neither does the subject learnt exist nor he who presides over that subject. But since we 39
undertook not only to construct a general argument against all the Professors but also a special one against each of them, let us assume that a subject of learning exists and that learning is possible, and consider whether the claims made regarding each subject are

μαθήματος ἐπάγγελμα δυνατόν ἐστι, λαμβάνοντες πρὸς τοὺς ἐλέγχους μὴ πάντα τὰ παρὰ τοῖς ἐλεγχομένοις λεγόμενα (τοῦτο γὰρ σὺν τῷ περισκελεῖ
40 καὶ ἀμέθοδον, ἐπεὶ τάχα καὶ ἀδύνατόν ἐστι) μήτε δ' ἐκ πάντων τὰ ὁποιαοῦν (τοῦτο γὰρ ἴσως οὐδὲ καθικνεῖται αὐτῶν), ἀλλὰ τὰ ὧν ἀναιρουμένων συναναιρεῖται πάντα. καὶ ὃν τρόπον οἱ πόλιν λαβεῖν σπεύδοντες ἐκείνων μάλιστα ἐγκρατεῖς γίνεσθαι σπουδάζουσιν ὧν ἁλόντων καὶ ἡ πόλις ἑάλωκεν, οἷον τείχη καθαιροῦντες ἢ στόλον ἐμπιπράντες ἢ τὰς εἰς τὸ ζῆν ἀφορμὰς ἀποκλείοντες οὕτω καὶ ἡμεῖς τοῖς ἀπὸ τῶν μαθημάτων διαγωνιζόμενοι ταῦτ' αὐτὰ[1] πειράζωμεν ἐξ ὧν αὐτοῖς σώζεται τὰ πάντα, οἷον ἢ ἀρχὰς ἢ τὰς ἐκ τῶν ἀρχῶν καθολικὰς μεθόδους ἢ τὰ τέλη· ἐν τούτοις γὰρ ἢ ἐκ τούτων συνίσταται πᾶν μάθημα.

Α'.—ΠΡΟΣ ΓΡΑΜΜΑΤΙΚΟΥΣ

41 Ἀρχέτω δὲ ἡμῖν εὐθὺς ἡ πρὸς τοὺς γραμματικοὺς ζήτησις, πρῶτον μὲν ἐπείπερ ἀπὸ νηπιότητος σχεδὸν καὶ ἐκ πρώτων σπαργάνων γραμματικῇ παραδιδόμεθα, ἔστι τε αὕτη οἷον ἀφετήριόν τι πρὸς τὴν τῶν ἄλλων μάθησιν, εἶθ' ὅτι παρὰ πάσας θρασύνεται τὰς ἐπιστήμας, σχεδόν τι τὴν τῶν
42 Σειρήνων ὑπόσχεσιν ὑπισχνουμένη. ἐκεῖναι μὲν γὰρ εἰδυῖαι ὅτι φύσει φιλοπευθής ἐστιν ἄνθρωπος καὶ πολὺς αὐτῷ κατὰ στέρνων τῆς ἀληθείας ἵμερος ἐντέτηκεν, οὐ μόνον θεσπεσίοις μέλεσι κηλήσειν

also possible ; and let us take as the matter for our
refutation not all the statements made by those
whom we are refuting (for this would be unsystematic
as well as inordinate, and probably impossible also),
nor yet any random selection out of them all (for this 40
perhaps would fail to reach all of them), but those
statements the destruction of which involves the
destruction of all. And just as those who are striving
to take a city are anxious most of all to make them-
selves masters of those things the capture of which
ensures also their capture of the city—such as break-
ing down their walls or setting fire to their equipment
or cutting off their food-supplies,—so let us, too, in
our conflict with the Professors attack those very
points upon which all their safety depends, such as
their principles, or the general arguments which are
derived from the principles, or their conclusions ; for
by these or from these every art and science is
formed.

Chapter I.—Against the Grammarians

Let us begin at once with our criticism of the 41
Professors of Letters (or Grammarians), firstly be-
cause we are handed over to the study of letters
well-nigh from infancy and our first swaddling clothes,
and this art is as it were the starting-point for our
learning of all the others, and next because it puffs
itself up above all the sciences, almost promising the
Sirens' promise. For they, being aware that man is 42
inquisitive by nature and that there is implanted in
his breast a great longing for the truth, promise not
only to charm those who are sailing by with wondrous

[1] ταῦτ' αὐτὰ cj. Bekk. : τὸ αὐτὸ MSS., Bekk.

τοὺς παραπλέοντας ὑπισχνοῦνται ἀλλὰ καὶ τὰ ὄντα αὐτοὺς διδάξειν. φασὶ γὰρ

δεῦρ' ἄγ' ἰών, πολύαιν' Ὀδυσεῦ, μέγα κῦδος Ἀχαιῶν,
νῆα κατάστησον, ἵνα νωιτέρην ὄπ' ἀκούσῃς.
οὐ γάρ πώ τις τῇδε παρέπλω νηὶ μελαίνῃ
πρίν γ' ἡμέων μελίγηρυν ἀπὸ στομάτων ὄπ' ἀκούσῃ,
ἀλλ' ὅ γε τερψάμενος νεῖται καὶ πλείονα εἰδώς.
ἴδμεν γάρ τοι πάνθ' ὅσ' ἐνὶ Τροίῃ εὐρείῃ
Ἀργεῖοι Τρῶές τε θεῶν ἰότητι μόγησαν,
ἴδμεν δ' ὅσσα γένηται ἐπὶ χθονὶ πουλυβοτείρῃ.

43 ἡ δὲ γραμματική, σὺν τῷ τὰ ἐκ τῶν μύθων τε καὶ ἱστοριῶν λόγῳ διορίζειν, καὶ τὸ περὶ τὰς διαλέκτους καὶ τεχνολογίας καὶ ἀναγνώσεις πραγματικὸν αὔχουσα πολὺν ἑαυτῆς ἐργάζεται τοῖς ἀκούουσι πόθον. ἀλλ' ἵνα μὴ παρὰ θύραν πλανᾶσθαι δοκῶμεν, ὑποδεικτέον πόσαι τέ εἰσι γραμματικαὶ καὶ περὶ τίνος αὐτῶν πρόκειται ζητεῖν.

Β'.—ΠΟΣΑΧΩΣ ΛΕΓΕΤΑΙ ΓΡΑΜΜΑΤΙΚΗ

44 Γραμματικὴ τοίνυν λέγεται κατὰ ὁμωνυμίαν κοινῶς τε καὶ ἰδίως, καὶ κοινῶς μὲν ἡ τῶν ὁποιωνδηποτοῦν γραμμάτων εἴδησις, ἐάν τε Ἑλληνικῶν ἐάν τε βαρβαρικῶν, ἣν συνήθως γραμματιστικὴν καλοῦμεν, ἰδιαίτερον δὲ ἡ ἐντελὴς καὶ τοῖς περὶ Κράτητα τὸν Μαλλώτην Ἀριστοφάνην τε καὶ
45 Ἀρίσταρχον ἐκπονηθεῖσα. δοκεῖ δὲ τούτων ἑκα-

ᵃ Homer, *Od.* xii. 184 ff.

music but also to teach them real happenings. Here is what they say [a]—

> Come hither now, Odysseus, far-famed, great pride of Achaeans:
> Stay thy ship on its course and hark to the words we utter.
> No man ever as yet has passed this point in his black ship
> Ere he has lent an ear to the honey-sweet sound of our singing;
> So he departs over-joy'd, and richer, to boot, in knowledge,
> Seeing that we know all that the Argive hosts and the Trojans
> Suffered in broad Troy-land by the dispensation of Heaven;
> Nothing, in sooth, is done on the bountiful Earth but we know it.

So the Art of Letters, by boasting of its dealing 43
systematically with dialects and rules of style and recitations, as well as the logical treatment of the contents of myths and histories, excites in the listeners a great desire for itself. But lest we should seem to be wandering past the door, we must indicate how many Arts of Letters there are and which of them we propose to investigate.

Chapter II.—The various Meanings of the Term "Art-of-Letters"

The term "Art-of-letters" is used, by homonym, 44
both in a general and in a special sense. In the general sense it means the knowledge of letters of any and every kind, either Greek or barbarian, which we customarily call "grammatistic"; in the special sense it is applied to the perfected art as elaborated by Crates of Mallus, Aristophanes and Aristarchus.[b]

[b] Aristarchus founded a School of Grammar at Alexandria (*circ.* 150 B.C.): he was a pupil of Aristophanes of Byzantium. Crates of Mallus, in Cilicia, founded the Pergamene School, in opposition to the Alexandrian.

τέρα καὶ ἀπό τινος ἐτύμου φερωνύμως προσηγορεῦσθαι. ἡ μὲν γὰρ πρώτη ἀπὸ τῶν γραμμάτων οἷς σημειούμεθα τὰς ἐνάρθρους φωνάς, ἡ δὲ δευτέρα τάχα μέν, ὥς τινες ἠξιώκασι, διατατικώτερον ἀπὸ τῆς πρώτης· μοῖρα γάρ ἐστιν ⟨αὕτη⟩[1] αὐτῆς, καὶ ὃν τρόπον ἰατρικὴ εἴρηται μὲν τὸ παλαιὸν ἀπὸ τῆς τῶν ἰῶν ἐξαιρέσεως, ἐπικατηγορεῖται δὲ νῦν καὶ τῆς τῶν ἄλλων παθῶν ἀνασκευῆς πολλῷ
46 τεχνικωτέρας οὔσης, καὶ ὡς γεωμετρία ἔσπακε μὲν τὴν κλῆσιν ἀρχικῶς ἀπὸ τῆς κατὰ τὴν γῆν καταμετρήσεως, τάττεται δὲ ἐπὶ τοῦ παρόντος καὶ κατὰ τῆς τῶν φυσικωτέρων θεωρίας, οὕτω καὶ ἡ τέλειος γραμματικὴ ἀπὸ τῆς τῶν γραμμάτων εἰδήσεως κατ᾽ ἀρχὰς ὀνομασθεῖσα διετάθη καὶ ἐπὶ τὴν ἐν τοῖς ποικιλωτέροις αὐτῶν καὶ τεχνικω-
47 τέροις θεωρήμασι γνῶσιν. τάχα δέ, ὡς φασὶν οἱ περὶ τὸν Ἀσκληπιάδην, καὶ αὐτὴ ἀπὸ μὲν γραμμάτων ὠνόμασται, οὐκ ἀπὸ τούτων δὲ ἀφ᾽ ὧν καὶ ἡ γραμματιστική, ἀλλ᾽ ἐκείνη μέν, ὡς ἔφην, ἀπὸ τῶν στοιχείων, αὕτη δὲ ἀπὸ τῶν συγγραμμάτων περὶ οἷς πονεῖται. γράμματα γὰρ καὶ ταῦτα προσηγορεύετο, καθὰ καὶ δημόσια καλοῦμεν γράμματα, καὶ πολλῶν τινὰ γραμμάτων ἔμπειρον ὑπάρχειν φαμέν, τουτέστιν οὐ τῶν στοιχείων ἀλλὰ
48 τῶν συγγραμμάτων. καὶ Καλλίμαχος δέ, ποτὲ μὲν τὸ ποίημα καλῶν γράμμα ποτὲ δὲ τὸ καταλογάδην σύγγραμμα, φησὶ

Κρεωφύλου πόνος εἰμί, δόμῳ ποτὲ θεῖον ἀοιδὸν
δεξαμένου, κλείω δ᾽ Εὔρυτον ὅσσ᾽ ἔπαθεν
καὶ ξανθὴν Ἰόλειαν, Ὁμήρειον δὲ καλοῦμαι
γράμμα. Κρεωφύλῳ, Ζεῦ φίλε, τοῦτο μέγα.

[1] ⟨αὕτη⟩ addidi.

In both cases the sense is evidently derived from the original meaning of the term " letters " ; for the first sense comes from the letters by which we represent the articulate sounds, and the second perhaps, as some have maintained, from the first by extension ; for the first is a part of it, and just as *iatrikê* (medical art) was so named in olden times from the extraction of *ioi* (poisons), but is now applied to the curative treatment of all other ailments as well, which is a much higher form of art,—and just as geometry 46
derived its name originally from the mensuration of *gê* (land), but is at present applied also to theory concerned with more fundamental entities,—so also the perfected Art of letters, after being named originally from the knowledge of letters, was extended so as to comprise acquaintance with the more complex and technical theories about them. But perhaps, as 47
Asclepiades says, this art, too, was named from letters, though not from the same letters as " grammatistic " ; for while the latter takes its name, as I said, from the elements, the former takes its name from the compositions with which it deals. For these, too, were called " letters," just as we speak of " public letters," and say that a man is skilled in many letters, meaning not the elements but the compositions. Callimachus,[a] too, calling the poem 48
in one place, and in another the prose composition, a letter, says—

45

Work am I of Creōphylus who welcomed once to his homestead
Homer, the singer divine. Woeful the tale I relate
How fared blond Ioleia and Eurytus. But they do call me
Homer's letter! By Jove! fame for the author is this!

[a] Callim. *Epigr.* 7 and 25.

καὶ πάλιν

εἴπας " ἥλιε χαῖρε " Κλεόμβροτος Ἀμπρακιώτης
ἥλατ' ἀφ' ὑψηλοῦ τείχεος εἰς ἀίδην,
ἄξιον οὐδὲν ἰδὼν θανάτου τέλος, ἀλλὰ Πλάτωνος
ἓν τὸ περὶ ψυχῆς γράμμ' ἀναλεξάμενος.

49 Πλὴν διττῆς οὔσης γραμματικῆς, τῆς μὲν τὰ στοιχεῖα καὶ τὰς τούτων συμπλοκὰς διδάξειν ἐπαγγελλομένης καὶ καθόλου τέχνης τινὸς οὔσης τοῦ γράφειν τε καὶ ἀναγινώσκειν, τῆς δὲ βαθυτέρας παρὰ ταύτην δυνάμεως, οὐκ ἐν ψιλῇ γραμμάτων γνώσει κειμένης ἀλλὰ κἀν τῷ ἐξετάζειν τὴν εὕρεσιν αὐτῶν καὶ τὴν φύσιν, ἔτι δὲ τὰ ἐκ τούτων συνεστῶτα λόγου μέρη καὶ εἴ τι τῆς αὐτῆς ἰδέας θεωρεῖται, πρόκειται νῦν ἀντιλέγειν οὐ τῇ προτέρᾳ· συμφώνως γὰρ κατὰ πάντας ἐστὶ χρειώδης. ἐν οἷς θετέον καὶ τὸν Ἐπίκουρον, εἰ καὶ δοκεῖ τοῖς ἀπὸ τῶν μαθημάτων διεχθραίνειν· ἐν γοῦν τῷ περὶ δώρων καὶ χάριτος ἱκανῶς πειρᾶται διδάσκειν ὅτι ἀναγκαῖόν ἐστι τοῖς σοφοῖς μανθάνειν γράμματα.
50 καὶ ἄλλως, εἴπαιμεν ἂν ἡμεῖς, οὐ σοφοῖς μόνον ἀλλὰ καὶ πᾶσιν ἀνθρώποις. ὅτι γὰρ πάσης τέχνης
51 τὸ τέλος εὔχρηστόν ἐστι τῷ βίῳ, φανερόν. τῶν δὲ τεχνῶν αἱ μὲν προηγουμένως ὑπὲρ τῆς τῶν ὀχληρῶν ἐκκλίσεως παρῆλθον, αἱ δὲ ὑπὲρ τῆς τῶν ὠφελίμων εὑρέσεως. καὶ ἔστι τῆς μὲν πρώτης ἰδέας ἰατρική, παιωνὶς οὖσα καὶ λυσίπονος τέχνη, τῆς δὲ δευτέρας κυβερνητική· τῆς γὰρ ἀπὸ τῶν ἄλλων ἐθνῶν χρείας μάλιστα δέονται πάντες
52 ἄνθρωποι. ἐπεὶ οὖν ἡ γραμματιστικὴ διὰ τῆς τῶν γραμμάτων ἐπινοίας ἰᾶται μὲν ἀργότατον πάθος, τὴν λήθην, συνέχει δὲ ἀναγκαιοτάτην

And again—

Crying "Farewell, O Sun!" Cleombrotus, born in Ambracia,
Leapt from the lofty wall, down to the house of the dead;
Naught had befallen him worthy of death, 'twas his reading of Plato
Mov'd him so—Plato's letter "Concerning the Soul."

However, as the Art of Letters is twofold,—the 49
one promising to teach the elements and their combinations and being in general an art of writing and reading, and the other being in comparison a more profound faculty and not consisting merely in the bare knowledge of letters but also in the investigation of their discovery and their nature, and in addition the parts of speech composed of letters and all other matters of the same kind—it is not our purpose now to controvert the former; for that it is useful is agreed by all men. And amongst them we must place Epicurus, although he seems to be bitterly hostile to the Professors of Arts and Sciences; certainly in his book *On Gifts and Gratitude* he definitely tries to prove that it is necessary for the wise to learn letters. Necessary moreover, as we 50
should say, not for the wise only but for all men. For it is plain that the end aimed at by every art is very useful for life. Some arts have been introduced 51
mainly with the object of averting things hurtful, others with that of discovering things beneficial; medicine is an example of the first kind, being a curative and pain-relieving art, and navigation of the second, for all men are very much in need of the assistance of the other nations. Since then "gramma- 52
tistic" by its comprehension of letters cures a most inactive disease, forgetfulness, and contains a most

ἐνέργειαν, τὴν μνήμην, τὰ πάντα ἐπ' αὐτῇ κεῖται σχεδόν, καὶ οὔτε ἄλλους τι ἔνεστι τῶν ἀναγκαίων διδάσκειν, οὔτε παρ' ἄλλου μαθεῖν τι τῶν λυσιτελῶν χωρὶς αὐτῆς δυνατὸν ἔσται. οὐκοῦν τῶν χρησι-
53 μωτάτων ἡ γραμματιστική. ἀμέλει γοῦν οὐδὲ θελήσαντες ταύτην δυνησόμεθα ἀπεριτρέπτως ἀνελεῖν· εἰ γὰρ αἱ ἄχρηστον διδάσκουσαι τὴν γραμματιστικὴν ἐπιχειρήσεις εἰσὶν εὔχρηστοι, οὔτε δὲ μνημονευθῆναι οὔτε τοῖς αὖθις παραδοθῆναι χωρὶς αὐτῆς δύνανται, χρειώδης ἐστὶν ἡ γραμματιστική. καίτοι δόξαι ἂν τισιν ἐπὶ τῆς ἐναντίας εἶναι προλήψεως ὁ προφήτης τῶν Πύρρωνος λόγων Τίμων ἐν οἷς φησὶ

γραμματική, τῆς οὔ τις ἀνασκοπὴ οὐδ' ἀνάθρησις
ἀνδρὶ διδασκομένῳ Φοινικικὰ σήματα Κάδμου.

54 οὐ μὴν οὕτως ἔχειν φαίνεται. τὸ γὰρ ὑπ' αὐτοῦ λεγόμενον οὐκ ἔστι τοιοῦτον κατ' αὐτῆς τῆς γραμματιστικῆς, καθ' ἣν διδάσκεται τὰ Φοινικικὰ σήματα Κάδμου, τὸ " οὐδεμία ἐστὶν ἀνασκοπὴ οὐδ' ἀνάθρησις "· πῶς γάρ, εἰ διδάσκεταί τις αὐτήν, οὐδεμίαν ἔσχηκεν ἐπιστροφὴν αὐτῆς; ἀλλὰ μᾶλλον τοιοῦτό φησι " διδαχθέντι τὰ Φοινικικὰ σήματα Κάδμου οὐδεμιᾶς ἄλλης παρὰ τοῦτό ἐστι γραμματικῆς ἐπιστροφή," ὅπερ καταστρέφει οὐκ εἰς τὸ ἀχρηστεῖν ταύτην τὴν ἐν τοῖς στοιχείοις καὶ τῷ δι' αὐτῶν γράφειν τε καὶ ἀναγινώσκειν θεωρου-
55 μένην, ἀλλὰ τὴν πέρπερον καὶ περιεργοτέραν. ἡ μὲν γὰρ τῶν στοιχείων χρῆσις ἤπειγεν εἰς τὴν τοῦ βίου διεξαγωγήν, τὸ δὲ μὴ ἀρκεῖσθαι τῇ ἐκ τῆς παρατηρήσεως τούτων παραδόσει, προσεπιδεικνύναι δὲ ὡς τάδε μέν ἐστι φωνάεντα τῇ

necessary activity, memory, almost everything depends upon it, and without it it is impossible to teach any necessary thing to others, and it will be impossible to learn anything profitable from another. Thus the "grammatistic" is one of the most useful arts. And in any case even if we wished we should 53
not be able to abolish it without upsetting ourselves; for if the arguments which show that "grammatistic" is useless are themselves useful but can neither be remembered nor passed on to posterity without it, then "grammatistic" is useful. Yet it might be thought by some that Timon, the expounder of Pyrrho's views, is of the contrary opinion when he says—

Grammar's an art that a man need neither heed nor consider
When he is still being taught the Punic symbols of Cadmus.

This, however, does not appear to be the case. For the 54
phrase he uses, "he need neither heed nor consider," is not aimed against the actual "grammatistic" by which the Punic symbols of Cadmus are taught; for if a man is being taught it, how can he have paid no attention to it? What he means is rather this,—"when a man has been taught the Punic symbols of Cadmus he need pay no attention to any further art of letters"; and this does not refer to the uselessness of the art which is found to deal with the elements and with employing them in writing and reading, but of that which is boastful and needlessly inquisitive. For 55
while the handling of the elements contributes to the conduct of life, not to be contented with what is given by the observation of the elements and attempting further to show that some of them are naturally

φύσει τάδε δὲ σύμφωνα, καὶ τῶν φωναέντων τὰ μὲν φύσει βραχέα τὰ δὲ μακρὰ τὰ δὲ δίχρονα καὶ κοινὰ μήκους τε καὶ συστολῆς, καὶ καθόλου τὰ λοιπὰ περὶ ὧν οἱ τετυφωμένοι τῶν γραμματικῶν
56 διδάσκουσιν ⟨ἄχρηστόν ἐστιν.⟩[1] ὥστε τῇ μὲν γραμματιστικῇ πρὸς τῷ μηδὲν ἐγκαλεῖν ἔτι καὶ τὰς ἀνωτάτω χάριτας ὀφείλομεν, τῇ δὲ λειπομένῃ προσάπτομεν τοὺς ἐλέγχους. τὸ δὲ εἴτε ὑγιῶς εἴτε τοὐναντίον μάθοιμεν ἂν προσεξαπλώσαντες αὐτῆς τὸν χαρακτῆρα.

Γ'.—ΤΙ ΕΣΤΙ ΓΡΑΜΜΑΤΙΚΗ

57 Ἐπεὶ οὔτε ζητεῖν οὔτε ἀπορεῖν ἔστι κατὰ τὸν σοφὸν Ἐπίκουρον ἄνευ προλήψεως, εὖ ἂν ἔχοι πρὸ τῶν ὅλων σκέψασθαι τί τ' ἔστιν ἡ γραμματική, καὶ εἰ κατὰ τὴν ἀποδιδομένην ὑπὸ τῶν γραμματικῶν ἔννοιαν δύναται συστατόν τι καὶ ὑπαρκτὸν νοεῖσθαι μάθημα. Διονύσιος μὲν οὖν ὁ Θρᾷξ ἐν τοῖς παραγγέλμασί φησι "γραμματική ἐστιν ἐμπειρία ὡς ἐπὶ τὸ πλεῖστον τῶν παρὰ ποιηταῖς τε καὶ συγγραφεῦσι λεγομένων," συγγραφεῖς καλῶν, ὡς ἔστιν ἐκ τῆς πρὸς τοὺς ποιητὰς ἀντεμφάσεως πρόδηλον, οὐκ ἄλλους τινὰς ἢ τοὺς καταλογάδην
58 πραγματευσαμένους. τά τε γὰρ παρὰ τοῖς ποιηταῖς ὁ γραμματικὸς ἑρμηνεύειν φαίνεται, καθάπερ Ὁμήρῳ τε καὶ Ἡσιόδῳ Πινδάρῳ τε καὶ Εὐριπίδῃ καὶ Μενάνδρῳ καὶ τοῖς ἄλλοις, τά τε παρὰ τοῖς συγγραφεῦσιν, οἷον Ἡροδότῳ καὶ Θουκυδίδῃ καὶ
59 Πλάτωνι, ὡς ἴδιον ἔργον μετέρχεται. παρὸ καὶ οἱ χαρίεντες ἐξ αὐτῶν περὶ πολλῶν ἐπραγματεύ-

[1] ⟨ἄχρηστόν ἐστιν.⟩ addidi.

vowels, others consonants, and that of the vowels some are naturally short, others long, others doubtful and indifferently long or short, and in general all the other rules that are taught by the conceited Grammarians ⟨are unprofitable proceedings⟩. Hence, 56
while we not only have no fault to find with "Grammatistic" but even owe it the warmest thanks, we bring our critical weapons to bear on the rest of the Art of Letters. Whether we do so rightly or wrongly, we shall learn when we have further explained its character.

Chapter III.—A Description of "The Art of Grammar"

Since, according to the sage Epicurus, it is not 57
possible either to inquire or to doubt without a preconception, it will be well first of all to consider what "Grammatic" (or the Art of Grammar) is and whether a systematic and real art or science can be conceived on the lines of the conception put forward by the Grammarians. Now Dionysius "the Thracian"[a] says in his *Directions* that "Grammar" is mainly expertness regarding the language of poets and composers," meaning by "composers" (as is plain from its contrast with "poets") none others than the writers in prose. For the Grammarian 58
appears to interpret the writings of the poets, such as Homer and Hesiod, Pindar and Euripides and Menander and the rest; and he also investigates those of the composers, such as Herodotus and Thucydides and Plato, as being his proper task. Accordingly, 59
some of their notable figures have dealt with many of

[a] Famous grammarian, who taught at Rome *circ.* 80 B.C.

σαντο συγγραφέων, τοῦτο μὲν ἱστορικῶν τοῦτο δὲ ῥητορικῶν καὶ ἤδη φιλοσόφων, ζητοῦντες τίνα τε δεόντως καὶ ἀκολούθως ταῖς διαλέκτοις εἴρηται καὶ τίνα παρέφθαρται, τί τε σημαίνει παρὰ μὲν Θουκυδίδῃ λόγου χάριν τὸ ζάγκλον καὶ[1] τορνεύοντες, παρὰ δὲ Δημοσθένει τὸ "ἐβόα ὥσπερ ἐξ ἁμάξης," ἢ πῶς ἀναγνωστέον παρὰ Πλάτωνι τὴν ηδος λέξιν, πότερον ψιλῶς ἐκφέροντα τὴν πρώτην συλλαβὴν ἢ δασέως, ἢ τὴν μὲν πρώτην ψιλῶς τὴν δὲ δευτέραν δασέως, ἢ ἀμφοτέρας ψιλῶς ἢ ἐν-
60 αλλάξ. διὰ γὰρ τὰ τοιαῦτα εἴρηται ἡ γραμματικὴ ἐμπειρία τῶν παρὰ ποιηταῖς τε καὶ συγγραφεῦσι λεγομένων.

Οὗτος μὲν οὖν οὕτως· ἐγκαλεῖ δὲ αὐτῷ Πτολεμαῖος ὁ περιπατητικὸς ὅτι οὐκ ἐχρῆν ἐμπειρίαν
61 εἰρηκέναι τὴν γραμματικήν (αὐτὴ μὲν γὰρ ἡ ἐμπειρία τριβή τίς ἐστι καὶ ἐργάτις ἄτεχνός τε καὶ ἄλογος, ἐν ψιλῇ παρατηρήσει καὶ συγγυμνασίᾳ κειμένη, ἡ δὲ γραμματικὴ τέχνη καθέστηκεν), οὐ συνορῶν ὅτι τάττεται μὲν καὶ ἐπὶ τέχνης τοὔνομα, καθὼς ἐν τοῖς ἐμπειρικοῖς ὑπομνήμασιν ἐδιδάξαμεν, ἀδιαφόρως τοῦ βίου τοὺς αὐτοὺς ἐμπείρους τε καὶ τεχνίτας καλοῦντος, ἀφ' ἧσπερ ἐννοίας καὶ ὁ Μητρόδωρος ἔφη μηδεμίαν ἄλλην πραγμάτων ἐμπειρίαν τὸ ἑαυτῆς τέλος συνορᾶν ἢ φιλοσοφίαν,
62 τουτέστι μηδεμίαν τέχνην, τάττεται δὲ ἐξόχως καὶ ἐπὶ τῆς τῶν πολλῶν καὶ ποικίλων πραγμάτων γνώσεως, καθὼς καὶ τοὺς πρεσβύτας πολλὰ μὲν

[1] ⟨παρὰ Πλάτωνι⟩ add. cj. Warmington.

[a] "*Zanklon*" occurs in Thucyd. vi. 4; but "*torneuontes*" is not found in Thucydides.

[b] *Cf.* Dem. *de Corona* 122. "From a waggon," *i.e.* like

the prose-writers, whether historians or orators or even philosophers, seeking to discover which of their writings are correctly and idiomatically expressed and which are faulty, and what, for example, is the meaning in Thucydides[a] of *zanklon* (" sickle ") and *torneuontes* (" rounding off "), and in Demosthenes[b] of " he shouted as though from a waggon " ; and how we should pronounce the word *êdos*[c] in Plato—whether we should utter the first syllable with or without the aspirate, or the first syllable without and the second with the aspirate, or both without the aspirate or both with it. It is because of such investi- 60
gations that Grammar has been called " expertness regarding the language of poets and composers."

Such, then, is the view of Dionysius. But Ptolemy the Peripatetic raises the objection that he ought not to have described Grammar as " expertness " (for 61
expertness as such is a kind of practice and operation, void of art and reasoning, which consists in mere observation and joint-exercise, whereas Grammar is an art) ; but he fails to notice that the term is also applied to art (as we have shown in our *Empiric Treatise*[d]), since ordinary folk call the same men " experts " and " artists " without distinction ; and it was with this notion that Metrodorus[e] said that " philosophy is the only expertness in affairs which perceives its own end," meaning " the only art " ; 62
and the term is applied especially to the knowledge of many and various matters, in the way that we speak of old men who have seen much and heard

women crying insults from the waggons in which they rode to the Eleusinian Mysteries.

[c] *i.e.* ἦ δ' ὅς, " said he."

[d] This work is no longer extant.

[e] An Epicurean philosopher.

ἰδόντας πολλὰ δὲ ἀκούσαντας ἐμπείρους τοῦ βίου φαμέν, ὡς καὶ ὁ Εὐριπίδης

> ὦ τέκνον, οὐχ ἅπαντα τῷ γήραι κακά,
> Ἐτεόκλεες, πρόσεστιν, ἀλλ' ἡμπειρία
> ἔχει τι λέξαι τῶν νέων σοφώτερον.

63 ἐφ' ὅπερ ἴσως ὁ Θρᾷξ φερόμενος σημαινόμενον, ἐπεὶ πολυειδήμονά τινα καὶ πολυμαθῆ βούλεται εἶναι τὸν γραμματικόν, ἔφη ἐμπειρίαν ὑπάρχειν τὴν γραμματικὴν τῶν παρὰ ποιηταῖς τε καὶ συγγραφεῦσι λεγομένων. ὥστε τοῦτο μὲν ὑπέλαφρον, ἐκεῖνο δὲ ἴσως τις πραγματικωτέρας[1]
64 ἐχόμενον ζητήσεως ἐρεῖ πρὸς αὐτόν· ἤτοι γὰρ τῶν παρὰ ποιηταῖς τε καὶ συγγραφεῦσι λεγομένων μόνον ἐμπειρίαν συμβέβηκεν εἶναι τὴν γραμματικήν, ἢ καὶ τῶν μήτε παρὰ ποιηταῖς μήτε παρὰ συγγραφεῦσι καθεστώτων. ἀλλὰ μόνων μὲν τῶν παρὰ ποιηταῖς καὶ συγγραφεῦσιν οὐκ ἂν εἴποιεν ἐμπειρίαν αὐτὴν ὑπάρχειν διὰ τὸ ποτὲ καὶ ταῖς ἀνὰ χεῖρα τῶν ἰδιωτῶν καὶ ἀνεπιστημόνων ὁμιλίαις[2] ἐφιστῶσαν ⟨θεωρεῖσθαι⟩,[3] καὶ τὸ βάρβαρον καὶ τὸ Ἑλληνικὸν τό τε σόλοικον καὶ τὸ μὴ τοιοῦ-
65 τον ἐξελέγχουσαν. εἰ δὲ καὶ τῶν μὴ παρὰ ποιηταῖς μηδὲ συγγραφεῦσι μόνον λεγομένων ἐμπειρία καθέστηκεν, οὐκ ἔδει αὐτὴν εἶναι λέγειν, ⟨ὃ⟩ ἀπὸ μέρους ἔχει[4] συμβεβηκός.

Ἀλλὰ παρέντες τὸ περὶ τῶν τοιούτων λεπτολογεῖν σκοπῶμεν, ὡς ὑπεσχόμεθα, εἰ δύναται τέλος, ὅσον ἐπὶ τῇ τοιαύτῃ ἐννοίᾳ, ὑποστῆναι ἡ γραμ-
66 ματική. ὅταν οὖν λέγωσιν αὐτὴν ἐμπειρίαν κατὰ τὸ πλεῖστον τῶν παρὰ ποιηταῖς καὶ συγγραφεῦσι

[1] πραγματικωτέρας Heintz : γραμματικωτέρας MSS., Bekk.

much as "experienced in life." So too Euripides [a]—

Not all is evil that besets old age,
My child Eteocles: experience
Can teach a wisdom which eludes the young.

And the Thracian [b] was probably brought to adopt 63
this meaning of the term through wishing the Grammarian to be a man of wide knowledge and wide learning, when he says that grammar is expertness in the language of poets and composers. So that this is rather a small point. But here is an argument more suited to a serious inquiry which one might perhaps bring against him: it belongs to grammar either to 64
be expertness in the language of poets and composers only, or also to be expertness in speech which is not found in either poets or composers. But they will not say that it is expertness in poets' and composers' language only, since it is sometimes seen to preside over the current conversation of ordinary and unlearned people, and also to examine and declare what is barbaric and what Hellenic, what is solecistic and what not. And if it is not expertness in the 65
speech of poets and composers only, they ought not to have identified it with a part of its contents.

But without proceeding to split hairs about such matters, let us consider, as we undertook to do, whether, if we are to go by the proposed definition, the Art of Grammar can in the end exist. Now 66
when they describe it as "expertness regarding most of the speech of poets and composers," they mean

[a] Eurip. *Phoen.* 528; *cf.* Ovid, *Met.* vi. 28.
[b] *i.e.* Dionysius.

---

[2] ταῖς . . . ὁμιλίαις cj. Bekk.: τὰς . . . ὁμιλίας MSS., Bekk.
[3] ⟨θεωρεῖσθαι⟩ add. cj. Bekk.
[4] ⟨ὃ⟩ . . . ἔχει cj. Bekk.: ἔχειν MSS., Bekk.

λεγομένων, φασὶ πάντων ἢ τινῶν. καὶ εἰ πάντων, πρῶτον μὲν οὐκέτι κατὰ τὸ πλεῖστον ἀλλὰ πάντων, καὶ εἰ πάντων, καὶ τῶν ἀπείρων· ἄπειρα γάρ ἐστι ταῦτα. τῶν δὲ ἀπείρων οὐκ ἔστιν ἐμπειρία· διόπερ οὐδὲ γραμματική τις γενήσεται. εἰ δὲ τινῶν, ἐπεὶ καὶ οἱ ἰδιῶταί τινα τῶν παρὰ ποιηταῖς καὶ συγγραφεῦσι λεγομένων εἰδότες οὐκ ἔχουσι γραμματικὴν ἐμπειρίαν, οὐδὲ ταύτῃ[1] εἶναι λεκτέον
67 γραμματικήν. ἐκτὸς εἰ μή τι διὰ τοῦτο ⟨τὸ⟩[2] "κατὰ τὸ πλεῖστον" εἰρῆσθαι φήσουσιν, ἵνα ἥ τε πρὸς τὴν πάντων ⟨ἐμπειρίαν⟩ ἐνιαχοῦ ἀπορία[3] ἥ τε πρὸς τὸν ἰδιωτισμὸν διαφορὰ ὑποβάλληται. τοῦ μὲν γὰρ ἰδιώτου διενήνοχεν ὁ γραμματικὸς παρόσον οὐκ ὀλίγων, ὡς ἐκεῖνος, ἀλλὰ πλείστων τῶν παρὰ ποιηταῖς τε καὶ συγγραφεῦσι λεγομένων ἔμπειρός ἐστι· τῆς δὲ τῶν πάντων γνώσεως ἀδυνάτου τάχα καθεστώσης κεχώρισται, ἐπεὶ οὐ πάντα τὰ δὲ πλεῖστα ἐξ αὐτῶν ἐπαγγέλλεται γινώσκειν.
68 ταῦτα δὲ οὐκ ἀπολογουμένου ἦν, ἀλλὰ κακοῖς ἐπιπληροῦντος κακὰ καὶ μηκέτι μετρίως ἀλλ' ἄρδην ἐπισπωμένου τὰς ἀπορίας. πρῶτον μὲν οὖν ὡς τὰ πολλὰ ἀόριστά ἐστι καὶ τὴν σωρικὴν γεννᾷ ἀπορίαν, οὕτω καὶ τὰ πλεῖστα. ὅθεν ἢ περιγραψάτωσαν ἡμῖν αὐτά, δείξαντες ἄχρι πόσων γνώσεως[4] τῶν παρὰ ποιηταῖς τε καὶ συγγραφεῦσι λεγομένων ῥητέον· ἢ εἴπερ ἐπὶ ἀορίστου μένουσιν ὑποσχέσεως, τὰ πλεῖστα γινώσκειν λέγοντες,
69 παραδεχέσθωσαν τὴν παρὰ μικρὸν ἐρώτησιν. τοῦ

[1] ταύτῃ scripsi : ταύτην MSS., Bekk.
[2] ⟨τὸ⟩ add. cj. Bekk.
[3] ⟨ἐμπειρίαν⟩ . . . ἀπορία cj. Bekk : ἀπορίαν MSS., Bekk.
[4] γνώσεως Heintz : γνώσεων MSS., Bekk.

either all or some of it. And if they mean " all," then, in the first place, it is no longer " regarding most of their speech " but all of it, and if all then endless (for their speech is endless): but of the endless there is no experience: hence no Art of Grammar will exist. But if they mean " some," then, since even ordinary folk understand some of the speech of the poets and composers though they possess no grammatical expertness, neither in this case can one say that an Art of Grammar exists.
Unless they shall say that the words " regarding 67
most " were expressly used for the purpose of suggesting their occasional lack of expertness regarding all their language and their difference from the ordinary man. For the Grammarian differs from the ordinary man in so far as he is expert not in regard to a little, like the other, but to most of the language of the poets and composers; and he is not committed to what is probably impossible, the knowledge of it all, since he does not pretend to know all but most of
it. But this is not the conduct of a man making his 68
defence, but of one who piles evils on evils and draws upon himself difficulties to an extent that is no longer moderate but complete. Now in the first place, just as " many " is indefinite and gives rise to the puzzle of " the Heap,"[a] so also is " most." Either, then, let them define for us this " most," showing how far we are to say that this knowledge of the speech of poets and composers extends; or if they abide by their indefinite claim and say that they know " most," let them submit to being questioned " little by little."

[a] *Cf. P.H.* ii. 253 n.

γὰρ πλείστου ὁρισθέντος ἀριθμοῦ ὁ ἑνὶ ἐλάσσων
πλεῖστος ἀκμήν ἐστιν, ἐπεὶ τελέως ἄτοπον μονάδος
προσθέσει τὸν μὲν πλεῖστον λέγειν τὸν δὲ μηδαμῶς.
διόπερ ἀεὶ μονάδι πλεονεκτούμενος ὁ κατ' αὐτοὺς
πλεῖστος ἀριθμὸς ἐλεύσεται πάντως εἰς τὸ μηκέτι
ἀριθμὸς πλεῖστος ὑπάρχειν, καὶ διὰ τοῦτο μηδὲ
γραμματική[1]· ὅπερ ἦν τῆς σωρικῆς ἀπορίας συμ-
70 πέρασμα. πῶς δ' οὐκ ὄντως γραμματικῆς παχύ-
τητος τὸ ἐν ἀπείρῳ πλήθει λέγειν πλεῖστα; ὡς
γὰρ τὸ ὀλιγώτερον πρός τί ἐστι καὶ κατὰ τὴν
ὡς πρὸς τὸ πλεῖστον σχέσιν νοεῖται, οὕτω καὶ
τὸ πλεῖστον κατὰ τὴν ὡς πρὸς τὸ ὀλίγον σχέ-
σιν θεωρήσεται. εἰ οὖν τῶν πλείστων τῶν παρὰ
ποιηταῖς καὶ συγγραφεῦσι λεγομένων ἐμπειρίαν
ἔχουσιν οἱ γραμματικοί, ὀλίγων τῶν λοιπῶν οὐκ
71 ἔχουσιν· εἰ δὲ καὶ τὸ ληφθέν ἐστι πλεῖστον καὶ
τὸ καταλειφθὲν ἔλασσον, οὐκέτι τὸ πᾶν γίνεται
ἄπειρον. ὅμως δ' οὖν, ἵνα μηδὲν περὶ τούτων
ἀκριβευώμεθα, ψεῦδός ἐστι τὸ τὰ πλεῖστα τῶν
παρὰ ποιηταῖς τε καὶ συγγραφεῦσι λεγομένων
γινώσκειν τὸν γραμματικόν· ἐλάχιστα γὰρ ἦν,
πολλαπλασιόνων ἀπολειπομένων ἃ οὐκ οἶδε, καθὼς
72 προβαινούσης τῆς ζητήσεως παραστήσω. τὰ νῦν
δὲ ἄλλην ἀπόδοσιν θεωρητέον.

Ἀσκληπιάδης τοίνυν μέμφεται τὸν Θρᾷκα Διονύσιον ἐμπειρίαν λέγοντα τὴν γραμματικήν, δι' ἣν αἰτίαν καὶ ὁ Πτολεμαῖος ἔφη, ἐγκαλεῖ δὲ αὐτῷ καὶ τὸ[2] κατὰ τὸ πλεῖστον ἐμπειρίαν αὐτὴν ἀποφαίνειν. τοῦτο μὲν γὰρ τῶν στοχαστικῶν καὶ

[1] γραμματική Fabr. : γραμματικήν Bekk.
[2] τὸ cj. Bekk : τῷ mss., Bekk.

For when a " very large " number is fixed, the number 69
that is less by one is still " very large," since it is perfectly absurd to call the one number, because of the addition of an unit, " very large " and the other not. Hence, if their " very large " number keeps on always being increased by one it will certainly come to being no longer a " very large " number, so that Grammar too is non-existent; and this is the conclusion of the puzzle of " the Heap." And surely it is 70
a piece of *grammatic* obtuseness to speak of " very many " in dealing with an infinite number; for just as " small " is a relative notion and is conceived as in relation to the condition " very large," so also the notion " very large " will be conceived as relative to the condition " small." If, then, the Grammarians are expert regarding " very much of the speech of poets and composers," they are not expert regarding the small quantity which remains over; and if what is in- 71
cluded is " very much," and what is omitted is " rather small," the total no longer makes up an infinite quantity.—But in any case, not to argue these points too closely, it is false to say that the Grammarian knows " very much of the speech of poets and composers"; for it is really " very little," since many times more of it remains which he does not know, as I shall prove in the course of our inquiry. But for the moment 72
we must consider another account of the matter.

Asclepiades blames Dionysius " the Thracian " for calling the Art of Grammar " expertness," and that for the reason stated by Ptolemy, and he also finds fault with his description of it as " expertness for the most part." For this is a feature of arts which are conjectural

ὑπὸ τὴν τύχην πιπτουσῶν ἐστὶ τεχνῶν, ὥσπερ κυβερνητικῆς καὶ ἰατρικῆς· γραμματικὴ δὲ οὐκ ἔστι στοχαστικὴ ἀλλὰ μουσικῇ τε καὶ φιλοσοφίᾳ
73 παραπλήσιος. εἰ μή τι δέδοικε, φησί, τὴν ὀλιγότητα τοῦ βίου ὡς οὐκ οὖσαν ἱκανὴν πρὸς τὸ πάντα περιλαβεῖν, ὅπερ ἐστὶν ἄτοπον, ⟨ὡς⟩[1] γραμματικοῦ ἀλλ' οὐ γραμματικῆς ποιήσεται τὸν ὅρον, ἐπείπερ οὗτος μὲν τυχὸν ἴσως ⟨ὀλίγων⟩[2] ἐπιστήμων ἐστὶ τῶν παρὰ ποιηταῖς τε καὶ συγγραφεῦσι λεγομένων, ὀλιγόβιον καθεστὼς ζῷον, ἡ δὲ γραμματικὴ
74 πάντων εἴδησις. ὅθεν τὸ μὲν ἀλλάξας τοῦ ὅρου τούτου τὸ δ' ἀνελών, οὕτως ἀποδίδωσι τῆς γραμματικῆς τὴν ἔννοιαν "γραμματική ἐστι τέχνη τῶν παρὰ ποιηταῖς καὶ συγγραφεῦσι λεγομένων." οὐκ ἀνεῖλε δὲ ὁ ἀνὴρ τὰς ἀπορίας ἀλλ' ἐπέτεινεν· καὶ ἐν οἷς θέλει τὴν γραμματικὴν αὔξειν, ἐν τούτοις αὐτῆς ἀφεῖλεν.[3] ἔστω γὰρ πάντων εἴδησις τῶν παρὰ ποιηταῖς καὶ συγγραφεῦσι λεγομένων. οὐκοῦν ἐπεὶ οὐδέν ἐστιν εἴδησις παρὰ τὸν εἰδότα, οὐδὲ γραμματικὴ παρὰ τὸν [εἰδότα] γραμματικόν, ὡς οὐδὲ περιπάτησις παρὰ τὸν περιπατοῦντα καὶ στάσις παρὰ τὸν ἑστῶτα καὶ κατάκλισις παρὰ
75 τὸν κατακείμενον. ὡμολόγηται δὲ ὁ γραμματικὸς μὴ ἔχειν πάντων εἴδησιν· οὐκ ἄρα ἔτι ἔστιν εἴδησις πάντων τῶν παρὰ ποιηταῖς καὶ συγγραφεῦσι λεγομένων, διὰ δὲ τοῦτο οὐδὲ γραμματική. καὶ ἄλλως, εἴπερ τέχνη ἐστὶν ἡ γραμματική, εἴδησις οὖσα πάντων τῶν παρὰ ποιηταῖς τε καὶ συγγραφεῦσι λεγομένων, ἡ δὲ τέχνη σύστημα ἐκ καταλήψεων [τῶν περὶ τὸν γραμματικόν],[4] ἐξ ἀνάγκης μηδενὸς ἔχοντος πάντων τῶν παρὰ ποιηταῖς καὶ

[1] ⟨ὡς⟩ add. cj. Bekk.
[2] ⟨ὀλίγων⟩ add. cj. Bekk.

and subject to accidents such as navigation and medicine; but Grammar is not a conjectural art but akin
to Music and Philosophy. "Unless," he says, "he has 73
some fear of the shortness of life as being insufficient for the comprehension of all, which is absurd; for in that case he will not be defining the Art of Grammar but the Grammarian, since he perchance, being a short-lived creature, is acquainted with a little of the speech of the poets and composers, whereas Grammar
is knowledge of it all. Hence, by altering one part and 74
deleting another part of this definition, he presents the concept of Grammar in this form: "Grammar is the art dealing with the speech of poets and composers." But instead of removing the difficulties, our critic has increased them, and by the very means whereby he intends to magnify Grammar, he makes it less. For let it be granted that it is knowledge of all the speech of poets and composers: then, since knowledge is nothing apart from him who knows, neither is Grammar anything apart from the Grammarian, just as walking is nothing apart from the walker, or standing apart from him who stands, or lying apart from him
who lies. But it has been agreed that the Grammarian 75
does not possess knowledge of all; therefore knowledge of all the speech of poets and composers no longer exists; neither, in consequence, does Grammar exist.—Again, if Grammar is an art, being knowledge of all the speech of poets and composers, and art is "a system composed of apprehensions," [a] then, since no one possesses an apprehension of all the speech of

[a] For this Stoic definition *cf. P.H.* iii. 188.

[3] *αὐτῆς ἀφεῖλεν* scripsi: *αὐτὴν ἀνεῖλεν* MSS., Bekk. (*ἔλυσε* cj. Bekk.). [4] [*τῶν . . . γραμματικόν*] del. Heintz.

συγγραφεῦσι λεγομένων κατάληψιν ἀνύπαρκτος γίνεται ἡ γραμματική.

76 Χάρης δὲ ἐν τῷ πρώτῳ περὶ γραμματικῆς τὴν τελείαν φησὶ γραμματικὴν ἕξιν εἶναι ἀπὸ τέχνης διαγνωστικὴν τῶν παρ' Ἕλλησι λεκτῶν καὶ νοητῶν ἐπὶ τὸ ἀκριβέστατον, πλὴν τῶν ὑπ' ἄλλαις τέχναις, τὸ τελευταῖον προσθεὶς οὐ παρέργως.

77 ἐπεὶ γὰρ τῶν παρ' Ἕλλησι λεκτῶν καὶ νοητῶν τὰ μέν ἐστιν ὑπὸ τέχναις τὰ δ' οὔ, τῶν μὲν ὑπὸ τέχναις οὐκ οἴεται τέχνην εἶναι καὶ ἕξιν τὴν γραμματικήν, οἷον ἐν μὲν μουσικῇ τῆς διὰ τεσσάρων συμφωνίας καὶ τῆς μεταβολῆς τῶν συστημάτων, ἐν δὲ μαθηματικῇ ἐκλείψεως ἢ τῆς τῶν κύκλων θέσεως. τὰ δὲ αὐτὰ καὶ ἐπὶ τῶν ἄλλων νοητέον τεχνῶν· οὐδενὸς γὰρ τῶν ὑπ' αὐταῖς εἴδησις ἡ γραμματική, ἀλλὰ μέθοδός τίς ἐστι τῶν

78 παρὰ ταύτας ἑτέρων λεκτῶν τε καὶ νοητῶν, νοητῶν μὲν ὡς ὅτι πίσυρες τέσσαρες καὶ βῆσσαι καὶ ἄγκεα οἱ βάσιμοι τόποι, λεκτῶν δὲ τῶν περὶ τὰς διαλέκτους, οἷον ὅτι τοῦτο μὲν εἴρηται Δωρικῶς τοῦτο δ' Αἰολικῶς, καὶ οὐχ ᾗπερ οἱ στωικοὶ τὸ σημαινόμενον, ἀλλ' ἀνάπαλιν τὸ σημαῖνον· τὸ γὰρ νοητὸν ἐπὶ τοῦ σημαινομένου μόνου παρείληπται.

79 ἔοικε δὲ καὶ Κρατήτειόν τινα κινεῖν λόγον. καὶ γὰρ ἐκεῖνος ἔλεγε διαφέρειν τὸν κριτικὸν τοῦ γραμματικοῦ· καὶ τὸν μὲν κριτικὸν πάσης, φησί, δεῖ λογικῆς ἐπιστήμης ἔμπειρον εἶναι, τὸν δὲ γραμματικὸν ἁπλῶς γλωσσῶν ἐξηγητικὸν καὶ προσῳδίας ἀποδοτικὸν καὶ τῶν τούτοις παρα-

poets and composers, Grammar is of necessity non-existent.

Chares in the first book of his treatise on the Art of 76
Grammar says that perfect Grammar is " a skill derived from art which distinguishes very precisely Greek language and thought, except in so far as these are dealt with by other arts,"—which last addition of his is not superfluous. For since part of the Greek 77
language and thought is dealt with by arts and part not, he holds that Grammar is not an art and skill concerned with what is treated by the arts,—as for instance in Music the *dia-tessaron* (or " Fourth ") concord and the change of the " systems," and in Mathematic the eclipse or the position of the spheres. And the same applies, one must suppose, to the rest of the arts : Grammar is not knowledge of any of the subjects the other arts deal with, but it is a methodical treatment of language and meanings not touched on by other arts,—of meanings, as that *pisures* means 78
tessares (four), and that *bēssai* and *ankea* mean accessible places ; and of language in so far as it concerns dialects, as for example, that such a phrase is Doric and such other Aeolic, but it does not (like the Stoics) regard the signification of the phrase but conversely the signifying phrase itself ; for the meaning is derived from the thing signified alone.—But Chares 79
seems to be disputing an argument of Crates.[a] For he used to say that the " Critic " differs from the Grammarian ; and, says he, the " Critic " must be expert in the whole range of linguistic science, but the Grammarian simply capable of expounding dialects and explaining prosodies and skilled in things

[a] For Crates see p. 27 n.

πλησίων εἰδήμονα· παρὸ καὶ ἐοικέναι ἐκεῖνον μὲν ἀρχιτέκτονι τὸν δὲ γραμματικὸν ὑπηρέτῃ.

80 Ἀλλὰ τὰ μὲν τῆς ἀποδόσεως τοιαῦτα, πῇ μὲν μετριώτερα τῶν Διονυσίου ἀτοπημάτων πῇ δὲ χείρονα. ὅτι μὲν γὰρ τῆς σωρικῆς ἀπορίας ἐξέλυσε τὴν γραμματικὴν καὶ τῶν ἀλλοτρίων κεχώρικε θεωρημάτων, μουσικῆς τε καὶ μαθηματικῆς, ὡς μὴ προσηκόντων, αὐτόθεν συμφανές· τοῦ δὲ μὴ ἀνυπόστατον ὑπάρχειν οὐδαμῶς αὐτὴν ἐρρύσατο, ἀλλὰ καὶ εἰς τὸ εἶναι τοιαύτην μᾶλλον
81 συνηγωνίσατο. ὁ μὲν γὰρ Διονύσιος κατά τι διώρισε τὸν τῆς γραμματικῆς ὅρον, ἐπὶ μόνων αὐτὴν ποιητῶν τε καὶ συγγραφέων στήσας· οὗτος δὲ περὶ πᾶσαν Ἑλληνικὴν φωνὴν καὶ περὶ πᾶν σημαινόμενον καταγίγνεσθαι ταύτην θέλει. ὅπερ, εἰ θεμιτὸν εἰπεῖν, οὐδὲ θεοῖς ἀνυτόν ἐστιν. ὡς γὰρ καὶ πρότερον ἐλέγομεν, οὐδεμία μέθοδος συνίσταται περί τι ἄπειρον, ἀλλὰ καὶ μάλιστα αὕτη τοῦτο περατοῖ· τῶν γὰρ ἀορίστων ἡ ἐπιστήμη
82 δεσμός ἐστιν· τὰ δὲ σημαίνοντα καὶ σημαινόμενα τῶν πραγμάτων ἐστὶν ἄπειρα· οὐκ ἄρα ἐστὶν ἡ γραμματικὴ τέχνη περὶ τὰ σημαίνοντα καὶ σημαινόμενα. καὶ μὴν παντοῖαι γίνονται τῶν φωνῶν μεταβολαὶ καὶ πρὸ τοῦ γεγόνασι καὶ εἰσαῦθις γενήσονται· φιλομετάβολον γάρ τί ἐστιν ὁ αἰών, οὐκ εἰς φυτὰ μόνον καὶ ζῶα ἀλλὰ καὶ εἰς ῥήματα.
83 περὶ ἑστῶσαν δὲ ἀπειρίαν, οὔ τοί γε καὶ μεταβάλλουσαν, ἀμήχανόν ἐστι γνῶσιν ἀνθρωπίνην εὑρεῖν. οὐδὲ ταύτῃ ἄρα ἡ γραμματικὴ συστήσεται. ἄλλως τε ἤτοι τεχνικὴν οἴεται εἶναι τὴν ἕξιν ἢ ἄτεχνον. καὶ εἰ μὲν τεχνικήν, πῶς οὐκ

of that sort; and accordingly the former may be compared to a head-craftsman, and the Grammarian to a subordinate.

Such, then, are the features of this definition, which 80
are in some ways more reasonable than the absurdities of Dionysius, but in others worse. For it is evident at once that he has freed Grammar from the "Heap" difficulty and has set it apart from the alien disciplines of Music and Mathematic, on the ground thay they have no connexion with it; but he has by no means rescued it from the charge of being non-existent, but rather has helped to confirm that it is so.
For whereas Dionysius in some degree limited the 81
scope of Grammar by confining it to the speech of poets and composers only, this man (Chares) proposes to make it cover all Greek language and every signification,—a thing which, if one may say so, is not feasible even for gods. For, as we have also said before, no scientific inquiry deals with anything unlimited, such inquiry itself being the chief agent of limitation; for science is what binds down the
undefined: but things signifying and things signified 82
are unlimited; therefore the Art of Grammar is not concerned with things signifying and signified. Moreover, as to words, changes of all sorts occur and have occurred in the past and will occur in the future; for Time is a lover of change and that not only in regard
to animals and plants but also to words.[a] But con- 83
cerning a stationary infinite, not to speak of a changing one, no human knowledge can be found. Neither, then, in this way will Grammar subsist.—Further, either he supposes that the "skill" is artistic or that it is without art. And if it is artistic, why did he

[a] *Cf.* Hor. *A.P.* 60 ff.

αὐτὴν εἶπε τέχνην ἀλλὰ τὸ ἀφ' οὗ ἔστιν; εἰ δὲ ἄτεχνον, ἐπεὶ οὐ δυνατὸν διὰ τοῦ ἀτέχνου τὸ τεχνικὸν ὁρᾶσθαι, οὐδὲ συστήσεταί τις γραμματικὴ ἕξις τεχνικῶς διαγινώσκουσα τὰ παρ' Ἕλλησι σημαίνοντά τε καὶ σημαινόμενα.

84 Δημήτριος δὲ ὁ ἐπικαλούμενος Χλωρὸς καὶ ἄλλοι τινὲς τῶν γραμματικῶν οὕτως ὡρίσαντο "γραμματική ἐστι τέχνη τῶν παρὰ ποιηταῖς τε καὶ τῶν κατὰ τὴν κοινὴν συνήθειαν λέξεων εἴδησις." μένουσι δὲ καὶ τούτους αἱ αὐταὶ ἀπορίαι· οὔτε γὰρ πάντων τῶν παρὰ ποιηταῖς λεγομένων δύναται
85 εἶναι τέχνη γραμματικὴ οὔτε τινῶν. καὶ πάντων μὲν αὐτόθεν ἀδύνατον, εἴγε καὶ περὶ θεῶν καὶ περὶ ἀρετῆς καὶ ψυχῆς λέγεται παρὰ τοῖς ποιηταῖς, ὧν ἀπείρως ἔχουσιν οἱ γραμματικοί· τινῶν δὲ διὰ τὸ μὴ εἰς τοὺς γραμματικοὺς τὸ τοιοῦτον πίπτειν μόνους ἀλλὰ καὶ ἄλλους τινάς, οἷον φιλοσόφους καὶ μουσικοὺς καὶ ἰατρούς· συνεώρων γὰρ
86 καὶ οἶδε τινὰ τῶν παρὰ ποιηταῖς. πάλιν τε ἐν τῷ λέγειν καὶ τῶν κατὰ τὴν κοινὴν συνήθειαν λέξεων εἴδησιν εἶναι τὴν γραμματικὴν εἰ μὲν τὸ καθολικὸν λαμβάνοιεν τὸ "εἴ τινές εἰσι κατὰ τὴν κοινὴν συνήθειαν λέξεις, ἐκείνων ἐστὶν εἴδησις ἡ γραμματική," ἁμαρτάνουσιν· ἄπειροι γὰρ αἱ κατὰ τὴν κοινὴν συνήθειαν λέξεις, καὶ τῶν ἀπείρων οὐκ
87 ἔστιν εἴδησις. εἰ δὲ ἐπὶ τὸ ἐπὶ μέρους φέροιντο, ὅπερ ἴσον ἐστὶ τῷ "εἰσί τινες λέξεις κατὰ τὴν συνήθειαν τὴν κοινὴν ὧν εἴδησίς ἐστιν ἡ γραμματική," οὐδ' οὕτω ποιήσουσί τι τὴν γραμματικήν· καὶ γὰρ ὁ Ἀθηναῖος εἴδησιν ἔχει τῶν κατὰ τὴν

not term it " art " instead of that wherefrom it is derived? But if it is without art, then, since the artistic cannot be perceived by means of the non-artistic, neither will there be formed any " grammatic skill " which distinguishes by art what things are signifying and what signified in the speech of the Greeks.

Demetrius, surnamed Chlorus, and certain other 84
Grammarians have given this definition : " The Art of Grammar is knowledge of the forms of speech in the poets and also those in common usage." But for these, too, the same difficulties remain ; the Art of Grammar cannot deal either with all the forms of
speech in the poets or with some of them. That it 85
should deal with them all is sheerly impossible, since in the poets there is discourse concerning the gods and concerning virtue and the soul, things whereof the Grammarians have no expert knowledge. Nor can it deal with some of them, since a subject such as this falls within the scope not of Grammarians only but also of certain others, such as philosophers and musicians and physicians ; for these too have in-
quired into some of the diction of the poets. Again, 86
in their statement that Grammar is " knowledge of the forms of speech in common usage " as well, if they are taking it as the universal proposition " whatsoever forms of speech are in common usage, of these the Art of Grammar is knowledge," then they are in error ; for the forms of speech in common usage are infinite, and of infinites there is no know-
ledge. But if they should tend to state it as a par- 87
ticular proposition, equivalent to " there are certain forms of speech in common usage whereof Grammar is knowledge," not even so will they make Grammar anything real ; for the Athenian has knowledge of

Ἀτθίδα συνήθων λέξεων, καὶ ὁ Δωριεὺς τῶν κατὰ τὴν Δώριον, καὶ ὁ ῥήτωρ τῶν κατὰ τὴν ῥητορικήν,
88 καὶ ὁ ἰατρὸς τῶν κατὰ τὴν ἰατρικήν. εἰ δὲ λέγοιεν πασῶν τῶν κατὰ τὴν κοινὴν συνήθειαν αὐτὴν λέξεων εἴδησιν οὐχ ὡς τῶν καθ' ἕκαστα καὶ ἐν μέρει πασῶν (τοῦτο γὰρ ὄντως ἀδύνατον) ἀλλὰ τῶν καθόλου πασῶν καὶ ἀνωτάτω ἐν ταῖς διαλέκτοις, οἷον ὅτι Δωριέων μέν ἐστι τοιούτῳ τόνῳ χρῆσθαι Ἰώνων δὲ ἄλλῳ, τάχα μέν τι πιθανὸν ἐροῦσιν, οὐ
89 μὴν ἀληθές· οὔτε γὰρ ἓν ἔθος ἐστὶ καθ' ἑκάστην διάλεκτον (πολλαὶ γὰρ Δωρίδες καὶ Ἀτθίδες), οὔτε οἱ κανόνες οὓς δοκοῦσι παραδιδόναι πρὸς πᾶσαν ἀποτείνονται λέξιν, ἀλλ' ἄχρι μὲν ποσῶν καὶ ὁμοτόνων, οἷον ὀξυτόνων ἢ βαρυτόνων, προκόπτουσιν, πάσας δὲ περιλαβεῖν ἀδυνατοῦσιν.

90 Δείγματος μὲν οὖν χάριν ταῦτ' εἰρήσθω εἰς τὸ ἀνυπόστατον εἶναι τὴν γραμματικὴν ὅσον ἐπὶ τῇ παρὰ τοῖς γραμματικοῖς αὐτῆς ἐπινοίᾳ· μετελθόντες δὲ ἀκολούθως καὶ τὰ κυριώτατα τῶν ἐν αὐτῇ θεωρημάτων, καὶ ἐξ ὧν μάλιστα λαμβάνει τὴν ὑπόστασιν, βασανίζωμεν.

Δ'.—ΤΙΝΑ ΜΕΡΗ ΓΡΑΜΜΑΤΙΚΗΣ

91 Πολλῆς οὔσης καὶ ἀνηνύτου παρὰ τοῖς γραμματικοῖς περὶ μερῶν γραμματικῆς διαστάσεως, ἵνα μήτε τὸ πάρεργον ἔργου χώραν ἐπέχειν ἡμῖν φαίνηται, μήτε εἰς ἀλλοτρίαν καὶ ὡς πρὸς τὸ παρὸν ἀνωφελῆ ἐμβαίνοντες ὕλην ἀπολειπώμεθα τῆς ἀναγκαιοτέρας ἀντιρρήσεως, ἀπαρκέσει λέγειν ἀσ-

the forms of speech commonly used in Attica, and the Dorian of those in Dorian lands, and the orator of those used in rhetoric, and the physician of those in
medicine. And if they should mean by this " know- 88
ledge of all the forms of speech in common usage," not that of all the forms taken separately and individually (for this is actually impossible), but knowledge of all those in the dialects which are universal and most widely used, as for instance that it is the practice of the Dorians to use one sort of pitch and the Ionians another, then perhaps what they say will
be plausible, yet it will not be true ; for neither is 89
there one customary practice in each dialect (for both the Dorian and the Attic dialects are numerous), nor do the rules which they suppose they are handing on extend to every form of speech but only go so far as to apply to a certain number and those with similar pitch, such as oxytones and barytones, and they are unable to include them all.

Let so much, then, be said by way of sample to 90
show the baselessness of the Art of Grammar, judging by the definition of it given by the Grammarians. Next let us investigate and test the most important of its theorems and those from which it mainly derives its constitution.

Chapter IV.—The Parts of Grammar

Since there exists amongst the Grammarians great 91
and endless dissension concerning the parts of the Art of Grammar, in order that we should not allow side-issues to seem to usurp the place of the main issue, nor be diverted from the more necessary mode of refutation by plunging into matter which is neither pertinent nor useful for our present purpose, it will

κοφαντητότερον ὡς ἄρα τῆς γραμματικῆς τὸ μέν ἐστιν ἱστορικὸν τὸ δὲ τεχνικὸν τὸ δὲ ἰδιαίτερον, δι' οὗ τὰ κατὰ τοὺς ποιητὰς καὶ συγγραφεῖς μεθ-
92 οδεύεται. ὧν τεχνικὸν μέν ἐστιν ἐν ᾧ περὶ τῶν στοιχείων καὶ τῶν τοῦ λόγου μερῶν ὀρθογραφίας τε καὶ ἑλληνισμοῦ καὶ τῶν ἀκολούθων διατάττονται, ἱστορικὸν δὲ ὅπου περὶ προσώπων οἱονεὶ θείων τε καὶ ἀνθρωπίνων καὶ ἡρωικῶν διδάσκουσιν, ἢ περὶ τόπων διηγοῦνται καθάπερ ὀρῶν ἢ ποταμῶν, ἢ περὶ πλασμάτων καὶ μύθων παραδιδόασιν ἢ εἴ τι
93 τῆς αὐτῆς ἰδέας ἐστίν. ἰδιαίτερον δ' ἐν ᾧ τὰ[1] κατὰ τοὺς ποιητὰς καὶ συγγραφεῖς ἐπισκοποῦσι, καθ' ὃ τὰ ἀσαφῶς λεγόμενα ἐξηγοῦνται, τά τε ὑγιῆ καὶ τὰ μὴ τοιαῦτα κρίνουσι, τά τε γνήσια ἀπὸ τῶν νόθων διορίζουσιν. ἀλλ' ὡς μὲν τύπῳ καὶ ὁλοσχερέστερον περιλαβεῖν, ταῦτά ἐστι τὰ τῆς
94 γραμματικῆς μέρη· νοητέον δὲ αὐτὰ οὐ κατ' εἰλικρίνειαν, οὐδ' ὡς ἄν τις εἴποι μέρη τοῦ ἀνθρώπου ψυχὴν[2] καὶ σῶμα. ταυτὶ μὲν γὰρ ὡς ἕτερα ὄντα ἀλλήλων νοεῖται, τὸ δὲ τεχνικὸν καὶ ἱστορικὸν καὶ τὸ περὶ τὰς ποιήσεις καὶ συγγραφὰς τῆς γραμματικῆς μέρη πολλὴν ἔχει συμπλοκὴν καὶ ἀνά-
95 κρασιν πρὸς τὰ λοιπά· καὶ γὰρ ἡ τῶν ποιητῶν ἐπίσκεψις οὐ χωρὶς τοῦ τεχνικοῦ καὶ ἱστορικοῦ γίνεται μέρους, καὶ ἑκάτερον τούτων οὐ δίχα τῆς τῶν ἄλλων παραπλοκῆς συνέστηκεν. ὥσπερ οὖν οἱ λέγοντες τῆς ἰατρικῆς μέρη δίαιταν χειρουργίαν φαρμακείαν οὕτω λέγουσιν ὡς πολλῆς οὔσης ἐν

[1] δ' ἐν ᾧ τὰ scripsi: δὲ τὸ MSS., Bekk. (δὲ ᾧ τὰ cj. Bekk.).
[2] ψυχὴν cj. Bekk.: ψυχὴ MSS., Bekk.

be sufficient to state without sophistry that one part of Grammar is "historical," another "technical," a third "special,"[a] in which last the writings of the poets and prose-writers are dealt with. Of these the 92
"technical" part is that in which they formulate the rules concerning the elements and the parts of speech and orthography and Greek idiom and consequential matters; and the "historical" is that wherein they give instruction regarding persons—divine persons, for example, and human, and heroic,—or else explain about places, such as mountains or rivers, or record fictions or legends, or anything else of that description. The "special" part is that in which they 93
examine the language of the poets and prose-writers, and in it they explain the obscurities in their language, and judge as to which of their expressions are sound and which unsound, and distinguish the genuine works from the spurious. Such then, to describe them in broad outline, are the parts of the Art of Grammar; but one should not conceive of them as 94
"parts" in the precise sense of that term, nor in the way that one speaks of the soul and body as "parts" of the man. For these latter are conceived as being distinct from each other, whereas the parts of Grammar,—the technical and historical and that which deals with poetry and prose,—are each closely interconnected and intermixed with the rest; for the 95
investigation of poetry is not separate from the "technical" and "historical" parts, nor is each of these devoid of mutual connexion with the other two. So, just as those who describe dieting, surgery and pharmacy as "parts" of the art of medicine, imply thereby that in their precepts there is much mutual

[a] Also termed "exegetical."

τοῖς θεωρήμασιν ἀλληλουχίας (καὶ γὰρ ἡ δίαιτα οὐ χωρὶς φαρμακείας καὶ χειρουργίας πρόεισι, καὶ ἡ φαρμακεία πάλιν περιείχετο καὶ τῇ τῶν ἄλλων δυνάμει), ὧδε καὶ ἐπὶ τοῦ παρόντος οὐκ ἀπότακτόν τί ἐστιν ἕκαστον μέρος, οὐδ' εἰλικρινὲς
96 ἀπὸ τῆς τῶν ἄλλων ἐπιμιξίας. τοῦτο δὲ προδιηρθρώσαμεν οὐ παρέργως, ἀλλ' ἵνα εἰδῶμεν ὡς ἂν ἕν τι ἐξ αὐτῶν δειχθῇ ἀσύστατον, δυνάμει καὶ τὰ λοιπὰ ἀνῄρηται, ὧν ἑκάτερον οὐ χωρὶς τοῦ ἀναιρεθέντος ὑφίσταται. ὅμως δὲ οὐ ποιήσομεν τοῦτο καίπερ ὂν σύντομον, ἀλλὰ πειρασόμεθα πρὸς ἕκαστον ἀντιλέγειν, ὡς εἰ καὶ μὴ ἔχρῃζε τῆς τῶν λοιπῶν παρουσίας. τάξει δὲ ἀρκτέον ἀπὸ τοῦ πρώτου.

Ε'.—ΠΕΡΙ ΤΟΥ ΤΕΧΝΙΚΟΥ ΤΗΣ ΓΡΑΜΜΑΤΙΚΗΣ ΜΕΡΟΥΣ

97 Διὰ πολλὰ μὲν καὶ ἄλλα δίκαιόν ἐστι μετὰ σπουδῆς ἐξετάζειν τὴν γραμματικὴν τεχνολογίαν, μάλιστα δὲ ἁπάντων διὰ τὸ ἐπ' αὐτῇ κομᾶν καὶ μέγα φρονεῖν τοὺς γραμματικούς, ἀεὶ δὲ τῶν κατὰ τὰ λοιπὰ κοσμουμένων μαθήματα κατατρέχειν ὡς μηδὲ τὴν κοινὴν τῶν Ἑλλήνων συνήθειαν ἐπισταμένων, καὶ ἔτι διὰ τό, εἴ ποτε θλίβοιντο ἐν ζητήσει, μὴ ἄλλην εὑρίσκειν πολλάκις ἀποφυγὴν εἰς τὸ περισπᾶν τοὺς συζητοῦντας αὐτοῖς ἢ τὸ ὅτι βάρβαρον ἢ σόλοικόν ἐστι τὸ ὑπ' αὐτῶν λεχθέν.
98 οὐκ ὀλίγην δὲ ἂν ἔχοι μοῖραν εἰς προτροπὴν καὶ ὅταν βλέπωμεν τοὺς μηδὲ δύο σχεδὸν ῥήματα δεξιῶς εἴρειν δυναμένους γραμματικοὺς θέλοντας

borrowing from one another (for dieting does not progress apart from pharmacy and surgery, and pharmacy in turn is involved in the functioning of the other " parts "), so also in the present instance each part of Grammar is not something quite separate and free from admixture with the others. This descrip- 96
tion we have given beforehand in full detail, not as a side-issue, but so that we may understand that if any one of these parts is proved to have no consistence the others too are virtually destroyed, since neither of them subsists when apart from the one which has been destroyed. We shall not, however, adopt this method, concise though it is, but shall try to refute each part as though it had no need of the presence of the rest. And, keeping the order, we must begin with the first part.

Chapter V.—Concerning the Technical Part of Grammar

While it is right for many other reasons to examine 97
with care the technology of Grammar, it is especially so because it is upon this that the Grammarians plume themselves and give themselves great airs, while they are always running down those who are highly reputed in other sciences as men who do not even understand the common usage of Greek speech ; and also for the further reason that whenever they are hard pressed in disputation they frequently find no other resort for the purpose of distracting their opponents than asserting that what they have said is a barbarism or a solecism. And what has no small 98
share in urging us on is our observing the Grammarians, who can hardly even join two sentences together skilfully, wishing to convict of barbarisms

ἕκαστον τῶν μέγα δυνηθέντων ἐν εὐφραδείᾳ καὶ ἑλληνισμῷ παλαιῶν, καθάπερ Θουκυδίδην Πλάτωνα Δημοσθένην, ὡς βάρβαρον ἐλέγχειν. μία γὰρ ἀντὶ πάντων ἄμυνα γενήσεται πρὸς αὐτούς, ἐὰν τὴν ψευδώνυμον αὐτῶν τεχνολογίαν ἄτεχνον ἀποδείξωμεν. τάξει δὲ λεκτέον ἡμῖν πρῶτον περὶ
99 τῶν στοιχείων, ἐξ ὧν τὰ πάντα κατ' αὐτοὺς συνέστηκεν καὶ ὧν ἀναιρεθέντων ἀγραμμάτους ἀνάγκη γίνεσθαι τοὺς γραμματικούς.

Καὶ δὴ τριχῶς λεγομένου τοῦ στοιχείου, τοῦ τε γραφομένου χαρακτῆρος καὶ τύπου καὶ τῆς τούτου δυνάμεως καὶ ἔτι τοῦ ὀνόματος, προαγέτω νῦν ἡ ζήτησις μάλιστα περὶ τῆς δυνάμεως· αὕτη γὰρ καὶ κυρίως στοιχεῖον παρ' αὐτοῖς προσηγόρευται. εἰκοσιτεσσάρων τοίνυν στοιχείων
100 ὄντων τῆς ἐγγραμμάτου φωνῆς, τούτων διττὴν ὑποτίθενται κατὰ τὸ ἀνωτάτω τὴν φύσιν. τὰ μὲν γὰρ αὐτῶν φωνάεντα προσαγορεύουσι, τὰ δὲ σύμφωνα, καὶ φωνάεντα μὲν ἑπτά, α ε η ι ο υ ω, σύμφωνα δὲ τὰ λοιπά. τῶν δὲ φωναέντων τρεῖς λέγουσι διαφοράς· δύο μὲν γὰρ αὐτῶν φύσει μακρὰ λέγουσι τυγχάνειν, τὸ η καὶ τὸ ω, ἰσάριθμα δὲ βραχέα, τὸ ε καὶ τὸ ο, τρία δὲ κοινὰ μήκους τε καὶ βραχύτητος, α ι υ, ἅπερ δίχρονα καὶ ὑγρὰ καὶ ἀμφίβολα καὶ μεταβολικὰ καλοῦσιν· ἕκαστον γὰρ
101 αὐτῶν πέφυκεν ὁτὲ μὲν ἐκτείνεσθαι ὁτὲ δὲ συστέλλεσθαι, οἷον τὸ μὲν α ἐπὶ τοῦ

Ἆρες Ἄρες βροτολοιγὲ μιαιφόνε τειχεσιπλῆτα,[a]

τὸ δὲ ι

Ἴλιον εἰς ἱερήν· τῇ δ' ἀντίος ὤρνυτ' Ἀπόλλων,[b]

---

[a] *Cf.* Hom. *Il.* v. 31. [b] *Cf.* Hom. *Il.* vii. 20.

every one of the ancients who were supreme masters
of correct language and Greek idiom, such as Thucy-
dides, Plato and Demosthenes. Against their attack
one defence will stand for all,—the proof, namely,
that their "technology" is non-technical. And first
in order we must discuss the "elements," wherefrom 99
their whole system is constructed, and the abolition
of which will necessarily bring it about that the
Grammarians are ungrammatic.

The term "element" being used in three senses,—
of the written character or form, and of its "func-
tion" (or sound), and also of the name,—let us
proceed now in our inquiry to deal mainly with the
function; for this is accounted by the Grammarians
to be the "element" proper. As, then, there are
twenty-four elements of the voice in written form, 100
they lay it down that the nature of these, in respect of
their *summa genera*, is twofold. The one class they
call vowels, the other consonants; and seven are
vowels—α ε η ι ο υ ω,—the rest being consonants.
And of the vowels there are, they say, three distinct
sorts: for they say that two of them are long, namely
η and ω, an equal number short, namely ε and ο, and
three indifferently long or short, namely α ι υ, which
they term "dichronous" (or having two quantities),
and "liquid" and "amphibolous" (or ambiguous)
and "metabolic" (or changeable); for each of them
is naturally capable of being at one time long, at 101
another short, as α in the example, Ἆρες Ἄρες—

Āres, Ăres, fort-wrecker and murderous slayer of mortals,[a]—

and ι in Ἴλιον εἰς ἱερήν,—

(So fared she)
Even to Holy Troy, and Apollo hastened to meet her,[b]—

τὸ δὲ υ

θεσπέσιον νεφέων ἐκ Διὸς ὗεν ὕδωρ.

102 τῶν δὲ συμφώνων τὰ μὲν ἡμίφωνά ἐστι κατ' αὐτοὺς τὰ δὲ ἄφωνα, καὶ ἡμίφωνα μὲν ὅσα δι' αὐτῶν ῥοῖζον ἢ σιγμὸν ἢ μυγμὸν ἤ τινα παραπλήσιον ἦχον κατὰ τὴν ἐκφώνησιν ἀποτελεῖν πεφυκότα, καθάπερ τὸ ζ θ λ μ ν ξ ρ σ φ χ ψ, ἢ ὥς τινές, χωρὶς τοῦ θ καὶ φ καὶ χ τὰ λειπόμενα ὀκτώ· ἄφωνα δέ ἐστι τὰ μήτε συλλαβὰς καθ' ἑαυτὰ ποιεῖν δυνάμενα μήτε ἤχων ἰδιότητας, αὐτὸ δὲ μόνον μετὰ τῶν ἄλλων συνεκφωνούμενα, καθάπερ
103 β γ δ κ π τ, ἢ ὥς ἔνιοι, καὶ τὸ θ φ χ. καὶ μὴν κοινῶς τῶν συμφώνων πάλιν τὰ μὲν φύσει δασέα λέγουσι τὰ δὲ ψιλά, καὶ δασέα μὲν θ φ χ, ψιλὰ δὲ κ π τ· μόνον δέ φασι τὸ ρ ἐπιδέχεσθαι ἑκάτερον, δασύτητα καὶ ψιλότητα. λέγουσι δέ τινα τῶν συμφώνων καὶ διπλᾶ, καθάπερ τὸ ζ ξ ψ· συνεστηκέναι γάρ φασι τὸ μὲν ζ ἐκ τοῦ σ καὶ δ, τὸ δὲ ξ ἐκ τοῦ κ καὶ σ, τὸ δὲ ψ ἐκ τοῦ π καὶ σ.

104 Τούτων δὴ προεστοιχειωμένων, φημὶ πρῶτον μὲν ἀτόπως αὐτοῖς λέγεσθαι τῶν στοιχείων τινὰ εἶναι διπλᾶ. τὸ γὰρ διπλοῦν σύστημά ἐστιν ἐκ δυοῖν, τὸ δὲ στοιχεῖον οὐκ ἔστι σύστημα ἐκ τινῶν· ἁπλοῦν γὰρ ὀφείλει τυγχάνειν καὶ οὐκ ἐξ ἑτέρων συστατόν. οὐκ ἄρα ἔστι διπλοῦν στοιχεῖον. ἄλλως τε, εἰ τὰ συστατικὰ τοῦ διπλοῦ στοιχείου στοιχεῖά ἐστι, τὸ διπλοῦν ἐκ τῶν στοιχείων συνεστὼς οὐκ ἔσται στοιχεῖον· ἀλλὰ μὴν τὰ συστατικὰ τοῦ διπλοῦ στοιχείου στοιχεῖα ἐστίν· οὐκ ἄρα τὸ διπλοῦν ἐστὶ στοιχεῖον.

---

[a] This pentameter comes, perhaps,—as Fabricius suggests—from a lost poem of Callimachus.

and the υ in ὗεν ὕδωρ—

Out of the Sky-god's clouds pourèd a marvellous flood.[a]

Of the consonants some, according to them, are "semi-vocal," others "non-vocal" (or mute): "semi-vocal" are all those which of themselves naturally produce at their utterance the *r*-sound or the *s*-sound or the *m*-sound, or some similar sound, for example ζ θ λ μ ν ξ ρ σ φ χ ψ, or, as some say, the eight which remain after the omission of θ φ and χ. "Non-vocal" are those which are not able by themselves to form syllables or sounds of their own, they being uttered only along with others, such as β γ δ κ π τ, or, as some say, with the addition of θ φ χ. Moreover, of the consonants again in general they say that some are naturally "rough" (or aspirated), others "smooth" (or without the aspirate), the "rough" being θ φ χ, and the "smooth" κ π τ; and ρ alone, they say, admits of either breathing, the "rough" or the "smooth." They declare also that some of the consonants are "twofold," such as ζ ξ ψ; for ζ, they say, is composed of σ and δ, ξ of κ and σ, ψ of π and σ. 102 103

104 These, then, being placed first as elements, I assert, firstly, that it is absurd of them to state that some of the elements are "twofold," for what is twofold is a compound of two things, but the element is not a compound of any things; for it must be simple and not compounded of other things. Therefore a twofold element does not exist. Again, if the components of the twofold element are elements, the twofold compound of those elements will not be an element; but in fact the components of the twofold element are elements; therefore the twofold (compound) is not an element.

105 Καὶ μὴν ὡς ταῦτα ἀναιρεῖται, οὕτω καὶ τὰ
δίχρονα, κοινὴν φύσιν μήκους τε καὶ βραχύτητος
ἀξιούμενα ἔχειν. εἴπερ γὰρ τοιαῦτα ἐστίν, ἤτοι
αὐτὸ τὸ γράμμα κατ' ἰδίαν καὶ ὁ ψιλός, εἰ τύχοι,
τοῦ α ι υ χαρακτὴρ ἐμφανιστικός ἐστι τῆς διχρόνου
φύσεως, καὶ νυνὶ μὲν συστέλλεσθαι νυνὶ δὲ ἐκτείνε-
106 σθαι δυνάμενον, ⟨ἢ σὺν[1]⟩ προσῳδίᾳ. ἀλλ' ὁ μὲν
χαρακτὴρ κατ' ἰδίαν οὐκ ἔστι κοινοῦ φύσει στοιχείου
μηνυτικός. οὔτε γὰρ ὅτι μηκύνεται οὔθ' ὅτι βρα-
χύνεται οὔθ' ὅτι τὸ συναμφότερον καὶ μηκύνεται
καὶ βραχύνεται ἐμφαίνει· ἀλλ' ὃν τρόπον ἡ δι'
αὐτοῦ συλλαβή, καθάπερ εἴρηται ἐπὶ τῆς Ἆρες λέ-
ξεως, οὐ[2] χωρὶς τῆς προστιθεμένης προσῳδίας οὔτ'
εἰ μακρά ἐστιν οὔτ' εἰ βραχεῖα γινώσκεται, οὕτω
καὶ τὸ α ι υ κατ' ἰδίαν λαμβανόμενα οὐ κοινὰ
107 ἑκατέρας ἔσται δυνάμεως ἀλλ' οὐδετέρας. λείπεται
οὖν σὺν προσῳδίᾳ λέγειν αὐτὸ κοινὸν ὑπάρχειν.
ὃ πάλιν ἐστὶν ἀμήχανον· προσλαμβάνον γὰρ
ταύτην ἢ μακρὸν γίνεται, ὅτε ἔστι μακρά, ἢ βραχύ,
ὅτε ἔστι βραχεῖα, κοινὸν δὲ οὐδέποτε. οὐκ ἄρα
108 ἔστι φύσει δίχρονα στοιχεῖα. εἰ δὲ λέγοιεν κοινὰ
φύσει ὑπάρχειν ταῦτα παρόσον ἐπιδεκτικά ἐστιν
ἑκατέρου, μήκους τε καὶ συστολῆς, λήσονται
σχεδὸν εἰς τὴν αὐτὴν ἐγκυλισθέντες ἀπορίαν. τὸ
γὰρ ἐπιδεκτικόν τινος οὐκ ἔσται ἐκεῖνο τὸ οὗπερ
ἐπιδεκτικόν ἐστιν· ὥσπερ γὰρ ὁ χαλκὸς ἐπιδεκτι-
κὸς μέν ἐστι τοῦ ἀνδριὰς γενέσθαι, οὐκ ἔστι δὲ
ἀνδριὰς ἐφ' ὅσον ἐπιδεκτικός ἐστι, καὶ ὃν τρόπον
τὰ ξύλα ἐπιτήδειον μὲν ἔχει φύσιν εἰς τὸ ναῦς

[1] ⟨ἢ σὺν⟩ addidi : τῇ mss., Bekk. (προσῳδία, ἢ σὺν προσῳδίᾳ κοινὸν ὑπάρχει cj. Bekk.).

[2] οὐ cj. Bekk. : ἢ mss., Bekk.

Moreover, just as these are destroyed, so also are 105
the "dichronous" elements which, as is claimed, are naturally "common" in respect of length and brevity. For if they are such, either the letter itself in itself and the mere character, of, say, the ε, ι, and υ, are indicative of their "dichronous" nature, and can be now shortened, now lengthened, or the "tension"[a]
must be added. But the character is not of itself indi- 106
cative of an element naturally "common," for it does not show clearly that it is long or that it is short or that it is at once both long and short; but just as it is not known about the syllable formed by it,—as we said in the case of the word Ἄρες—whether it is long or short apart from the added tension, so also the letters α ι and υ taken by themselves will not be "common"
and having either quantity, but will have neither. It 107
remains, then, to say that the letter is "common" when conjoined with tension. But this again is impossible; for when the letter takes the tension it becomes either long, when the tension is long, or short when it is short, but never "common." Hence,
elements "dichronous" by nature do not exist.—And 108
if they should say that these elements are "common" by nature inasmuch as they are susceptible of either quantity, length or shortness, they will unwittingly be involving themselves in much the same difficulty. For what is susceptible of a thing will not be that thing whereof it is susceptible; for just as the bronze is susceptible of becoming the statue, but inasmuch as it is susceptible is not the statue, and just as the timber possesses a nature suitable for becoming a

[a] See note on § 113.

γενέσθαι, οὔπω δὲ ἔστι ναῦς, οὕτω καὶ τὰ τοιαῦτα τῶν στοιχείων ἐπιδεκτικὰ μέν ἐστι μήκους τε καὶ συστολῆς, οὔτε δὲ μακρά ἐστιν οὔτε βραχέα οὔθ' ἑκάτερον πρὶν ἀπὸ προσῳδίας ποιωθῆναι.
109 πρός γε μὴν τοῖς λεχθεῖσιν ἐναντίον ἐστὶν ἥ τε βραχύτης καὶ ἡ ἐπέκτασις, καὶ οὐ συνυφίσταται· ἀναιρέσει γὰρ τῆς βραχύτητος ἡ ἐπέκτασις συνίσταται, καὶ ἀναιρουμένης μακρᾶς βραχεῖα γίνεται. παρ' ἣν αἰτίαν ἀδύνατον περισπωμένην βραχεῖαν γενέσθαι, διὰ τὸ τῷ περισπασμῷ κατ' ἀνάγκην
110 συνυφίστασθαι τὴν ἐπέκτασιν. διόπερ εἰ φύσει τι δίχρονόν ἐστι στοιχεῖον, ἤτοι ὑφ' ἓν περὶ αὐτὸ ἥ τε τῆς βραχύτητος καὶ ἡ τῆς ἐπεκτάσεως ὑποστήσεται δύναμις ἢ παρὰ μέρος. ἀλλ' ὑφ' ἓν μὲν ἀμήχανον· περὶ γὰρ τὴν αὐτὴν ἐκφώνησιν κατὰ τὸ αὐτὸ ἀναιρετικαὶ ἀλλήλων δυνάμεις οὐκ ἂν ὑποσταῖεν. λείπεται ἄρα παρὰ μέρος. ὃ πάλιν ἐστὶν ἀπίθανον· ὅτε γὰρ ἔστι ⟨βραχὺ καὶ ὅτε ἔστι⟩ μακρόν, τότε οὐκ ἔστι κοινὸν στοιχεῖον βραχύτητος καὶ μήκους, ἀλλὰ βραχὺ μόνον ⟨καὶ μακρὸν μόνον⟩.[1]
111 Ὁ δ' αὐτὸς τῆς ἐπιχειρήσεως τρόπος γινέσθω καὶ ἐπὶ τῶν φύσει ψιλῶν ἢ δασέων ἢ καθ' ἑκάτερον κοινῶν· ἡμῖν δὲ ἀπόχρη τὸ γένος τῆς ἐπιχειρήσεως ὑποδεῖξαι.

Καὶ μὴν ἐπεὶ ἀνῄρηται τὰ κοινὰ καὶ δέδεικται τὸ ἐκτείνεσθαι μόνον αὐτὰ ἢ συστέλλεσθαι, ἀκολουθήσει καὶ τὸ δισσὸν ὑπάρχειν ἕκαστον, τὸ
112 μὲν φύσει μακρὸν τὸ δ' αὖ φύσει βραχύ. δισσοῦ οὖν ὄντος τοῦ α καὶ ι καὶ υ οὐκέτι ἑπτὰ γενήσεται μόνον στοιχεῖα φωνάεντα, ὧν δύο μὲν μακρά, τό

[1] ⟨βραχὺ . . . ἔστι⟩ et ⟨καὶ μ. μόνον⟩ addidi (Bekk. cj. ⟨καὶ ὅτι βραχύ⟩ post μακρὸν et ⟨μακρὸν μόνον καὶ⟩ post ἀλλὰ).

ship, but is not as yet a ship, so also elements of this kind are susceptible of both length and shortness but are neither long nor short, nor either of the two, before they have been qualified by tension.—Moreover, in 109
addition to what has been said, shortness and length are contraries and do not exist together; for length exists by the abolition of shortness, and when a long is abolished a short comes into existence. And for this reason it is impossible for a circumflexed letter to be short, since the lengthening necessarily co-exists with the circumflex. Consequently, if an 110
element is "dichronous" by nature the quality of both shortness and length will subsist in it either both together or one at a time. But it is impossible for both qualities to subsist together, for qualities which are destructive of each other will not subsist in the same utterance simultaneously. It remains, then, to say that they subsist "one at a time." But this too is incredible; for when the element is short, or when it is long, it is not, at that time, "common" as regards shortness and length, but is solely short or solely long.

Let the same method of attack be employed in the 111
case of the elements which are "smooth" or "rough," or "common," as taking either breathing. We content ourselves here with merely indicating the kind of the attack.

And now that the "common" elements have been destroyed and it has been proved that they are only either shortened or lengthened, it will follow that each of them is twofold, the one sort naturally long, the other naturally short. As then α ι and υ are each 112
twofold, there will no longer be only seven "vocal" elements (or vowels), of which two are long, namely

τε η καὶ τὸ ω, δύο δὲ βραχέα, τό τε ε καὶ τὸ ο, τρία δὲ δίχρονα, τό τε α καὶ ι καὶ υ, ἀλλὰ τὰ σύμπαντα δέκα, καὶ τούτων τὰ πέντε μὲν μακρά, τό τε η καὶ τὸ ω καὶ τὸ μακρὸν α καὶ ι καὶ υ, ἰσάριθμα δὲ τὰ βραχέα, τὸ ο καὶ τὸ ε καὶ τὸ βραχὺ
113 α καὶ ι καὶ υ. ἀλλ' ἐπεὶ οὐ δύο μόνον ὑπειλήφασιν εἶναι προσῳδίας γραμματικῶν παῖδες, τήν τε μακρὰν καὶ βραχεῖαν, ἀλλὰ καὶ ὀξεῖαν βαρεῖαν περισπωμένην δασεῖαν ψιλήν, ἕκαστον τῶν ὑποδεδειγμένων φωναέντων ἔχον τινὰ τούτων κατ' ἰδίαν προσῳδίαν γενήσεται στοιχεῖον· καὶ ᾧ λόγῳ οὐδὲν ἦν κοινὸν μήκους τε καὶ βραχύτητος στοιχεῖον, ἀλλ' ἢ μακρὸν μόνον, ὅτ' εἶχε τὴν μακράν, ἢ βραχύ, ὅτ' εἶχε τὴν βραχεῖαν, τῷ αὐτῷ λόγῳ οὐδὲν ἔσται κοινὸν ὀξύτητος καὶ βαρύτητος, ἀλλ' ἢ ὀξὺ μόνον, ὅτε προσειλήφει τὴν ὀξεῖαν, ἢ βαρύ, ὅτε τὴν βαρεῖαν. καὶ ἐπὶ τῶν ἄλλων τὸ ἀνάλογον. ἐπεὶ οὖν τὰ μὲν βραχέα δύ' ὄντα ἀνὰ πέντε ἐπιδέχεται προσῳδίας, βραχεῖαν ὀξεῖαν βαρεῖαν δασεῖαν
114 ψιλήν, δέκα γενήσεται. τὰ δὲ μακρὰ πάλιν δύο ὄντα εἰ ἐκ περισσοῦ προσλαμβάνει καὶ τὴν προσῳδίαν τὴν περισπωμένην (μηκύνεται γὰρ ταῦτα καὶ ὀξύνεται καὶ βαρύνεται καὶ δασύνεται καὶ ψιλοῦται καὶ ἰδιαίτερον περισπᾶσθαι πέφυκε), γενήσεται δώδεκα. τὰ δὲ κοινὰ τρία καθεστῶτα ⟨εἰ⟩[1] τὰς ἑπτὰ προσῳδίας καθ' ἕκαστον ἐπιδέχεται,

[1] ⟨εἰ⟩ add. cj. Bekk.

[a] προσῳδία (*cf.* § 119) "comprises accent, breathing, and quantity—all the elements in the spoken word which in the ancient mode of writing were left to be supplied by the reader" (Bywater, *Aristotle on Poetry*, p. 336). The symbols were invented to help foreigners. See Rhys Roberts, *Demetrius on Literary Composition*, glossary. The τόνος,

η and ω, and two short, namely ε and ο, and three "dichronous," namely α ι and υ, but the total will be ten; and of these five will be long, namely η and ω and the long α and ι and υ, and the short will be equal in number, namely ο and ε and the short α and ι and υ. But since the School of Grammarians have assumed 113 that there are not only two tensions,[a] the long and the short, but also the acute, the grave,[b] the circumflex, the rough and the smooth, each of the vowels indicated having some one of these tensions peculiar to itself will become an element; and just as we found that an element was never "common" in respect of length and brevity but was either long only, when it had the long tension added, or short when it had the short, so, by the same reasoning, an element will never be "common" in respect of acuteness and gravity, but will be acute only, when it has received the acute added, or grave, when it has received the grave. So also with the others. Since, then, the short elements, being two, admit of five tensions apiece,—the short, the acute, the grave, the rough and the smooth,—they will become ten. The 114
long ones, again, which are two, if they receive the further addition of the circumflex,—for these elements can naturally take the circumflex as their own special tension as well as the long, the acute, the grave, the rough and the smooth,—will become twelve. And if the "common," being three, each admit of the

which we call accent, is explained by Demetrius as a musical pitch of about a musical fifth, what we call the acute, all the rest is what we call grave accent. The circumflex is a drawl when a vowel with acute slides into one with an unaccented vowel (*e.g.* when όο-μαι becomes οῦ-μαι).

[b] The grave is, then, the general monotone of the speaker's voice.

καὶ ταῦτα γίνεται εἴκοσι καὶ ἕν· ὥστε πάντα τεσσαράκοντα καὶ τρία τυγχάνειν· οἷς τῶν δεκαεπτὰ συμφώνων προστιθεμένων ἑξήκοντα γίνεται[1] στοιχεῖα, ἀλλ' οὐκ εἰκοσιτέσσαρα.

115 Ἔστι δὲ καὶ ἕτερος λόγος καθ' ὃν ἀξιοῦται διαφόρως τὰ φωνάεντα πάλιν στοιχεῖα ἐλάσσονα εἶναι τῶν παρὰ τοῖς γραμματικοῖς θρυλουμένων ἑπτά. εἰ γὰρ τὸ α κατ' αὐτοὺς ἐκτεινόμενον καὶ συστελλόμενον οὐχ ἕτερόν ἐστι στοιχεῖον ἀλλ' ἓν κοινόν, ὡσαύτως δὲ καὶ τὸ ι καὶ τὸ υ, ἀκολουθήσει καὶ τὸ ε καὶ τὸ η ἓν εἶναι στοιχεῖον κατὰ τὴν αὐτὴν δύναμιν κοινόν· ἡ γὰρ αὐτὴ δύναμις ἐπ' ἀμφοτέρων ἐστί, καὶ συσταλὲν μὲν τὸ η γίνεται ε, ἐκταθὲν δὲ τὸ ε γίνεται η. κατὰ δὲ τὸν αὐτὸν τρόπον καὶ τὸ ο καὶ τὸ ω μία στοιχείου γενήσεται φύσις κοινή, ἐκτάσει καὶ συστολῇ διαφέρουσα, ἐπείπερ τὸ μὲν ω μακρόν ἐστιν ο, τὸ δὲ ο βραχύ
116 ἐστιν ω. τυφλώττουσιν οὖν οἱ γραμματικοὶ καὶ τὸ ἀκόλουθον αὐτοῖς οὐ συνορῶσι, λέγοντες ἑπτὰ φωνάεντα τυγχάνειν, πέντε μόνων ὄντων πρὸς τὴν φύσιν.

Καὶ ἀναστρόφως ἔσεσθαί τινά φασιν ἔνιοι τῶν φιλοσόφων πλείονα στοιχεῖα, διάφορον ἔχοντα δύναμιν τῶν συνήθως παραδιδομένων, οἷον καὶ τὸ αι καὶ τὸ ου καὶ πᾶν ὃ τῆς ὁμοίας ἐστὶ φύσεως. τὸ γὰρ στοιχεῖον κριτέον μάλιστα, ὅτι στοιχεῖόν ἐστιν, ἐκ τοῦ ἀσύνθετον καὶ μονόποιον ἔχειν φθόγγον, οἷός ἐστιν ὁ τοῦ α καὶ ε καὶ ο καὶ τῶν
117 λοιπῶν. ἐπεὶ οὖν ὁ τοῦ αι καὶ ει φθόγγος ἁπλοῦς ἐστὶ καὶ μονοειδής, ἔσται καὶ ταῦτα στοιχεῖα. τεκμήριον δὲ τῆς ἁπλότητος καὶ μονοειδείας τὸ

[1] γίνεται cj. Bekk. : γίνεσθαι mss., Bekk.

seven tensions, they become twenty-one. Hence their total is forty-three; and when to these are added the seventeen consonants, the elements turn out to be sixty and not twenty-four.

There is, on the other hand, a different argument 115
by which it is made out that the vowels are fewer in number than the seven which are so much talked about by the Grammarians. For if the α when long and when short is not, according to them, a different element but one "common" element, and so likewise both the ι and the υ, then it will also follow that ε and η are one "common" element as having the same potency; for in the case of both there is the same potency, and η when shortened becomes ε, while ε when lengthened becomes η. In the same way ο and ω will also become one "common" element, differing in length and brevity, since ω is long ο, and ο is short
ω. The Grammarians, then, are blind and fail to 116
perceive the consequence, when they assert that there are seven vowels, whereas in actual fact there are only five.

Contrariwise, some philosophers declare that there will be some more elements which possess a potency different from those which are usually handed down, —αι, for instance, ου and every one which is of similar nature. For the "element" must be judged to be an element mainly on the ground that it has a sound which is non-composite and of single quality,
such as that of α and ε and ο and the rest. Since, then, 117
the sound of αι and of ει is simple and of one kind only, these, too, will be elements. And a proof of the simplicity of their sounds and that they are of one kind only is to be found in the following state-

λεχθησόμενον. ὁ μὲν γὰρ σύνθετος φθόγγος οὐχ οἷος ἀπ' ἀρχῆς προσπίπτει τῇ αἰσθήσει, τοιοῦτος ἄχρι τέλους παραμένειν πέφυκεν, ἀλλὰ κατὰ παράτασιν ἑτεροιοῦται, ὁ δὲ ἁπλοῦς καὶ ὄντως τοῦ στοιχείου λόγον ἔχων τοὐναντίον ἀπ' ἀρχῆς μέχρι τέλους ἀμετάβολός ἐστιν. οἷον τοῦ μὲν ρα φθόγγου ἐν παρατάσει προφερομένου, δῆλον ὡς οὐχ ὡσαύτως αὐτοῦ κατὰ τὴν πρώτην πρόσπτωσιν ἀντιλήψεται ἡ αἴσθησις καὶ κατὰ τὴν τελευταίαν, ἀλλὰ κατ' ἀρχὰς μὲν ὑπὸ τῆς τοῦ ρ ἐκφωνήσεως κινηθήσεται, μεταῦθις δὲ ἐξαφανισθείσης αὐτῆς εἰλικρινοῦς τῆς τοῦ α δυνάμεως ποιήσεται τὴν ἀντίληψιν. ὅθεν οὐκ ἂν εἴη στοιχεῖον τὸ ρα καὶ
118 πᾶν τὸ ἐοικὸς αὐτῷ. εἰ δὲ τὸν τοῦ αι φθόγγον λέγοιεν, οὐδὲν ἔσται τοιοῦτον, ἀλλ' οἷον ἀπ' ἀρχῆς ἐξακούεται τῆς φωνῆς ἰδίωμα, τοιοῦτον καὶ ἐπὶ τέλει, ὥστε στοιχεῖον ἔσται τὸ αι. τούτου δὲ οὕτως ἔχοντος, ἐπεὶ καὶ ὁ τοῦ ει φθόγγος καὶ ὁ τοῦ οι[1] μονοειδὴς καὶ ἀσύνθετος καὶ ἀμετάβολος ἐξ ἀρχῆς ἄχρι τέλους λαμβάνεται, ἔσται καὶ οὗτος στοιχεῖον.

119 Ἀλλὰ ἀφέμενοί γε ταύτης τῆς ζητήσεως ἐκεῖνο ἂν λέγοιμεν, ὃ μᾶλλον δύναται θλίβειν τοὺς γραμματικούς. εἰ γὰρ κοινὰ λέγεται στοιχεῖα τρία, α ι υ, διὰ τὸ ἐπιδεκτικὰ τυγχάνειν μήκους τε καὶ συστολῆς, ἀκολουθήσει πᾶν στοιχεῖον κοινὸν εἶναι λέγειν· ἐπιδεκτικὸν γάρ ἐστι τῶν τεσσάρων προσῳδιῶν, βαρύτητος ὀξύτητος ψιλότητος δασύτητος. ἢ εἴπερ οὐχ ὑπομένουσι πᾶν στοιχεῖον κοινὸν εἶναι λέγειν, μηδ' ἐκεῖνα λεγέτωσαν κοινὰ παρόσον ἐκτάσεως καὶ συστολῆς ἐστὶν ἐπιδεκτικά.

120 Ἤρκει μὲν οὖν ἠπορημένων τῶν στοιχείων τῆς

ment: the composite sound does not naturally re-
main to the end just the same in quality as when it
first strikes the sense-organ, but is altered through
its prolongation, whereas the simple sound, which
really has the character of "element," is on the
contrary unchanged from beginning to end. Thus,
when the sound *ρα* is prolonged in utterance, it is
plain that the sense will not perceive it alike at the
first impression and at the last, but will be stirred at
first by the utterance of the *ρ* and afterwards, when
this sound has vanished, it will gain the perception of
the *α* sound in its purity. Hence, *ρα* and all similar
vocables will not be elements. But if they should 118
pronounce the sound *αι*, the result will not be at all
similar, but its peculiar tone, as heard at the begin-
ning, is just the same at the end, so that *αι* will be an
element. And such being the case in respect of *αι*,
since the sound *ει* and the sound *οι* are received from
beginning to end as of single quality and non-com-
posite and unaltered, these too will be elements.

But letting go this line of inquiry we may state 119
another which is capable of pressing the Grammarians
still harder. If the three elements *α*, *ι*, and *υ* are
termed "common" because they are receptive of
length and brevity, it will follow that we must say
that every element is common, for it is receptive of
the four tensions, gravity, acuteness, smoothness,
roughness. Else, if they cannot endure to say that
every element is common, neither let them term
those three common on the ground that they are
receptive of length and brevity.

It would have been sufficient to have put an end to 120

---

[1] *οι* Coraes: *ου* MSS., Bekk.; ⟨*καὶ ὁ τοῦ αυ*⟩ add. cj. Warmington.

γραμματικῆς πέρας ἐπιτεθεικέναι τῇ ζητήσει· τίς γὰρ ἀπολείπεται λόγος περὶ τῶν μετὰ τὰς ἀρχὰς τοῖς τὰς ἀρχὰς οὐκ ἔχουσι γραμματικοῖς; ὅμως δ' οὖν ἐνδοτέρω προχωροῦντας οὐκ ἔστιν ἀλλότριον κἀκείνων δείγματος χάριν ἀποπειραθῆναι. καὶ ἐπεὶ ἐκ στοιχείων αἱ συλλαβαί εἰσι, ⟨τὰ⟩[1] περὶ τούτων ἐπισυνάπτωμεν.

ϛʹ.—ΠΕΡΙ ΣΥΛΛΑΒΗΣ

121 Πᾶσα οὖν συλλαβὴ ἢ μακρά ἐστιν ἢ βραχεῖα. μακρὰ δὲ γίνεται, φασί, διχῶς, φύσει τε καὶ θέσει, φύσει μὲν τριχῶς, ἢ ὅταν ἔχῃ στοιχεῖον φύσει μακρὸν ὡς ἐπὶ τῆς ἠώς λέξεως, ἑκατέρα γὰρ τούτων τῶν συλλαβῶν ἐστὶ μακρὰ διὰ τὸ τὴν μὲν τὸ η τὴν δὲ τὸ ω φύσει ἔχειν μακρόν, ἢ ὅταν ἐκ δυοῖν φωναέντων συνεστήκῃ ὡς ἐπὶ τῆς αἰεί λέξεως, αἱ γὰρ δύο συλλαβαὶ μακραὶ τῷ ἑκατέραν ἐκ δυοῖν φωναέντων ὑπάρχειν, ἢ ὅταν κοινὸν ἔχῃ τὸ στοιχεῖον μακροτόνως παρειλημμένον ὡς ἐπὶ τῆς Ἄρης· τὸ γὰρ α δίχρονον νῦν μακροτόνως ἐκ-
122 φέρεται. οὐκοῦν φύσει τριχῶς μηκύνεται συλλαβή, θέσει δὲ πενταχῶς, ἤτοι ὅταν εἰς σύμφωνα τῶν ἁπλῶν λήγῃ δύο, ἢ ὅταν ἡ μετ' αὐτὴν συλλαβὴ ἀπὸ συμφώνων δύο ἄρχηται, ἢ ὅταν εἰς σύμφωνον λήγῃ καὶ ἀπὸ συμφώνου ἡ ἑξῆς ἄρχηται, ἢ ὅταν εἰς διπλοῦν λήγῃ στοιχεῖον, ἢ ὅταν μετ' αὐτὴν
123 διπλοῦν ἐπιφέρηται. εἰ δὴ πᾶσα συλλαβὴ ἤτοι μακρά ἐστιν ἢ βραχεῖα κατὰ τοὺς ὑποδεδειγμένους τῆς τεχνολογίας τρόπους, ἐὰν παραστήσωμεν μηδετέραν οὖσαν αὐτῶν, δῆλον ὡς οὐδὲ λέξιν

[1] ⟨τὰ⟩ add. Heintz.

our investigation at this point, now that the " elements " of the Art of Grammar have been shown to be doubtful ; for what is left for the Grammarians to say about what comes after their principles when they do not possess those principles ? None the less, it is not irrelevant to penetrate further and, as a sample, to test also what follows the principles. And since the " syllables " are formed from the " elements," let us subjoin a discussion of these.

Chapter VI.—Concerning the Syllable

Every syllable is either long or short. It is long, 121
they say, in two ways, by nature and by position : by nature it is long in three cases, either when it has an element which is naturally long, as in the word ἠώς, for each of these syllables is long since the one contains η, the other ω, which are naturally long ; or when it is composed of two vowels, as in the word αἰεί, for the two syllables are long since each is made up of two vowels ; or when it has its element " common " and pronounced with a long tension, as in the case of Ἄρης ; for the " dichronous " α is now uttered
with a long tension. Thus the syllable is long by 122
nature in three ways, but by position in five ways, either when it ends with two of the simple consonants ; or when the next syllable begins with two consonants ; or when it ends with a consonant and the next begins with a consonant ; or when it ends with a double element ; or when a double element is inserted after
it.—If every syllable is either long or short in the ways 123
indicated in their " technology," then if we shall establish that neither of these syllables exists, it is plain that the Grammarians will not have a " word "

ἕξουσιν οἱ γραμματικοί· καθὰ γὰρ τῶν στοιχείων ἀναιρουμένων συναναιροῦνται καὶ αἱ συλλαβαί, οὕτω καὶ τῶν συλλαβῶν μὴ οὐσῶν οὔτε αἱ λέξεις γενήσονται οὔτε κοινῶς τὰ τοῦ λόγου μέρη, διὰ δὲ τοῦτ᾽ οὐδὲ λόγος.

124 Ἵν᾽ οὖν ᾖ τις βραχεῖα συλλαβή, δεῖ προωμολογῆσθαι ὅτι ἐλάχιστος καὶ βραχύς ἐστι χρόνος, ἐν ᾧ ὑφίσταται. οὐκ ἔστι δὲ ἐλάχιστος χρόνος· πᾶς γὰρ εἰς ἄπειρον τέμνεται, ὡς ἐν τοῖς ὕστερον δείξομεν· εἰ δὲ εἰς ἄπειρον τέμνεται, οὐκ ἔστιν ἐλάχιστος. οὐκ ἄρα ἔσται βραχεῖα συλλαβὴ βραχὺν ἔχουσα χρόνον. εἰ δὲ λέγοιεν νῦν βραχεῖαν καλεῖν καὶ ἐλαχίστην συλλαβὴν οὐ τὴν πρὸς φύσιν ἐλαχίστην οὖσαν ἀλλὰ τὴν πρὸς αἴσθησιν, ἑαυτοῖς
125 προσαύξουσι τὴν ἀπορίαν. τὰς γὰρ λεγομένας παρ᾽ αὐτοῖς βραχείας συλλαβὰς εὑρήσομεν ὡς πρὸς αἴσθησιν μεριστάς, οἷον τὴν ερ. αἰσθητῶς γὰρ ἐπιβάλλομεν ἐπ᾽ αὐτῆς ὅτι προεκφωνεῖται τῆς τοῦ ρ δυνάμεως ἡ τοῦ ε δύναμις. καὶ ἐναλλάξαντες εἰ λέγοιμεν ρε, πάλιν ἀντιληψόμεθα ὅτι πρώτη μέν ἐστι κατὰ τὴν τάξιν ἡ τοῦ ρ δύναμις, δευτέρα
126 δὲ ἡ τοῦ ε. ἐπεὶ οὖν πᾶν ὃ πρῶτον καὶ δεύτερον μέρος ἔχει πρὸς αἴσθησιν, οὐκ ἔστιν ἐλάχιστον πρὸς αἴσθησιν, φαίνεται δὲ ἡ κατὰ τοὺς γραμματικοὺς βραχεῖα συλλαβὴ πρῶτον καὶ δεύτερον ἔχουσα, οὐκ ἂν εἴη πρὸς αἴσθησιν ἐλαχίστη[1] καὶ βραχεῖα συλλαβή. μουσικοὶ μὲν γὰρ ἴσως ἀλόγους τινὰς χρόνους καὶ φωνῶν παραυξήσεις δυνήσονται ἀπολιπεῖν· τοῖς δὲ μὴ χωροῦσι τὸ τοιοῦτον βάθος [γραμματικοῖς] τῆς ἀπειρίας,[2] ἀλλ᾽ αὐτὸ μόνον

[1] ἐλαχίστη cj. Bekk. : ἐλάχιστος MSS., Bekk.
[2] ἀπειρίας Fabr. : ἀπορίας Bekk.

either ; for just as when the elements are destroyed the syllables, too, are destroyed with them, so also, if the syllables are non-existent, neither the words will exist nor the parts of speech in general, nor, consequently, speech itself.

In order that a short syllable may exist, it must 124
be settled beforehand that there is a minimal and a short time wherein it exists : but there is not a minimal time, for every period of time is divided to infinity,[a] as we shall prove later on ; and if it is divided to infinity it is not minimal. Hence, there will not exist a short syllable occupying a short time.—And if they should now say that what they are calling a short and minimal syllable is not that which is minimal by nature but that which is minimal for sense, they will be adding to their difficulty. For we 125
shall find that what are termed by them short syllables are divisible for sense,—for example the syllable ερ. For in its case we perceive by sense that the sound ε is uttered before the sound ρ. And conversely, if we were to say ρε, we shall again perceive that the sound ρ comes first in order and the sound ε second. Since, then, everything which has 126
for sense a first part and a second is not a minimal for sense, and since the short syllable of the Grammarians evidently has a first part and a second, there will not exist for sense a minimal and short syllable. Musicians perhaps will be able to leave us certain irrational times and extensions of tones ; but to those who have no room for such a depth of infinitude, but

[a] *Cf. Adv. Mus.* 62.

εἰς βραχεῖαν καὶ μακρὰν διαιρουμένοις τὴν γενικὴν συλλαβήν, οὐκ ἔστι συγγνωμονεῖν δίκαιον. οὐκοῦν ἀνυπόστατός ἐστιν ἡ βραχεῖα συλλαβή.

127 Καὶ μὴν ἡ μακρὰ πάλιν ἔσται ἀνύπαρκτος· δίχρονον μὲν γὰρ αὐτὴν εἶναι λέγουσι, δύο δὲ χρόνοι οὐ συνυπάρχουσιν ἀλλήλοις. εἰ γὰρ δύο εἰσί, τούτῳ διορίζονται ὅτι εἰσὶ δύο, τῷ τὸν μὲν ἐνεστηκέναι τὸν δὲ μή· εἰ δὲ ὃς μὲν ἐνέστηκεν ὃς δὲ οὐκ ἐνέστηκεν, οὐ συνυπάρχουσιν ἀλλήλοις.
128 διόπερ καὶ ἡ μακρὰ συλλαβὴ εἴπερ ἐστὶ δίχρονος, ὀφείλει, ὅτε μὲν αὐτῆς ἐνέστηκεν ὁ πρῶτος χρόνος, τότε ὁ δεύτερος μὴ ἐνεστηκέναι, καὶ ὅτε ὁ δεύτερος ἐνίσταται, τότε ὁ πρῶτος μηκέτι εἶναι. ἀσυνυπάρκτων δὲ αὐτῆς ὄντων τῶν μερῶν ὅλη μὲν οὐχ ὑφέστηκε, μέρος δέ τι αὐτῆς. ἀλλὰ τό γε μέρος αὐτῆς οὐκ ἦν αὐτή, ἐπεὶ οὐ διοίσει τῆς βραχείας
129 ἡ μακρὰ συλλαβή. οὐκ ἄρα οὐδὲ μακρά τις ἔστι συλλαβή. εἰ δὲ λέγοιεν κατὰ συμμνημόνευσιν νοεῖσθαι μακρὰν συλλαβήν (τοῦ γὰρ προλεχθέντος φθόγγου μνημονεύοντες καὶ τοῦ νῦν λεγομένου ἀντιλαμβανόμενοι τὸ ἐξ ἀμφοτέρων συντιθέμενον μακρὰν ἐνενοήσαμεν συλλαβήν)—εἰ δὴ τοῦτο λέγοιεν, οὐδὲν ἄλλο ἢ ἀνυπόστατον ὁμολογήσουσιν
130 εἶναι τὴν τοιαύτην συλλαβήν. εἰ γὰρ ὑφέστηκεν, ἤτοι ἐν τῷ προαναφωνουμένῳ φθόγγῳ ὑφέστηκεν ἢ ἐν τῷ ἐπαναφωνουμένῳ. οὔτε δὲ ἐν τῷ προαναφωνουμένῳ οὔτε ἐν τῷ ἐπαναφωνουμένῳ· ἑκάτερος γὰρ αὐτῶν κατ' ἰδίαν μὴ ὑφεστὼς οὐδὲ τὴν ἀρχὴν συλλαβή ἐστιν· ὥστε οὐδὲ ὑφέστηκεν. ὑφεστῶσα[1] δὲ βραχεῖά ἐστιν ἀλλ' οὐ μακρὰ συλλαβή. οὔτε

[1] ὑφεστῶσα cj. Bekk.: ὑφεστὼς MSS., Bekk.

merely divide the class of syllables into short and long, it is not right to extend pardon. Wherefore the short syllable is non-existent.

Moreover, the long syllable, too, will be non- 127
existent; for they say that it is "dichronous" (or two-timed), but two times do not co-exist with each other. For if they are two, they are distinguished as being two by the fact that the one is in present existence, the other not; but if one is present and the other is not present, they do not co-exist with each other. Consequently, if the long syllable is "di- 128
chronous," it must follow that when its first time is present, then its second is not present, and when the second becomes present, then the first no longer exists. And as its parts are not co-existent the whole syllable does not exist but only a part of it; but the part of it is not itself, since otherwise the long syllable will not differ from the short. Therefore no 129
long syllable exists. And if they should say that the long syllable is perceived through "joint memory"[a] (that is to say, that it is by remembering the sound already uttered and sensing that which is now being uttered that we have perceived the combination of both as a long syllable),—if they should say this, they will simply be confessing that a syllable such as this is non-existent. For if it exists, it exists either in the 130
sound pronounced first or in that pronounced later; but it does not exist either in the sound pronounced first or in that pronounced later; for as neither of these exists by itself, it is not a syllable at all; so that the long syllable does not exist. But if it exists, it is a short, and not a long, syllable. Nor, again, does it

[a] Or "concurrent recollection," a Stoic term, *cf. P.H.* iii. 108.

δὲ ἐν ἀμφοτέροις· ὁ γὰρ ἕτερος αὐτῶν τοῦ ἑτέρου ὑφεστῶτος οὐχ ὑφέστηκεν, ἐκ δὲ τοῦ ὑπάρχοντος καὶ μὴ ὑπάρχοντος οὐδὲν ἔστιν ἐπινοῆσαι συγκείμενον ὡς ἐκ μερῶν. οὐκ ἄρα ἔστι τις μακρὰ συλλαβή.

131 Ἀνάλογον δὲ τούτοις ἐστὶ καὶ τὰ περὶ λέξεως καὶ τῶν τοῦ λόγου μερῶν ἀπορεῖσθαι ὀφείλοντα. πρῶτον μὲν γάρ, ὡς μικρῷ πρότερον ὑπεδείκνυμεν, μὴ οὔσης συλλαβῆς μηδὲ λέξιν εἶναι δυνατόν· ἐκ συλλαβῶν γὰρ αἱ λέξεις λαμβάνουσι τὴν ὑπόστασιν. εἶτα καὶ κατὰ τὴν αὐτὴν ἔφοδον ἐξέσται προηγουμένως ἐπ' αὐτῆς τῆς λέξεως τὰς αὐτὰς χειρίζειν ἀπορίας. ἢ γὰρ συλλαβή ἐστιν ἢ ἐκ συλλαβῶν συνέστηκεν· ὅπως δ' ἂν ἔχῃ, τὰς ἐκκειμένας ἡμῖν
132 ἐπὶ τῆς συλλαβῆς ἀπορίας ἐπιδέξεται. ἀλλ' ὅμως ἵνα μὴ καινοτέρων ἐλέγχων ἀπορεῖν δοκῶμεν, προσφωνητέον τι κἀνταῦθα τοῖς γραμματικοῖς.

Ὅταν γὰρ μέρη τινὰ λόγου καλῶσιν, οἷον ὄνομα ῥῆμα ἄρθρον καὶ τὰ λοιπά, πόθεν λαβόντες; ἤτοι γὰρ ταῦτα ὅλον[1] τὸν λόγον καλοῦσιν, ἢ ταῦτα μέρη ἐκείνου, μήτε δ' ἐκείνου ὡς ὅλου νοεῖσθαι δυναμένου μήτε τούτων ὡς μερῶν ἐκείνου. λαμβανέσθω δὲ τὰ εἰς τὴν ὑπόθεσιν παραδείγματα, ὡς μὴ ἀπίωμεν[2] τῶν τῆς γραμματικῆς θεωρημάτων.
133 ἔστω τοίνυν ὑποθέσεως χάριν λόγος μὲν ὁ σύμπας οὗτος ὁ στίχος,

μῆνιν ἄειδε θεὰ Πηληιάδεω Ἀχιλῆος,

μέρη δὲ αὐτοῦ ταῦτα, τό τε μῆνιν προσηγορία

[1] ὅλον cj. Bekk. : ὅλα MSS., Bekk.
[2] ἀπίωμεν scripsi : ἀπῶμεν MSS., Bekk. (ἀποστῶμεν cj. Bekk.).

[a] Cf. § 123.

exist in both the sounds ; for when the one of them exists the other does not exist, and from an existent and a non-existent it is impossible to conceive anything compounded as though from parts. Hence no long syllable exists.

Analogous to these are the difficulties which ought 131
to be raised regarding the " word " and the " parts of speech." For in the first place, as we showed a short while ago,[a] if the syllable does not exist, the " word " cannot exist either ; for words derive their substance from syllables. And next, in the same line of attack, it will be possible to employ the same objections directly against the " word " itself. For either it is a syllable or it is composed of syllables ; but in either case it will be open to the objections we set forth against the syllable. All the same, lest we 132
should appear to be at a loss for fresh arguments, we must have something to say on this point also to the Grammarians.

When they term certain things " parts of speech," —such as the name (or noun), the verb, the article, and the rest,—whence do they derive the designation ? For they designate them thus either as being the whole of speech or as being parts of it, when neither can speech be conceived as a whole nor these as its parts. Let us take examples which bear on 133
their assumption, so that we may not depart from the theorems of Grammar. Let us suppose, then, that the whole of this verse is a " speech " (or sentence)—

μῆνιν ἄειδε θεὰ Πηληιάδεω Ἀχιλῆος

(Sing, O goddess, the wrath of the son of Peleus, Achilles),[b]—

and that its parts are these,—*μῆνιν*, which is a noun,

[b] This is the first verse of the *Iliad* : *cf. Adv. Phys.* i. 350.

καθεστώς, καὶ τὸ ἄειδε ῥῆμα προστακτικόν, καὶ τὸ θεά προσηγορία πάλιν θηλυκή, καὶ τὸ Πηληιάδεω ὄνομα πατρωνυμικόν, πρὸς δὲ τούτοις καὶ τὸ
134 Ἀχιλῆος ὄνομα κύριον. οὐκοῦν ἤτοι ἄλλο τί ἐστιν ὁ λόγος παρὰ τὰ μέρη αὐτοῦ καὶ ἄλλα τὰ μέρη παρὰ τὸν λόγον, ἢ τὸ ἄθροισμα τῶν μερῶν ὁ λόγος ὑπείληπται. καὶ εἰ μὲν ἄλλο τι τῶν μερῶν ἐστὶν ὁ λόγος, αἰρομένων δηλονότι τῶν ἐκκειμένων τοῦ λόγου μερῶν ὑπολειφθήσεται ὁ λόγος. τοσοῦτον δὲ ἀπέχει ὁ προειρημένος στίχος τοῦ μένειν πάντων αἰρομένων αὐτοῦ τῶν μερῶν, ὡς κἂν ἓν ὁδηποτοῦν αὐτοῦ μέρος ἀνέλωμεν, οἷον τὸ μῆνιν ἢ τὸ ἄειδε,
135 μηκέτι στίχος ὑπάρχειν. εἰ δὲ τὸ ἄθροισμα τῶν τοῦ λόγου μερῶν νοεῖται λόγος, τῷ μηδὲν εἶναι τὸν ἀθροισμὸν παρὰ τὰ ἠθροισμένα καθάπερ καὶ τὸ διάστημα παρὰ τὰ διεστηκότα, οὐδὲν ὑπάρξει ὁ λόγος οὗ νοηθήσεταί τινα μέρη. μηδενὸς δὲ ὄντος ὅλου λόγου οὐδὲ μέρη τινὰ τούτου γενήσεται. ὥσπερ οὖν εἰ μηδέν ἐστιν ἀριστερόν, οὐδὲ δεξιὸν ἔστιν, οὕτως εἰ μὴ ἔστι τι ὅλον λόγος, οὐδὲ τὰ
136 μέρη ὑπάρξει. καθόλου τε, εἰ τὸν ἀθροισμὸν τῶν τοῦ λόγου μερῶν ὅλον ἡγήσονται λόγον, ἀκολουθήσει αὐτοῖς τὰ μέρη τοῦ λόγου ἀλλήλων λέγειν εἶναι μέρη. εἰ γὰρ μηδὲν ὑπόκειται ὅλον παρ' αὐτὰ οὗ γενήσεται μέρη, ἀλλήλων ἔσται μέρη.
137 τοῦτο δὲ ὡς ἔστιν ἀλογώτατον, σκοπῶμεν. τὰ γὰρ μέρη πάντως ἐμπεριέχεται ἐκείνοις τοῖς ⟨ὅλοις⟩[1] ὧν λέγεται μέρη, ἴδιον τόπον ἐπέχοντα καὶ ἰδίαν ὑπόστασιν ἔχοντα, ἐν ἀλλήλοις δὲ οὐκ ἐμπεριέχεται. οἷον ἀνθρώπου μὲν μέρη χεῖρες

[1] ⟨ὅλοις⟩ addidi.

and ἄειδε an imperative verb, and θεά also a noun and feminine, and Πηληιάδεω a patronymic noun, and besides these Ἀχιλῆος a proper noun. Then either 134
the sentence is something else than its parts and the parts something else than the sentence, or the aggregate of the parts is assumed to be the sentence. And if the sentence is something else than its parts, then it is plain that when the expressed parts of the sentence are removed the sentence will be left. But the verse we have given above is so far from remaining when all its parts are removed that, if we should remove even any one single part, such as μῆνιν or ἄειδε, the verse ceases to exist. But if the aggre- 135
gate of the parts of the sentence is conceived to be a sentence, then because the aggregation is nothing apart from the parts aggregated, just as distance is nothing apart from the objects which are distant, the sentence of which any parts shall be conceived will not be anything. And when the whole sentence is nothing, neither will any parts of it exist. Thus, just as, if there is no left-hand, neither is there a right-hand, so too if a sentence does not exist as a whole, neither will its parts exist.—If, then, they shall regard 136
the aggregation of the parts of the sentence as the whole sentence, the general consequence will be that they will have to say that the parts of the sentence are parts of one another. For if no whole subsists other than themselves whereof they may be parts, they will be parts of one another. But let us consider how extremely irrational this is. For the parts are 137
certainly included in the wholes of which they are said to be parts, each occupying its own place and having its own reality, but they are not included in one another. For example, the hands are parts of the

καθεστᾶσι, χειρῶν δὲ δάκτυλοι καὶ δακτύλων ὄνυχες. διόπερ ἐν μὲν ἀνθρώπῳ χεῖρες περιέχονται, ἐν χερσὶ δὲ δάκτυλοι, ἐν δακτύλοις δὲ ὄνυχες, καὶ οὐχ ἡ μὲν δεξιὰ χεὶρ τὴν ἀριστερὰν συμπληροῖ, ὁ δὲ λιχανὸς τὸν ἀντίχειρα δάκτυλον ἀπαρτίζει, ἡ δὲ κεφαλὴ τοὺς πόδας συντίθησι καὶ οἱ πόδες
138 τὸν θώρακα. ὅθεν καὶ τὰ μέρη τοῦ λόγου οὐ ῥητέον ἀλλήλων εἶναι μέρη, ἐπεὶ ἐν ἀλλήλοις αὐτὰ δεήσει περιέχεσθαι, τὸ μὲν μῆνιν ἐν τῷ ἄειδε, τὸ δὲ ἄειδε ἐν τῷ θεά, καὶ καθόλου πάντα ἐν πᾶσιν, ὅπερ ἀδύνατον. οὐ τοίνυν ἀλλήλων τῶν τοῦ λόγου μερῶν δυναμένων εἶναι μερῶν διὰ τὸ ἀνόητον τοῦ πράγματος, οὔτε ὅλου τινὸς εὑρισκομένου λόγου παρὰ τὰ ἑαυτοῦ μέρη, μηδενός τε εὑρισκομένου πράγματος παρ᾽ αὐτὰ τὰ μέρη οὗ λέξομεν εἶναι τὰ μέρη, λείπεται λέγειν ὡς οὐδὲν ἔστι μέρος λόγου. διὰ δὲ τοῦτ᾽ οὐδὲ λόγος.

139 Ἐπακτέον δὲ καὶ οὕτως. εἴπερ τὸ μῆνιν μέρος ἐστὶ τοῦ στίχου, ἤτοι ὅλου τοῦ στίχου μέρος ἐστὶν ἢ τοῦ "ἄειδε θεὰ Πηληιάδεω Ἀχιλῆος." ἀλλ᾽ εἰ μὲν τοῦ ὅλου στίχου μέρος ἐστίν, ἐπεὶ ὅλος σὺν αὐτῷ τῷ μῆνιν νοεῖται, καὶ ἑαυτοῦ μέρος συμπληρωτικὸν γενήσεται τὸ μῆνιν, διὰ δὲ τοῦτο καὶ μεῖζον ἑαυτοῦ καὶ ἧττον, μεῖζον μὲν ἑαυτοῦ ᾗ συμπληροῦται ὑφ᾽ ἑαυτοῦ (τὸ γὰρ ὑπό τινος συμπληρούμενον μεῖζόν ἐστι τοῦ συμπληροῦντος αὐτό), ἔλασσον δὲ ᾗ συμπληροῖ ἑαυτό· τὸ γὰρ τινος συμπληρωτικὸν ἔλασσόν ἐστι τοῦ συμπληρουμένου. οὐ πάνυ δὲ ταῦτα πιθανά· οὐκ ἄρα τοῦ
140 ὅλου στίχου μέρος ἐστὶ τὸ μῆνιν. καὶ μὴν οὐδὲ τοῦ λειπομένου, φημὶ δὲ τοῦ "ἄειδε θεὰ Πηληιάδεω Ἀχιλῆος." πρῶτον μὲν γὰρ τὸ μέρος

man, the fingers of the hand, and the nails of the fingers. Hence the hands are included in the man, the fingers in the hand, and the nails in the fingers; but the right hand does not help to fill up the left, nor does the forefinger complete the thumb, nor the head compose the feet, nor the feet the chest. Hence also 138
we must not say that the parts of the sentence are parts of one another, for if so they will have to be included in one another,—*μῆνιν* in *ἄειδε*, *ἄειδε* in *θεά*, and in short all in all,—which is impossible. Since, then, the parts of the sentence cannot be parts of one another, because the thing is inconceivable, and since no whole sentence is found other than its own parts, and nothing is found other than the parts themselves whereof we can say that they are the parts, it remains to say that no part of the sentence exists. Consequently, neither does the sentence exist.

We may also make the following attack. If *μῆνιν* 139
is a part of the verse, it is either a part of the whole verse or of *ἄειδε θεὰ Πηληιάδεω Ἀχιλῆος*. But if it is a part of the whole verse, then, since it is with the inclusion of *μῆνιν* itself that the verse is conceived as a whole, *μῆνιν* will come to be a supplementary part of itself, and therefore both greater and less than itself; it will be greater than itself inasmuch as it is supplemented by itself (for that which is supplemented by anything is greater than the thing which supplements it), and it will be less inasmuch as it supplements itself (for that which supplements anything is less than the thing supplemented). But these results are by no means credible; therefore *μῆνιν* is not a part of the whole verse. Nor, in fact, is it a part of the re- 140
mainder of the verse,—I mean of *ἄειδε θεὰ Πηληιάδεω Ἀχιλῆος*. For, firstly, the part is included in that

περιέχεται ἐν τῷ οὗ ἔστι μέρος, τὸ δὲ μῆνιν οὐκ ἐμπεριέχεται ἐν τῷ "ἄειδε θεὰ Πηληιάδεω Ἀχιλῆος," ὥστε οὐκ ἂν εἴη μέρος τούτου. εἶτα δὲ οὐδὲ τὸ "ἄειδε θεὰ Πηληιάδεω Ἀχιλῆος" χρῄζει συμπληρώσεως· κατὰ γὰρ τὸν ἴδιον λόγον συμπεπλήρωται. ἀλλὰ ὅλος ὁ λόγος, λέγω δὲ ὁ στίχος, οὐκ ἔστι τὸ "ἄειδε θεὰ Πηληιάδεω Ἀχιλῆος." τοίνυν οὐδὲ τούτου μέρος ἐστὶ τὸ μῆνιν. ἀλλ' εἰ μήτε τοῦ ὅλου στίχου μέρος ἐστὶ τὸ μῆνιν μήτε τοῦ ἀπολειπομένου μέρους, παρὰ δὲ ταῦτα οὐδὲν ἄλλο ὑπόκειται, οὐδενὸς λόγου μέρος ἐστὶ τὸ μῆνιν.

141 Ταῦτα μὲν οὖν καθολικώτερον πρὸς τὰ μέρη τοῦ λόγου ῥητέον· ἐμβάντες δὲ εἰς τὰς κατὰ μέρος παρ' αὐτοῖς περὶ τούτων τεχνολογίας πολὺν λῆρον εὑρήσομεν. καὶ τοῦτο πάρεστι μαθεῖν οὐκ ἐπὶ τὴν πᾶσαν ὕλην φοιτήσαντας (ἀδόλεσχον γάρ ἐστι καὶ γραμματικῆς γραολογίας πλῆρες), ἀλλ' ὅμοιόν τι τοῖς οἰνοκαπήλοις ποιήσαντας,[1] καὶ ὃν τρόπον ἐκεῖνοι ἐξ ὀλίγου γεύματος τὸν ὅλον δοκιμάζουσι φόρτον, οὕτω καὶ αὐτοὶ ἓν λόγου μέρος προχειρισάμενοι, καθάπερ τὸ ὄνομα, ἐκ τῆς περὶ τούτου τεχνολογίας συνοψόμεθα καὶ τὴν ἐν τοῖς ἄλλοις τῶν γραμματικῶν ἐντρέχειαν.

Ζ'.—ΠΕΡΙ ΟΝΟΜΑΤΟΣ

142 Αὐτίκα τοίνυν ὅταν τῶν ὀνομάτων τὰ μὲν ἀρσενικὰ φύσει λέγωσι τὰ δὲ θηλυκὰ τὰ δὲ οὐδέτερα, καὶ τὰ μὲν ἑνικὰ τῷ ἀριθμῷ τὰ δὲ δυϊκὰ τὰ δὲ πληθυντικά, καὶ ἤδη τὰς ἄλλας ἐπισυνείρωσι

whereof it is a part, but μῆνιν is not included in ἄειδε θεὰ Πηληιάδεω Ἀχιλῆος, so that it will not be a part thereof. And next, the phrase ἄειδε θεὰ Πηληιάδεω Ἀχιλῆος does not need supplementing; for it is a complete sentence in itself. But the whole sentence,—I mean the verse,—is not ἄειδε θεὰ Πηληιάδεω Ἀχιλῆος. Thus μῆνιν is not a part of this ⟨remainder⟩ either. But if μῆνιν is a part neither of the whole verse nor of its remaining part, and nothing else besides these exists, then μῆνιν is not a part of any sentence.

These are the objections to be stated against the 141
parts of speech (or of the sentence) in general; and when we go on to their particular technical theories regarding them we shall find much idle talk. This we may learn not by roving over the whole field (for it is empty verbiage and full of the old wives' prattle of the Grammarians), but rather by doing much as the wine-merchants do; for just as they judge of the whole cargo from a little taste, so too we, when we have first dealt with a single part of speech, such as the noun, will gain, from their "technology" concerning this, a general view of the skill of the Grammarians in treating of the other parts.

Chapter VII.—Concerning the Noun

When, for instance, they say that some nouns are 142
"naturally" masculine, some feminine and some neuter, and some singular in number, some dual and some plural, and link on to these all their other

---

[1] ποιήσαντας Fabr.: ποιήσαντες MSS., Bekk.

διαιρέσεις, ἐπιζητήσωμεν τί ποτε ἔστι τὸ ἐπι-
143 φωνούμενον τοῦτο φύσει. ἢ γὰρ ὅτι οἱ πρῶτοι ἀναφθεγξάμενοι τὰ ὀνόματα φυσικὴν ἐποιήσαντο τὴν ἀναφώνησιν αὐτῶν ὡς καὶ τὴν ἐπὶ τῷ ἀλγεῖν κραυγὴν καὶ τὴν ἐπὶ τῷ ἥδεσθαι ἢ τῷ θαυμάζειν ἐκβόησιν, οὕτω λέγουσι φύσει τὰ μὲν τοιαῦτα εἶναι τῶν ὀνομάτων τὰ δὲ τοιάδε· ἢ ὅτι καὶ ἐπὶ τοῦ παρόντος ἕκαστον αὐτῶν φυσικῶς ἡμᾶς κινεῖ ὅτι ἀρρενικόν, κἂν ἡμεῖς μὴ νομίζωμεν αὐτὸ ἀρρενικὸν εἶναι, καὶ πάλιν φυσικῶς αὐτὸ ἐνδείκνυται ὅτι
144 θηλυκόν ἐστι, κἂν ἡμεῖς μὴ θέλωμεν. ἀλλὰ τὸ μὲν πρῶτον οὐκ ἂν εἴποιεν. πόθεν γὰρ γραμματικῇ παχύτητι διαγινώσκειν πότερον φύσει ἢ θέσει τὰ ὀνόματα, ἢ τινὰ μὲν οὕτως τινὰ δὲ ἐκείνως; ὅτε οὐδὲ τοῖς ἐπ᾿ ἄκρον ἥκουσι φυσιολογίας εὐ-
145 μαρὲς εἰπεῖν διὰ τὰς ἑκατέρωθεν ἰσολογίας. ἄλλως τε καὶ ἰσχυρὸς ἀντικάθηται τούτῳ λόγος, πρὸς ὃν οὐδ᾿ εἰ καταπέλτην ὑπομένοιεν, φασίν, οἱ γραμματικοὶ δυνήσονταί τι συνιδεῖν ἱκνούμενον. εἴπερ γὰρ φύσει τὰ ὀνόματα ἦν καὶ μὴ τῇ καθ᾿ ἕκαστον θέσει σημαίνει, ἐχρῆν πάντας πάντων ἀκούειν, Ἕλληνας βαρβάρων καὶ βαρβάρους Ἑλλήνων καὶ βαρβάρους βαρβάρων. οὐχὶ δέ γε τοῦτο· οὐκ ἄρα φύσει σημαίνει τὰ ὀνόματα. ὥστε τοῦτο μὲν οὐκ
146 ἐροῦσιν· εἰ δ᾿ ὅτι φυσικῶς διαδείκνυσιν ἕκαστον ὄνομα ὅτι ἀρρενικόν ἐστιν ἢ θηλυκὸν ἢ οὐδέτερον, φασὶ τὰ μὲν τοιάδε τὰ δὲ τοιαῦτα τυγχάνειν, ἴστωσαν λειότερον αὐτοῖς τρίβοντες τὸν κλοιόν.
147 πάλιν γὰρ φήσομεν ὅτι τὸ φύσει κινοῦν ἡμᾶς

[a] *Cf. P.H.* ii. 214, iii. 267.

distinctions, let us inquire what exactly is meant by the designation " naturally." Either it means that 143
those who first gave utterance to the nouns produced their " natural " sound ; and just as crying is natural in pain and shouting in joy or in admiration, so some nouns, they say, are " naturally " of this sort and others of that sort. Or else it means that at the present moment each of them naturally affects us as being masculine, even if we do not consider that it is masculine, or again naturally indicates that it is feminine, even if we are not willing. But they will 144
not declare for the first alternative; for how could the Grammarians' stupidity decide whether names are due to nature or to convention, or some to the one and some to the other, when even for those who have attained the summit of natural science it is no easy matter to settle because of the equipollence of the arguments on either side ? Moreover, this view is 145
confronted by a strong argument to which the Grammarians—even if they could (as men say) stand up against a bolt from a cross-bow—will be unable to discover any fitting reply. If nouns exist " by nature" and are not significant in each instance by reason of convention, then all men ought to understand the speech of all, Greeks that of barbarians and barbarians that of Greeks and barbarians that of (other) barbarians.[a] But this is not the case; therefore nouns are not "naturally" significant. This, then, they will not assert.—But if it is because each noun 146
"naturally" indicates that it is masculine or feminine or neuter that they assert that some are of this sort and others of that sort, let them know that they are rubbing their collar smoother. For we shall tell them 147
again that what affects us "naturally" affects all men

ὁμοίως πάντας κινεῖ, καὶ οὐχ οὓς μὲν οὕτως οὓς δὲ ἐναντίως. οἷον φύσει τὸ πῦρ ἀλεαίνει, βαρβάρους Ἕλληνας, ἰδιώτας ἐμπείρους, καὶ οὐχ Ἕλληνας μὲν ἀλεαίνει βαρβάρους δὲ ψύχει· καὶ ἡ χιὼν φύσει ψύχει, καὶ οὐ τινὰς μὲν ψύχει τινὰς δὲ θερμαίνει. ὥστε τὸ φύσει κινοῦν ὁμοίως τοὺς ἀπαραποδίστους
148 ἔχοντας τὰς αἰσθήσεις κινεῖ. τὰ δὲ αὐτὰ ὀνόματα οὐ πᾶσίν ἐστι τὰ αὐτά, ἀλλὰ τοῖς μὲν ἀρρενικὰ τοῖς δὲ θηλυκὰ τοῖς δὲ οὐδέτερα. οἷον Ἀθηναῖοι μὲν τὴν στάμνον λέγουσι θηλυκῶς, Πελοποννήσιοι δὲ τὸν στάμνον ἀρρενικῶς, καὶ οἱ μὲν τὴν θόλον οἱ δὲ τὸν θόλον, καὶ οἱ μὲν τὴν βῶλον οἱ δὲ τὸν
149 βῶλον. καὶ οὐ διὰ τοῦτο οὗτοι ἢ ἐκεῖνοι λέγονται ἁμαρτάνειν· ἕκαστος γάρ, ὡς τεθεμάτικεν, οὕτω χρῆται. καὶ οἱ αὐτοὶ δὲ διαφόρως ταὐτὰ ὁτὲ μὲν ἀρρενικῶς ἐκφέρουσιν ὁτὲ δὲ θηλυκῶς, λέγοντες τὸν λιμόν καὶ τὴν λιμόν. οὐκ ἄρα φύσει τῶν ὀνομάτων τὰ μὲν ἀρρενικὰ τὰ δὲ θηλυκά, ἀλλὰ κατὰ θεματισμὸν τὰ μὲν τοιαῦτα γίνεται τὰ δὲ
150 τοιαῦτα. καὶ μὴν εἴπερ φύσει τῶν ὀνομάτων τὰ μὲν ἦν ἀρρενικὰ τὰ δὲ θηλυκά, ὤφειλον αἱ ἀρρενικαὶ φύσεις ἀεί ποτε ἀρρενικοῖς ὀνόμασι προσαγορεύεσθαι καὶ αἱ θηλυκαὶ θηλυκοῖς καὶ μήτε αἱ ἀρρενικαὶ φύσεις μήτε αἱ θηλυκαὶ οὐδετέρως.
151 οὐχὶ δὲ τοῦτο, ἀλλὰ καὶ τὰς ἀρρενικὰς φύσεις θηλυκῶς καλοῦμεν καὶ τὰς θηλυκὰς ἀρρενικῶς καὶ τὰς οὔτε ἀρρενικὰς οὔτε θηλυκὰς ἤτοι ἀρρενικῶς ἢ θηλυκῶς, οὐχὶ δὲ οὐδετέρως. οἷον κόραξ

alike, and not some in one way and others in an opposite way.[a] Fire, for instance, " naturally " warms barbarians and Greeks, unskilled and skilled, and does not warm Greeks but chill barbarians ; and snow " naturally " chills, and does not chill some but heat others. Thus, that which affects us " naturally " affects in a similar way those who have their senses unimpaired. But the same nouns are not the same 148
for all, but are masculine for some, feminine for others, and for others neuter. The Athenians, for example, speak of ἡ στάμνος (" the jar ") in the feminine,[b] but the Peloponnesians of ὁ στάμνος in the masculine, and some of ἡ θόλος (" the rotunda "), others of ὁ θόλος, and some of ἡ βῶλος (" the lump "), others of ὁ βῶλος ; and neither the 149
one set nor the other is said to be wrong because of this ; for each treats the word as laid down by custom. And even the same people utter the same nouns differently, at one time as masculine, at another as feminine, saying both ὁ λιμός (" the hunger ") and ἡ λιμός. So it is not " by nature " that some nouns are masculine, others feminine, but it is by conventional usage that some are of one sort and some of the other. Moreover, if some nouns had 150
been masculine and others feminine by nature, male objects ought to have been called always by masculine names and female by feminine names, and objects neither male nor female by neuters. This, however, 151
is not so, but we call male objects by feminine names, and female by masculine, and objects which are neither male nor female by names which are either masculine or feminine, and not by neuters. Thus, even in the case of the female, the masculine name is

[a] *Cf. P.H.* iii. 179.

[b] *Cf.* § 187, *infra.*

μὲν λέγεται ἀετός κώνωψ κάνθαρος σκορπίος μῦς ἀρρενικῶς καὶ ἐπὶ τοῦ θήλεος, καὶ πάλιν χελιδών χελώνη κορώνη ἀκρίς μυγαλῆ ἐμπίς καὶ ἐπὶ τοῦ
152 ἄρρενος τὴν φύσιν θηλυκῶς. ὡσαύτως δὲ κλίνη θηλυκῶς ἐπὶ τῆς μήτε ἄρρενος μήτε θηλείας τὴν φύσιν, καὶ στῦλος ἀρρενικῶς ἐπὶ τοῦ οὐδετέρου. τοίνυν εἰ φύσει οὐδέν ἐστιν ἀρρενικὸν ἢ θηλυκὸν ὄνομα, ζητῶ πῶς ὁ γραμματικὸς ἐπιλήψεται τοῦ διαστρόφως λέγοντος ὁ χελιδών καὶ ἡ ἀετός. ἤτοι γὰρ ὡς φύσει τοῦ ὀνόματος τῆς χελιδόνος θηλυκοῦ ὄντος, ἐκείνου δὲ ἀρρενικὸν αὐτὸ τῷ ἄρθρῳ βιαζομένου γενέσθαι, ἢ ὡς τῆς κοινῆς συνηθείας θηλυκὸν αὐτὸ θεματισάσης ἀλλ' οὐκ
153 ἀρρενικόν. ἀλλ' εἰ μὲν ὡς φύσει θηλυκοῦ καθεστῶτος, ἐπεὶ οὐδὲν φύσει θηλυκόν ἐστι καθὼς παρεστήσαμεν, ἀδιάφορον τὸ ὅσον ἐπὶ τούτῳ ἐάν τε οὕτως ἐάν τε ἐκείνως ἐκφέρηται· εἰ δ' ὡς ὑπὸ τῆς κοινῆς συνηθείας ἀντὶ θηλυκοῦ θεματισθέν, γενήσεται τοῦ τε εὖ λεγομένου καὶ μὴ κριτήριον οὐχὶ τεχνικός τις καὶ γραμματικὸς λόγος ἀλλ' ἡ ἄτεχνος καὶ ἀφελὴς τῆς συνηθείας παρατήρησις.

154 Τὰ δὲ αὐτὰ ταῦτα μετακτέον καὶ ἐπὶ τὰ ἑνικὰ καὶ πληθυντικὰ τῶν ὀνομάτων. Ἀθῆναι γὰρ λέγονται πληθυντικῶς ἡ μία πόλις καὶ Πλαταιαί, καὶ πάλιν Θήβη ἑνικῶς καὶ Θῆβαι πληθυντικῶς, καὶ Μυκήνη καὶ Μυκῆναι. ῥηθήσεται δὲ ἐπιμελέστερον περὶ τῆς ἐν τούτοις ἀνωμαλίας προβαινούσης τῆς ζητήσεως.[a]

Τὰ νῦν δὲ ἐπεὶ καὶ ὑποδειγματικῶς κατωπτεύ-

[a] *Cf.* §§ 195 ff.

given to κόραξ ("raven"), ἀετός ("eagle"), κώνωψ ("gnat"), κάνθαρος ("beetle"), σκόρπιος ("scorpion"), μῦς ("mouse"), and conversely the feminine, even when they are of the male sex, to χελιδών ("swallow"), χελώνη ("tortoise"), κορώνη ("crow"), ἀκρίς ("locust"), μυγαλῆ ("shrew-mouse"), ἐμπίς ("mos-
quito"). So likewise κλίνη ("couch") is feminine, 152
though the object itself is neither of the male nor of the female sex, and στῦλος ("pillar") is masculine, though applied to what is neuter. If, then, no noun is masculine or feminine by nature, I ask how the Grammarian will censure the man who perversely says ὁ χελιδών and ἡ ἀετός. Either (he will do so) on the ground that the noun χελιδών is feminine by nature, whereas the man forces it to become masculine by means of the article, or else on the ground that common usage has ruled it to be feminine and
not masculine. But if he takes the ground that it is 153
feminine by nature, then, since none is feminine by nature, as we have established, it is, so far as this goes, a matter of indifference whether it is expressed in this way or in that. If, on the other hand, he assumes that the noun is ruled to be feminine by common usage, the criterion of correct and incorrect speech will not be any technical and grammatical theory but untechnical and simple adherence to actual usage.

We may use the same arguments against singular 154
and plural names. Thus Ἀθῆναι in the plural is the name given to one city, as is Πλαταιαί, and Θήβη conversely in the singular as well as Θῆβαι in the plural, and both Μυκήνη and Μυκῆναι. But with these anomalies we shall deal more carefully as our inquiry proceeds.[a]

For the present, now that by means of examples

καμεν τὴν ἐν τούτοις τῶν γραμματικῶν ἀκρίβειαν,
φέρε κἀκεῖνο, πρὶν ἐπ' ἄλλον τρόπον ἀπελθεῖν,
155 ἐξετάσωμεν, φημὶ δὲ τίνα λόγον καλοῦσιν ἢ μέρη
λόγου. ἤτοι γὰρ αὐτὴν τὴν σωματικὴν φωνὴν
ἐροῦσιν ἢ ἀσώματον λεκτόν, διαφέρον ταύτης.
οὔτε δὲ τὴν φωνὴν ἐροῦσιν· ταύτης μὲν γὰρ
ῥηθείσης πάντες ἀκούουσιν, Ἕλληνές τε καὶ
βάρβαροι, καὶ ἰδιῶται καὶ οἱ παιδείας ἐντός, τοῦ
δὲ λόγου καὶ τῶν τούτου μερῶν Ἕλληνες μόνοι
καὶ οἱ τούτου ἔμπειροι. τοίνυν οὐχ ἡ φωνή ἐστιν
156 ὁ λόγος καὶ μέρη λόγου. καὶ μὴν οὐδὲ τὸ ἀσώ-
ματον λεκτόν. πῶς γὰρ ἀσώματον ἔτι ἔστι τι
ἄλλο τοιοῦτο παρὰ τὸ σῶμα καὶ τὸ κενόν, πολλῆς
καὶ ἀνηνύτου γενομένης παρὰ τοῖς φιλοσόφοις
περὶ αὐτοῦ διαμάχης; εἰ μὲν γὰρ κινεῖται, σῶμά
ἐστιν· τὸ γὰρ κινούμενον σῶμα· εἰ δὲ μένει, δεχό-
μενον μὲν τὰ εἰς αὐτὸ φερόμενα σώματα καὶ μὴ
ἀντιτυποῦν κενὸν γενήσεται, κενοῦ γὰρ ἴδιον τὸ
μὴ ἀντιτυπεῖν, ἀντιτυποῦν δὲ τοῖς εἰς αὐτὸ φερο-
μένοις σῶμά ἐστιν, ἰδίωμα γὰρ σώματος τὸ ἀντι-
157 τυπεῖν. ἄλλως τε ὁ λέγων ἀσώματόν τι λεκτὸν
ὑπάρχειν ἤτοι φάσει μόνον ἀρκούμενος λέγει ἢ
ἀπόδειξιν παραλαμβάνων. ἀλλὰ φάσει μὲν ἀρκού-
μενος ἐν ἀντιφάσει ἐπισχεθήσεται· ἀπόδειξιν δὲ
παραλαμβάνων, ἐπεὶ καὶ αὐτὴ δι' ἀναμφισβητήτων
ὀφείλει λημμάτων προάγειν, τὰ δὲ λήμματά ἐστι
λεκτά, προαρπάζων τὸ ζητούμενον ὡς ὁμολογού-
158 μενον ἄπιστος ἔσται. παρ' ἣν αἰτίαν λοιπόν, εἰ
μήτε ἡ φωνὴ λόγος ἐστὶ μήτε τὸ σημαινόμενον

---

[a] *Cf.* § 78.

[b] The only two "reals" admitted by the Epicureans.

we have gained an insight into the precision of the Grammarians in regard to these matters, come and before we turn away to another subject let us investigate this point too,—I mean what, in their view, 155
is speech or the parts of speech. Either they will say that it is the corporeal sound of the voice, or that it is the incorporeal "lekton" (or meaning) [a] which differs from the sound. But they will not say that it is the sound; for when this is uttered all men hear it, Greeks and barbarians, plain people and people of culture, but speech and the parts of speech are heard (and understood) only by the Greeks and those who are skilled therein. So, then, speech and the parts of 156
speech are not the sound.—Neither indeed are they the incorporeal "lekton." For how can there still exist any other incorporeal of the sort beside body and void,[b] when there has been a great and endless dispute about it amongst the philosophers? For if it moves it is a body; for what moves is corporeal; and if it is stationary, it will be void if it admits the bodies which move into it without resistance (for not to resist is the property of void), but, if it resists the bodies moving into it, it is body, for resistance is the property of body.—Moreover, he who says that an incorporeal 157
"lekton" exists, in saying so relies only on assertion or else adduces a proof.[c] But if he relies on assertion he will be rebutted by a counter-assertion; and if he adduces proof, then, since the proof itself must proceed by means of unquestioned premisses, and the premisses are "lekta," by begging of the question, as though it were settled, he will be discredited. And 158
for this reason, if neither the sound of the voice is speech, nor yet the incorporeal "lekton" signified by

[a] *Cf.* § 188.

ὑπ' αὐτῆς ἀσώματον λεκτόν, παρὰ δὲ ταῦτα νοεῖν οὐδὲν ἐνδέχεται, οὐδέν ἐστι λόγος.

Ἔστω δὲ νῦν καὶ ὁ λόγος καὶ μέρη τούτου ὁπόσα θέλουσιν οἱ γραμματικοὶ ὑπάρχειν. ἀλλ' εἰπάτωσάν γε ἡμῖν πῶς τὸν λόγον μερίζουσιν.

Η'.—ΠΕΡΙ ΜΕΡΙΣΜΟΥ

159 Ἐπεὶ γὰρ τὸν μερισμὸν τὸν τῶν μέτρων ἐν δυσὶ μάλιστα τοῖς ἀναγκαιοτάτοις κεῖσθαι συμβέβηκεν, ἔν τε τῷ βαίνειν, τουτέστι τῇ εἰς τοὺς πόδας διανομῇ, καὶ ἐν τῇ εἰς τὰ τοῦ λόγου μέρη διαιρέσει, ἀκόλουθον μὲν ἦν τοῖς τελέως πρὸς αὐτοὺς ἀντιλέγουσιν ἑκάτερον κινεῖν, τόν τε τρόπον τοῦ βαίνειν, σκελίσαντας αὐτῶν ἅπαντας τοὺς οἷς βαίνουσι πόδας ὡς ἀνυπάρκτους, καὶ ἔτι τὸν τρόπον τῆς τῶν τοῦ λόγου μερῶν διανομῆς, δείξαντας τὸ
160 ἀδύνατον τῆς διαιρέσεως. ἀλλ' ἐπεὶ κἂν τοῖς πρὸς τοὺς μουσικοὺς προηγουμένως περὶ ποδῶν ζητοῦμεν, ἵνα μὴ προλαμβάνωμεν τὰ μελλήσοντα πρὸς ἐκείνους λέγεσθαι ἢ μὴ δὶς τὰ αὐτὰ λέγωμεν, ταύτην μὲν τὴν ἀπορίαν εἰς τὸν δέοντα καιρὸν ὑπερθησόμεθα, περὶ δὲ τῆς διαιρέσεως τῶν τοῦ λόγου μερῶν σκεψώμεθα.

161 Ὁ οὖν μερίζων τινὰ στίχον τὰ μὲν ἀφαιρεῖ τὰ δὲ προστίθησι, καὶ ἀφαιρεῖ μὲν τὸ μῆνιν, εἰ τύχοι, χωρίζων τοῦ παντὸς στίχου, καὶ πάλιν τὸ ἄειδε καὶ τὰ λοιπὰ μέρη, προστίθησι δὲ τοῖς κατὰ συναλοιφὴν ἐκφερομένοις, οἷον τῷ "αἷμ' ἐμέων" τὸ α, τὸ γὰρ πλῆρες ἦν "αἷμα ἐμέων," καὶ πάλιν τῷ "βῆ δ' ἀκέων" τὸ ε, κατὰ γὰρ ἐκπλήρωσιν

the sound, and no alternative other than these can be conceived, nothing is speech.

But for the present let it be granted that both speech exists and as many parts of it as the Grammarians desire. Yet let them tell us how they divide speech into parts.

Chapter VIII.—Concerning Division of Speech into Parts

Since it happens that the division of metres is 159
mainly into the two most necessary parts, namely, scansion,—that is, the division into feet,—and distinguishing the parts of the sentence, it follows that those who are completely refuting the Grammarians attack each of these,—both the method of scansion, by upsetting all the feet with which they scan as being non-existent, and also their method of distinguishing the parts of the sentence, by showing the impossibility of their distinctions. But seeing 160
that in our treatise *Against the Musicians* we are specially discussing "feet," in order to avoid forestalling what will then be said against them, or saying the same things twice, we shall postpone this difficulty to the proper time and consider now the question of the division of the parts of the sentence.

Now he who divides a verse subtracts some words 161
and adds others: he subtracts, let us say, *μῆνιν*, separating it from the whole verse, and likewise *ἄειδε* and the rest of the parts[a]; but he adds to words uttered with elision,—*α*, for example, to *αἶμ' ἐμέων* ("vomiting blood"), for the full expression is *αἶμα ἐμέων*,[b] and likewise *ε* to *βῆ δ' ἀκέων* ("he went

[a] *Cf.* §§ 133 ff.

[b] *Cf.* Hom. *Il.* xv. 11.

οὕτως εἶχε "βῆ δὲ ἀκέων." μηδενὸς μέντοι μήτε ἀφαιρεῖσθαι δυναμένου ἀπό τινος μήτε προστίθεσθαί τινι πεφυκότος ἀδύνατος γίνεται ὁ κατὰ
162 γραμματικὴν μερισμός. τὸ δὲ ὅτι οὐδὲν οὐδενὸς ἀφαιρεῖται μάθοιμεν ἂν τόνδε τὸν τρόπον. εἰ γὰρ ἀφαιρεῖταί τι ἀπό τινος, ἢ ὅλον ἀφ' ὅλου ἀφαιρεῖται ἢ μέρος ἀπὸ μέρους ἢ ὅλον ἀπὸ μέρους ἢ μέρος ἀπὸ ὅλου. ὅλον μὲν οὖν ἀπὸ ὅλου οὐκ ἀφαιρεῖται· ἑνὸς γὰρ ὑποκειμένου στίχου, εἰ ὅλον ἐστὶ τὸ ἀφαιρούμενον, ὅλον στίχον ἀφελοῦμεν. καὶ οὕτως εἰ μὲν ἔτι μένει ὁ στίχος ἀφ' οὗ ἡ ἀφαίρεσις, οὐδὲ ὅλως ἔσται γεγονυῖά τις ἀφαίρεσις ἀπ' αὐτοῦ· πῶς γὰρ ἔτι μένειν οἷόν τέ ἐστι τὸ ὅλον, εἴπερ ἀφῄρηται; εἰ δὲ μὴ μένει, δῆλον ὡς ἐκ τοῦ μὴ ὄντος οὐκ ἔστι πάλιν γεγονυῖα ἀφαίρεσις.
163 ὥστε ὅλον ἀπὸ ὅλου οὐκ ἀφαιρεῖται. καὶ μὴν οὐδὲ ὅλον ἀπὸ μέρους· ἐν μὲν γὰρ τῷ μέρει οὐκ ἐμπεριέχεται τὸ ὅλον, οἷον τῷ μῆνιν τὸ "⟨μῆνιν⟩[1] ἄειδε θεὰ Πηληιάδεω Ἀχιλῆος," τὸ δὲ ἀφαιρούμενον ὀφείλει ἐμπεριέχεσθαι τῷ τὴν ἀφαίρεσιν ἐπιδεχομένῳ. λείπεται ἄρα ἢ μέρος ἀφ' ὅλου ἢ μέρος ἀπὸ μέρους ἀφαιρεῖσθαι. ἀλλὰ καὶ τοῦτ' ἄπορον. τὸ γὰρ μῆνιν εἰ μὲν ἀφ' ὅλου ἀφαιρεῖται τοῦ στίχου, καὶ ἀπὸ αὑτοῦ ἀφαιρεῖται· σὺν αὐτῷ γὰρ ὅλος ὁ στίχος ἐνοεῖτο. καὶ ἄλλως, εἰ ἀφ' ὅλου ἀφαιρεῖται, τὸ δ' ὅλον ἦν "μῆνιν ἄειδε θεὰ Πηληιάδεω Ἀχιλῆος," ὤφειλεν ἠλαττῶσθαι καὶ τὸ "ἄειδε θεὰ Πηληιάδεω Ἀχιλῆος" καὶ μὴ μένειν ἐν τῷ αὐτῷ, παντὸς τοῦ ἀφαίρεσιν ἐπιδεξαμένου
164 μὴ μένοντος ἐν ταὐτῷ. ἐχρῆν δὲ καὶ αὐτὸ τὸ

[1] ⟨μῆνιν⟩ add. Heintz.

[a] *Cf.* Hom. *Il.* i. 34.

in silence"), for when fully expressed it is βῆ δὲ ἀκέων.[a] Since, however, nothing can really be subtracted from anything or added to anything,[b] grammatical division becomes impossible. And we may 162
learn the fact that nothing is subtracted from anything in this way: if anything is subtracted from anything, either a whole is subtracted from a whole, or a part from a part, or a whole from a part, or a part from a whole. But a whole is not subtracted from a whole; for when a single verse is set down, if it is the whole that is subtracted we shall be subtracting the whole verse. And thus, if the verse from which the subtraction is made still remains, no subtraction at all will have been made from it; for how is it possible for the whole still to remain when it has been subtracted? And if it does not remain, clearly no subtraction has taken place from what does not exist.
Consequently a whole is not subtracted from a whole. 163
Nor, again, is a whole subtracted from a part. For the whole is not included in the part,—μῆνιν ἄειδε θεὰ Πηληιάδεω Ἀχιλῆος, for instance, in μῆνιν,—but what is subtracted ought to be included in that which suffers the subtraction. It remains, then, to say that either a part is subtracted from a whole or a part from a part. But this too is questionable. For if μῆνιν is subtracted from the whole verse, it is also subtracted from itself; for the whole verse was conceived as including it. Besides, if it is taken from the whole, and the whole was μῆνιν ἄειδε θεὰ Πηληιάδεω Ἀχιλῆος, then ἄειδε θεὰ Πηληιάδεω Ἀχιλῆος ought also to have been diminished and not to have remained unaltered, since nothing which suffers subtraction remains unaltered.
And μῆνιν itself, too, when being sub- 164

[b] *Cf. P.H.* iii. 85, *Adv. Phys.* i. 280 ff., 308 ff.

μῆνιν, ἀφ' ὅλου ἐκείνου λαμβάνον τὴν ἀφαίρεσιν, ἔχειν τι ἐξ ἑκάστου τῶν ἐν ἐκείνῳ, ὃ πάλιν ἐστὶ ψεῦδος. εἰ οὖν μήτε ὅλον στίχον ἀπὸ στίχου δυνατὸν μερίζειν μήτε μέρος στίχου ἀπὸ μέρους μήτε ὅλον ἀπὸ μέρους μήτε μέρος ἀφ' ὅλου, καὶ παρὰ ταῦτα οὐδὲν ἐνδέχεται ποιεῖν, ἀδύνατος τῷ γραμματικῷ ὁ μερισμός.

165 Οὐ μὴν ἀλλὰ καὶ ἡ κατὰ τὰς συναλειπτικῶς ἐκφερομένας λέξεις τινῶν πρόσθεσις οὐκ ἔσται. καὶ τοῦτ' ἔσται σαφές, ἂν μὴ ἐπὶ συλλαβῶν ἢ στοιχείων χειρίζηται ὁ λόγος, ὧν μάλιστα τὰς προσθέσεις ἐν τοῖς μερισμοῖς ποιοῦνται οἱ γραμματικοί, ἀλλ' ἐπὶ ὅλων λέξεων. ὑποκειμένου τοίνυν ἡμιστιχίου τοῦ "ἄειδε θεὰ Πηληιάδεω Ἀχιλῆος" (ἔστω γὰρ πρὸς τὸ παρὸν τουτὶ ἡμιστίχιον, καὶ προσλαμβανέτω τὸ μῆνιν, ὥστε τὸ ἐξ ἀμφοτέρων ἡρωικὸν γενέσθαι μέτρον) ζητοῦμεν
166 τίνι ἡ πρόσθεσις γίνεται; ἤτοι γὰρ ἑαυτῷ τὸ μῆνιν προστίθεται ἢ τῷ προϋποκειμένῳ ἡμιστιχίῳ ἢ τῷ ἐξ ἀμφοτέρων ἀποτελεσθέντι ἡρωικῷ μέτρῳ. καὶ ἑαυτῷ μὲν οὐκ ἂν προστεθείη· μὴ ὂν γὰρ ἕτερον ἑαυτοῦ καὶ μὴ διπλασιάζον ἑαυτὸ οὐκ ἂν λέγοιτο ἑαυτῷ προστίθεσθαι. τῷ δὲ προϋποκειμένῳ ἡμιστιχίῳ πῶς ἐνδέχεται; ὅλῳ μὲν γὰρ αὐτῷ προστιθέμενον καὶ αὐτὸ παρισαζόμενον ἐκείνῳ
167 ἡμιστίχιον γενήσεται, ταύτῃ τε ἀκολουθήσει καὶ τὸ μέγα ἡμιστίχιον λέγειν εἶναι βραχύ, βραχεῖ συνεξισούμενον τῷ μῆνιν, καὶ τὸ βραχὺ μέγα, μείζονι ἀντιπαρῆκον τῷ ἡμιστιχίῳ, εἴπερ τῷ παντὶ ἡμιστιχίῳ προστίθοιτο τὸ μῆνιν, οἷον τῷ ἄειδε ⟨καὶ τὰ λοιπά⟩.[1] καὶ εἰ μὲν μόνον αὐξήσει τὸ ἄειδε, τὸ ὅλον οὐ ποιήσει στίχον. λείπεται οὖν φάσκειν

tracted from that whole verse ought to have extracted something from each of the parts of that whole; and that again is false. If, then, it is not possible to divide the whole verse from the whole or a part of the verse from a part or the whole from a part or a part from the whole, and no other operation besides these is feasible, division is impossible for the Grammarian.

Moreover, even in phrases which contain elisions 165
there will be no addition of any letters. And this will
be clear if language be treated not as a matter of
syllables or elements,—and it is mainly these which
the Grammarians use as additions in their divisions,—
but as a matter of whole phrases. So if the half-verse
ἄειδε θεὰ Πηληιάδεω Ἀχιλῆος is set down (for let it be
granted for the present that this is a half-verse, and
let it have μῆνιν added to it so that the combination
forms a hexameter), to what, we ask, is the addition
made? For μῆνιν is added either to itself or to the 166
half-verse already set down or to the hexameter made
up of the two together. But it will not be added to
itself; for as it is not other than itself and does not
double itself it will not be said to be added to itself.
And how can it be added to the half-verse already
set down? For as it is added to the whole of it, it too
being made equal to the half-verse will become a half-
verse, and thus it will follow that we must say that the 167
large half-verse is short, being made equal to the
short one, μῆνιν, and the short one large, as being
equal in extent to the larger half-verse, if in fact
μῆνιν is added to the whole half-verse, such as ἄειδε
etc. But if it shall increase ἄειδε only, the whole will
not make a verse. It remains then to assert that

[1] ⟨καὶ τὰ λοιπά⟩ addidi.

τῷ ἐξ ἀμφοῖν, αὐτοῦ τε τοῦ μῆνιν καὶ τοῦ προϋποκειμένου ἡμιστιχίου, ἀποτελουμένῳ ἑξαμέτρῳ καὶ ἡρωικῷ στίχῳ προστίθεσθαι. ὃ τελέως ἦν
168 ἀπίθανον· τὸ γὰρ ἐπιδεχόμενον πρόσθεσιν προϋπόκειται τῆς προσθέσεως, οὐ μὴν τὸ γινόμενον ἐκ τῆς προσθέσεως προϋπόκειται ταύτης. οὐκ ἄρα οὖν τῷ γινομένῳ ἐκ τῆς προσθέσεως τοῦ μῆνιν ἑξαμέτρῳ στίχῳ προστίθεται τὸ μῆνιν· ὅτε μὲν γὰρ γίνεται ἡ πρόσθεσις, οὔπω ἑξάμετρός ἐστιν, ὅτε δὲ ἔστιν ἑξάμετρος, οὐκέτι γίνεται ἡ πρόσθεσις. πλὴν συνῆκται τὸ προκείμενον, καὶ μήτε προσθέσεως μήτε ἀφαιρέσεως οὔσης ἀναιρεῖται ὁ προειρημένος τοῦ μερισμοῦ τρόπος.

Ἀλλὰ δὴ πάλιν τὴν ἐν τούτοις τῶν γραμματικῶν ἀκρίβειαν κατανοήσαντες, φέρε καὶ τῆς ἐν τῷ γράφειν αὐτῶν δυνάμεως ἀποπειραθῶμεν.

Θʹ.—ΠΕΡΙ ΟΡΘΟΓΡΑΦΙΑΣ

169 Τὴν γὰρ ὀρθογραφίαν φασὶν ἐν τρισὶ κεῖσθαι τρόποις, ποσότητι ποιότητι μερισμῷ. ποσότητι μὲν οὖν, ὅταν ζητῶμεν εἰ ταῖς δοτικαῖς προσθετέον τὸ ι, καὶ εὐχάλινον καὶ εὐώδινας τῷ ι μόνον γραπτέον ἢ τῇ ει· ποιότητι δέ, ὅταν σκεπτώμεθα πότερον διὰ τοῦ ζ γραπτέον ἐστὶ τὸ σμιλίον καὶ τὴν Σμύρναν ἢ διὰ τοῦ σ· μερισμῷ δέ, ἐπειδὰν διαπορῶμεν περὶ τῆς ὄβριμος λέξεως, πότερόν ποτε τὸ β τῆς δευτέρας ἐστὶ συλλαβῆς ἀρχὴ ἢ τῆς προηγουμένης πέρας, καὶ ἐπὶ τοῦ Ἀριστίων
170 ὀνόματος ποῦ τακτέον τὸ σ. πάλιν δ' ἡ τοιαύτη τεχνολογία, ἵνα μηδὲν τῶν ἀπορωτέρων κινῶμεν, μάταιος εἶναι φαίνεται, πρῶτον μὲν ἐκ τῆς δια-

μῆνιν is added to the combination of the two,—of μῆνιν itself and the previously stated half-verse,—which is the complete hexameter and heroic verse. 168
But this is wholly incredible ; for that which receives addition subsists before the addition, but that which results from the addition certainly does not subsist before it. Therefore μῆνιν is not added to that which results from the addition of μῆνιν to the hexameter verse ; for when the addition is being made, it is not as yet a hexameter, and when it is a hexameter the addition is no longer being made. Howbeit the task we proposed is now concluded, and, as neither addition exists nor subtraction, the method of division stated above is destroyed.

But now that we have again had an insight into the precision of the Grammarians in these matters, come and let us make trial of their power in the matter of writing.

Chapter IX.—Concerning Orthography

Orthography, they say, is to be found in three 169
modes,—quantity, quality, division. In quantity when we inquire if ι should be added to datives and whether εὐχάλινον and εὐώδινας should be written with an ι only or with ει. And in quality when we consider whether σμιλίον and Σμύρναν are to be written with a ζ or with a σ. And in division when we question regarding the word ὄβριμος whether the β is the beginning of the second syllable or the end of the first, and in the case of the name Ἀριστίων
where we should place the σ. But here again, not to 170
raise any worse objections, technology of this sort appears to be useless, firstly because of the disagree-

φωνίας, ἔπειτα δὲ καὶ ἐξ αὐτῶν τῶν ἀποτελεσμάτων. καὶ ἐκ μὲν τῆς διαφωνίας, ἐπείπερ οἱ τεχνικοὶ μάχονταί τε καὶ εἰς αἰῶνα μαχήσονται πρὸς ἀλλήλους, τῶν μὲν οὕτως τῶν δὲ ἐκείνως
171 τὸ αὐτὸ γράφειν ἀξιούντων. ὅθεν καὶ οὕτως αὐτοὺς ἐρωτητέον. εἰ χρειώδης ἐστὶν ἡ περὶ ὀρθογραφίας τεχνολογία τῷ βίῳ, ἐχρῆν ἡμᾶς τε καὶ ἕκαστον τῶν διαφωνούντων περὶ αὐτῆς γραμματικῶν, ἀνεπικρίτου ἀκμὴν καθεστώσης τῆς κατὰ ταύτην διαφωνίας, παραποδίζεσθαι εἰς ἃ ἂν δέῃ
172 γράφειν. οὔτε δὲ ἡμῶν οὔτε τούτων ἕκαστος παραποδίζεται, ἀλλὰ συμφώνως πάντες τυγχάνουσι τῆς προθέσεως, ἅτε δὴ μὴ ἀπ᾽ ἐκείνης ἀλλ᾽ ἀπὸ κοινοτέρας τινὸς καὶ συμφώνου ὁρμώμενοι τριβῆς, καθ᾽ ἣν τὰ μὲν κατ᾽ ἀνάγκην ὀφείλοντα παραλαμβάνεσθαι στοιχεῖα πρὸς τὴν μήνυσιν τοῦ ὀνόματος πάντες παραλαμβάνουσι, καὶ οἱ γραμματικοὶ καὶ οἱ μὴ γραμματικοί, περὶ δὲ τῶν μὴ κατ᾽ ἀνάγκην ἀδιαφοροῦσιν. οὐκ ἄρα χρειώδης ἐστὶν ἡ περὶ ὀρθογραφίας παρὰ τοῖς γραμματικοῖς ὑφήγησις.

173 Ἀλλ᾽ ὁ μὲν ἀπὸ τῆς διαφωνίας ἔλεγχος τοιοῦτος, ὁ δὲ ἀπὸ τῶν ἀποτελεσμάτων ἐμφανής. οὐδὲν γὰρ βλαπτόμεθα, ἐάν τε σὺν τῷ ι γράφωμεν τὴν δοτικὴν πτῶσιν ἐάν τε μή, καὶ ἐάν τε διὰ τοῦ σ τὸ σμιλίον καὶ τὴν Σμύρναν ἐάν τε διὰ τοῦ ζ, καὶ ἐπὶ τοῦ Ἀριστίων ὀνόματος ἐάν τε τῇ προηγουμένῃ συλλαβῇ τὸ σ προσμερίζωμεν ἐάν τε τῇ ἐπιφερο-
174 μένῃ τοῦτο συντάττωμεν. εἰ μὲν γὰρ παρὰ τὸ διὰ τοῦ σ ἀλλὰ μὴ διὰ τοῦ ζ γράφειν τὸ σμιλίον οὐκέτι σμιλίον γίνεται ἀλλὰ δρέπανον, καὶ εἰ παρὰ τὸ τοῦ Ἀριστίων ὀνόματος οὕτως ἀλλὰ μὴ ἐκείνως συντάσσεσθαι τὸ σ ὁ Ἀριστίων, καθώς φησί τις τῶν

ments about it, and next because of its actual results themselves. Because of the disagreements, inasmuch as the technicians fight and will fight to eternity with one another, some insisting on writing the same word in this way and some in that. Hence 171
we must question them in this wise : If the technology which deals with orthography is profitable for life, both we and each of the Grammarians who dispute about it ought to have been in a tangle as to what we ought to write, seeing that the dispute about it is still unsettled. But neither any of us nor 172
any of them is in a tangle, but we all achieve our purpose without dispute, inasmuch as we set out not from this technology but from more general and undisputed practice, in accordance with which we all —Grammarians and non-Grammarians alike—adopt the elements which necessarily must be adopted for the indication of the noun, while as to such as are not necessary we are indifferent. Therefore the instruction given by the Grammarians regarding orthography is not profitable.

173 Such, then, is the refutation based on their disagreement, and that based on the actual results is obvious. For we are in no wise injured whether we write the dative case with an ι or without one, and whether we write σμιλίον and Σμύρναν with a σ or with a ζ, and whether, in the case of the name Ἀριστίων, we attach the σ to the preceding syllable or assign it to that which follows. For if σμιλίον, because it is written 174
with a σ and not with a ζ, no longer is σμιλίον ("a scalpel") but δρέπανον ("a sickle"), and if, because the σ in the name Ἀριστίων is assigned to this syllable rather than to that, Ἀριστίων ("the breakfaster")—as one of the witty fellows says—becomes Δειπνίων

χαριεντιζομένων, Δειπνίων γίνεται, ἥρμοζε μὴ ἀδιαφορεῖν. εἰ δ' ὅπως ἂν ἔχῃ τὰ τῆς γραφῆς, τὸ σμιλίον ⟨ἐστὶ σμιλίον⟩,[1] ἐάν τε διὰ τοῦ σ ἐάν τε διὰ τοῦ ζ κατάρχηται, ὅ τε Ἀριστίων ἀεί ποτέ ἐστιν Ἀριστίων, ἐάν τε τῷ ι ἐάν τε τῷ τ τὸ σ προσμερίζωμεν, τίς χρεία τῆς πολλῆς καὶ ματαίας παρὰ τοῖς γραμματικοῖς περὶ τούτων μωρολογίας;

175 Κεφαλαιωδέστερον δὴ καὶ περὶ ὀρθογραφίας διεξιόντες, ἴδωμεν εἰς συμπλήρωσιν τῆς πρὸς τὸ τεχνικὸν μέρος αὐτῶν ἀντιρρήσεως πότερον ἔχουσί τινα πρὸς τὸ ἑλληνίζειν συνεστῶσαν μέθοδον ἢ οὐδαμῶς.

Ι'.—ΠΕΡΙ ΕΛΛΗΝΙΣΜΟΝ

176 Ὅτι μὲν δεῖ τινὰ φειδὼ ποιεῖσθαι τῆς περὶ τὰς διαλέκτους καθαριότητος, αὐτόθεν συμφανές· ὅ τε γὰρ ἑκάστοτε βαρβαρίζων καὶ σολοικίζων ὡς ἀπαίδευτος χλευάζεται, ὅ τε ἑλληνίζων ἱκανός ἐστι πρὸς τὸ σαφῶς ἅμα καὶ ἀκριβῶς παραστῆσαι τὰ νοηθέντα τῶν πραγμάτων. ἤδη δὲ τοῦ ἑλληνισμοῦ δύο εἰσὶ διαφοραί· ὃς μὲν γάρ ἐστι κεχωρισμένος τῆς κοινῆς ἡμῶν συνηθείας καὶ κατὰ γραμματικὴν ἀναλογίαν δοκεῖ προκόπτειν, ὃς δὲ κατὰ τὴν ἑκάστου τῶν Ἑλλήνων συνήθειαν ἐκ παραπλασμοῦ καὶ τῆς ἐν ταῖς ὁμιλίαις παρατηρή-
177 σεως ἀναγόμενος. οἷον ὁ μὲν τῆς Ζεύς ὀρθῆς πτώσεως τὰς πλαγίους σχηματίζων τοῦ Ζεός τῷ Ζεΐ τὸν Ζέα κατὰ τὸν πρότερον τοῦ ἑλληνισμοῦ χαρακτῆρα διαλέκεται, ὁ δὲ ἀφελῶς τοῦ Ζηνός λέγων καὶ τῷ Ζηνί καὶ τὸν Ζῆνα κατὰ τὸν δεύτερον καὶ συνηθέστερον ἡμῖν. πλὴν δυοῖν ὄντων τῶν ἑλληνισμῶν εὔχρηστον μὲν εἶναί φαμεν τὸν

("the diner"), then it would be fitting not to be indifferent. But if, whatever be the form of writing, σμιλίον is σμιλίον, whether it begins with a σ or with a ζ, and Ἀριστίων is always Ἀριστίων, whether we attach the σ to the ι or to the τ, what is the use of the long, vain and stupid disputation about these points which is carried on by the Grammarians?

And now that we are discussing orthography in 175
rather a summary fashion, in order to complete our confutation of their technical section let us consider whether or not they have any systematic method of dealing with "hellenism" (or "Greek idiom").

Chapter X.—Concerning Greek Idiom

That one must preserve with some care purity of 176
speech is at once plain; for the man who is constantly using barbarisms and solecisms is jeered at as one of no culture, whereas he who speaks good Greek is capable of presenting his ideas both clearly and exactly. But now there are two distinct kinds of "hellenism": one stands apart from our common usage and seems to proceed in accordance with grammatical analogy; the other conforms to the common usage of each of the Greeks and is derived from framing words and from observation in ordinary converse. For 177
example, the man who forms from the nominative Ζεύς the oblique cases Ζεός, Ζεΐ, Ζέα, frames the declension in accordance with the first kind of "hellenism," but he who simply says Ζηνός, Ζηνί, Ζῆνα frames it in accordance with the second, the one more usual with us. Only, as two forms of "hellenism" exist, we declare, for the reasons already stated, that the

[1] ⟨ἐστὶ σμιλίον⟩ addidi (σμιλίον post γραφῆς add. cj. Bekk.).

δεύτερον διὰ τὰς προειρημένας αἰτίας, ἄχρηστον
178 δὲ τὸν πρῶτον διὰ τὰς λεχθησομένας. ὥσπερ γὰρ
ἐν πόλει νομίσματός τινος προχωροῦντος κατὰ τὸ
ἐγχώριον ὁ μὲν τούτῳ στοιχῶν δύναται καὶ τὰς
ἐν ἐκείνῃ τῇ πόλει διεξαγωγὰς ἀπαραποδίστως
ποιεῖσθαι, ὁ δὲ τοῦτο μὲν μὴ παραδεχόμενος ἄλλο
δέ τι καινὸν χαράσσων ἑαυτῷ καὶ τούτῳ νομι-
στεύεσθαι θέλων μάταιος καθέστηκεν, οὕτω κἀν
τῷ βίῳ ὁ μὴ βουλόμενος τῇ συνήθως παραδεχθείσῃ,
καθάπερ νομίσματι, ὁμιλίᾳ κατακολουθεῖν ἀλλ'
179 ἰδίαν αὐτῷ τέμνειν μανίας ἐγγὺς ἐστίν. διόπερ εἰ
οἱ γραμματικοὶ ὑπισχνοῦνται τέχνην τινὰ τὴν καλου-
μένην ἀναλογίαν παραδώσειν, δι' ἧς κατ' ἐκεῖνον
ἡμᾶς τὸν ἑλληνισμὸν ἀναγκάζουσι διαλέγεσθαι,
ὑποδεικτέον ὅτι ἀσύστατός ἐστιν αὕτη ἡ τέχνη, δεῖ
δὲ τοὺς ὀρθῶς βουλομένους διαλέγεσθαι τῇ ἀτέχνῳ
καὶ ἀφελεῖ κατὰ τὸν βίον καὶ τῇ κατὰ τὴν κοινὴν
τῶν πολλῶν συνήθειαν παρατηρήσει προσανέχειν.
180 Εἴπερ οὖν ἔστι τις περὶ ἑλληνισμὸν τέχνη, ἤτοι
ἔχει ἀρχὰς ἐφ' αἷς συνέστηκεν ἢ οὐκ ἔχει. καὶ
μὴ ἔχειν μὲν οὐκ ἂν φαῖεν οἱ γραμματικοί· πᾶσα
γὰρ τέχνη ἀπό τινος ἀρχῆς ὀφείλει συνίστασθαι.
εἰ δὲ ἔχει, ἤτοι τεχνικὰς ταύτας ἔχει ἢ ἀτέχνους.
καὶ εἰ μὲν τεχνικάς, πάντως ἢ ἀφ' ἑαυτῶν ἢ
ἀπ' ἄλλης τέχνης συνέστησαν, κἀκείνη πάλιν ἀπὸ
τρίτης, καὶ ἡ τρίτη ἀπὸ τετάρτης, καὶ τοῦτ' εἰς
ἄπειρον, ὥστε ἄναρχον γιγνομένην τὴν περὶ ἑλλη-
181 νισμὸν τέχνην μηδὲ τέχνην ὑπάρχειν· εἰ δὲ ἀτέχνους,
οὐκ ἄλλαι τινὲς εὑρεθήσονται παρὰ τὴν συνήθειαν.
ἡ ἄρα συνήθεια τοῦ τί τέ ἐστιν ἑλληνικὸν καὶ τί
ἀνελλήνιστον γίνεται κριτήριον, καὶ οὐκ ἄλλη τις
182 περὶ τὸν ἑλληνισμὸν τέχνη. ἄλλως τε, ἐπεὶ τῶν

second is of great use, but the first useless, for reasons now to be stated. For just as in a city where a certain 178
local coinage is current, he who makes use of this is able to carry on his business in that city without hindrances, but he who does not adopt it, but coins for himself some new money and desires to have this passed, is a fool, so also in ordinary intercourse the man who refuses to follow the mode of converse—like the coinage—which is usually adopted, and cuts out a new way of his own, is near to madness. Where- 179
fore, if the Grammarians promise to present us with an art described as "analogy," by which they compel us to discourse in accordance with that kind of "hellenism," one must point out that this art has no foundation, and that those who wish to discourse correctly must cleave to the non-technical, simple and ordinary style and to the observing of the rules which accord with the usage of the majority.

If there does exist any art concerned with "hellen- 180
ism," it either has or has not principles upon which it is based. That it has none the Grammarians would deny, for every art must be based on some principle. But if it has some, the principles it has are either technical or non-technical. If they are technical, they certainly are constructed either from themselves or from another art, and that again from a third, and the third from a fourth, and so on to infinity, so that the art dealing with "hellenism" is found to have no beginning (or principle) and thus to be no art. And 181
if its principles are untechnical, none will be found other than common usage. Thus common usage becomes the criterion of what is "hellenic" and what is not "hellenic," and not some other art concerning "hellenism."—Again, since some arts—such as 182

τεχνῶν αἱ μὲν τῷ ὄντι εἰσὶ τέχναι, ὡς ἡ ἀνδριαντοποιικὴ καὶ ζωγραφία, αἱ δὲ ἐπαγγέλματι μέν εἰσι τέχναι, οὐ πάντως δὲ καὶ κατ' ἀλήθειαν, ὡς Χαλδαϊκή τε καὶ θυτική, ἵνα μάθωμεν πότερόν ποτε καὶ ἡ περὶ τὸν ἑλληνισμὸν λεγομένη τέχνη ὑπόσχεσις μόνον ἐστὶν ἢ καὶ ὑποκειμένη δύναμις, δεήσει κριτήριόν τι ἡμᾶς ἔχειν εἰς τὴν ταύτης
183 δοκιμασίαν. τοῦτ' οὖν τὸ κριτήριον πάλιν ἤτοι τεχνικόν τί ἐστι καὶ περὶ ἑλληνισμόν, εἴγε καὶ τῆς περὶ τὸν ἑλληνισμὸν κρινούσης, εἰ ὑγιῶς κρίνει, δοκιμαστικὸν καθέστηκεν, ἢ ἄτεχνον. ἀλλὰ τεχνικὸν μὲν περὶ ἑλληνισμὸν οὐκ ἂν εἴη διὰ τὴν προειρημένην εἰς ἄπειρον ἔκπτωσιν· ἄτεχνον δ' εἰ λαμβάνοιτο τὸ κριτήριον, οὐκ ἄλλο τι εὑρήσομεν ἢ τὴν συνήθειαν. ἡ ἄρα συνήθεια καὶ αὐτὴν τὴν περὶ ἑλληνισμὸν τέχνην κρίνουσα οὐ δεήσεται τέχνης.

184 Εἴπερ δὲ οὐκ ἄλλως ἔστιν ἑλληνίζειν ἐὰν μὴ παρὰ γραμματικῆς μάθωμεν τὸ ἑλληνικόν, ἤτοι ἐναργές ἐστι τοῦτο καὶ ἐξ αὐτοῦ βλεπόμενον ἢ ἀδηλότερον. ἀλλ' ἐναργὲς μὲν οὐκ ἔστιν, ἐπεὶ σύμφωνον ἂν ἦν παρὰ πᾶσιν ὡς τὰ λοιπὰ τῶν
185 ἐναργῶν. καὶ ἄλλως πρὸς μὲν τὴν τοῦ ἐναργοῦς ἀντίληψιν οὐδεμιᾶς τέχνης ἐστὶ χρεία, καθάπερ οὐδὲ πρὸς τὸ λευκὸν ὁρᾶν ἢ γλυκέος γεύεσθαι ἢ θερμοῦ θιγγάνειν· πρὸς δὲ τὸ ἑλληνίζειν μεθόδου τινὸς καὶ τέχνης κατὰ τοὺς γραμματικούς ἐστι χρεία. οὐκ ἄρα ἐναργές ἐστι τὸ ἑλληνίζειν.
186 ἄδηλον δὲ εἴπερ ἐστί, πάλιν ἐπεὶ τὸ ἄδηλον ἔκ τινος ἑτέρου γνωρίζεται, ἤτοι φυσικῷ τινὶ κατακολουθητέον κριτηρίῳ, ἐξ οὗ διαγιγνώσκεται τί τὸ ἑλληνικὸν καὶ τί τὸ ἀνελλήνιστον, ἢ τῇ ἑνὸς

statuary and painting—are really arts, but others which are claimed as arts—such as astrology and the art of the haruspex—are not wholly and truly arts, in order that we may learn whether the so-called art of "hellenism" is merely a profession or a substantial power, it will be necessary for us to possess some criterion by which to test it. Then this criterion, 183
too, is either a technical one and deals with "hellenism"—since it is to test whether the art which judges "hellenism" judges soundly—or it is non-technical. But it will not be a technical one concerning "hellenism" because of the regress *ad infinitum* already stated[a]; and if the criterion is to be taken as non-technical, we shall find no other criterion than common usage. Common usage, then, as judging the art of "hellenism" itself, will not need art.

But if it is not possible to speak good Greek other- 184
wise than by learning good Greek from the Art of Grammar, either it is something evident and seen of itself or it is obscure. But it is not evident, since then it would have been agreed upon by all men, like all other evident things. And besides, there is need of 185
no art for the perception of what is evident, any more than for seeing the white object, tasting the sweet, or feeling the hot; but for speaking good Greek there is need, according to the Grammarians, of a certain method and art. Therefore, speaking good Greek is not evident.—But if it is obscure, then again, since 186
what is obscure is revealed by something else, we must either be guided by some natural criterion by means of which it is determined what is good Greek and what is not, or we must employ, in order to grasp

[a] *Cf.* § 180.

συνηθείᾳ ὡς ἄκρως ἑλληνίζοντος χρηστέον πρὸς
187 τὴν τούτου κατάληψιν, ἢ τῇ πάντων. ἀλλὰ
φυσικὸν μὲν κριτήριον εἰς τὸ ἑλληνικὸν καὶ τὸ
μὴ τοιοῦτον οὐδὲν ἔχομεν· τοῦ γὰρ Ἀττικοῦ τὸ
τάριχος λέγοντος ὡς ἑλληνικὸν καὶ τοῦ Πελοπον-
νησίου ὁ τάριχος προφερομένου ὡς ἀδιάστροφον,
καὶ τοῦ μὲν τὴν στάμνον ὀνομάζοντος τοῦ δὲ τὸν
στάμνον, οὐδὲν ἔχει ἐξ ἑαυτοῦ κριτήριον πιστὸν
ὁ γραμματικὸς εἰς τὸ οὕτως ἀλλὰ μὴ οὕτως δεῖν
λέγειν, εἰ μὴ ἄρα τὴν ἑκάστου συνήθειαν, ἥτις
188 οὔτε τεχνικὴ οὔτε φυσική ἐστιν. τῇ δὲ τινὸς
συνηθείᾳ δεῖν ἀκολουθεῖν εἴπερ ἐροῦσιν, ἤτοι
φάσει μόνον ἐροῦσιν ἢ ἐμμεθόδοις ἀποδείξεσι
χρησάμενοι. ἀλλὰ φάσιν μὲν λέγουσι φάσιν ἀντι-
θήσομεν περὶ τοῦ τοῖς πολλοῖς μᾶλλον ἢ τῷ ἑνὶ
δεῖν ἀκολουθεῖν· ἐμμεθόδως δὲ ἀποδεικνύντες ὅτι
οὗτος ἑλληνίζει, ἀναγκασθήσονται ἐκείνην τὴν
μέθοδον κριτήριον ἑλληνισμοῦ λέγειν δι' ἣν καὶ
οὗτος ἑλληνίζων δέδεικται, ἀλλ' οὐχὶ τοῦτον.
189 λείπεται οὖν τῇ πάντων συνηθείᾳ προσέχειν. εἰ
δὲ τοῦτο, οὐ χρεία τῆς ἀναλογίας ἀλλὰ παρατηρή-
σεως τοῦ πῶς οἱ πολλοὶ διαλέγονται καὶ τί ὡς
ἑλληνικὸν παραδέχονται ἢ ὡς οὐ τοιοῦτον ἐκκλί-
νουσιν. τό γε μὴν ἑλληνικὸν ἤτοι φύσει ἐστὶν ἢ
θέσει. καὶ φύσει μὲν οὐκ ἔστιν, ἐπεὶ οὐκ ἄν ποτε
ταὐτὸν τοῖς μὲν ἑλληνικὸν ἐδόκει τυγχάνειν τοῖς
190 δὲ οὐχ ἑλληνικόν· θέσει δὲ εἴπερ ἔστι καὶ νόμῳ
τῶν ἀνθρώπων, ὁ συνασκηθεὶς μάλιστα καὶ τριβεὶς
ἐν τῇ συνηθείᾳ, οὗτος ἑλληνίζει, καὶ οὐχ ὁ τὴν
ἀναλογίαν ἐπιστάμενος. καὶ γὰρ ἄλλως ἔνεστι

it, either the habit of speech of one man, as being pre-eminent in " hellenism," or that of all men. But 187
we possess no natural criterion regarding what is good Greek and what is not; for when the man of Attica uses τὸ τάριχος (" the dried fish ") as being good Greek,[a] while the Peloponnesian enounces ὁ τάριχος as being correct, and the one speaks of ἡ στάμνος (" the jar "), the other of ὁ στάμνος, the Grammarian possesses no criterion which is of itself reliable for deciding that one ought to speak in this way rather than in that way, unless indeed it be each man's practice, and this is neither technical nor natural. And if 188
they shall declare that we must follow the customary practice of a certain person, they will declare this either by mere assertion or by employing methodical proofs. But to their assertion we shall reply by the counter-assertion that one ought to follow the many rather than the one; and if they try to prove methodically that this man speaks good Greek, they will be forced to say that that method by which this man has been proved to speak good Greek is the criterion, and not this man himself. It remains, then, to hold fast 189
by the common usage of all men. And if so, there is no need of analogy but of observing how most men converse and what they adopt as good Greek or avoid as not good.—However, good Greek exists either by nature or by convention. But it does not exist by nature,[b] since then the same phrase would never have seemed good Greek to some and not good Greek to others; and if it exists by convention and human 190
enactment, the man who speaks good Greek is he who is most practised and versed in common usage, and not he who understands analogy.—By another

[a] *Cf.* § 148.

[b] *Cf.* § 145.

παραστῆσαι ὅτι οὐ δεόμεθα πρὸς τὸ ἑλληνίζειν
191 τῆς γραμματικῆς. ἐν γὰρ ταῖς ἀνὰ χεῖρα ὁμιλίαις ἤτοι ἀντικόψουσιν ἡμῖν οἱ πολλοὶ ἐπί τισι λέξεσιν ἢ οὐκ ἀντικόψουσιν. καὶ εἰ μὲν ἀντικόψουσιν, εὐθὺς καὶ διορθώσονται ἡμᾶς, ὥστε παρὰ τῶν ἐκ τοῦ βίου καθεστώτων ἀλλ' οὐχὶ παρὰ γραμματικῶν
192 ἔχειν τὸ ἑλληνίζειν. εἰ δ' οὐ δυσχεραίνουσιν ἀλλ' ὡς σαφέσι καὶ ὀρθῶς ἔχουσι συμπεριφέροιντο τοῖς λεγομένοις, καὶ ἡμεῖς αὐτοῖς ἐπιμενοῦμεν. κατά τε ταύτην τὴν ἀναλογίαν ἤτοι πάντες ἢ οἱ πλεῖστοι ἢ οἱ πολλοὶ διαλέγονται· οὔτε δὲ πάντες οὔθ' οἱ πλεῖστοι οὔθ' οἱ πολλοί· μόλις γὰρ δύο ἢ τρεῖς τοιοῦτοι εὑρίσκονται, οἱ δὲ πλεῖστοι οὐδὲ ἴσασιν
193 αὐτήν. τοίνυν ἐπεὶ τῇ τῶν πολλῶν συνηθείᾳ καὶ οὐ τῇ τῶν δυοῖν ἀναγκαῖον κατακολουθεῖν, ῥητέον τὴν παρατήρησιν τῆς κοινῆς συνηθείας χρησιμεύειν πρὸς τὸ ἑλληνίζειν, ἀλλὰ μὴ τὴν ἀναλογίαν. ἐπὶ πάντων γε μὴν σχεδὸν τῶν χρησιμευόντων τῷ βίῳ μέτρον ἐστὶν ἱκανὸν τὸ μὴ παραποδίζεσθαι
194 πρὸς τὰς χρείας. διόπερ εἰ καὶ ὁ ἑλληνισμὸς διὰ δύο μάλιστα προηγούμενα ἔτυχεν ἀποδοχῆς, τήν τε σαφήνειαν καὶ τὴν προσήνειαν τῶν δηλουμένων (τούτοις γὰρ ἔξωθεν κατ' ἐπακολούθησιν συνέζευκται τὸ μεταφορικῶς καὶ ἐμφατικῶς καὶ κατὰ τοὺς ἄλλους τρόπους φράζειν), ζητήσομεν οὖν ἐκ ποτέρας ταῦτα μᾶλλον περιγίνεται, ἆρά γε τῆς κοινῆς συνηθείας ἢ τῆς ἀναλογίας, ἵνα ἐκείνῃ
195 προσθώμεθα. βλέπομεν δέ γε ὡς ἐκ τῆς κοινῆς συνηθείας μᾶλλον ἢ ὅτι ἐκ τῆς ἀναλογίας. ἐκείνῃ ἄρα ἀλλ' οὐ ταύτῃ χρηστέον. τὸ μὲν γὰρ τῆς ὀρθῆς πτώσεως ὁ Ζεύς οὔσης τὰς πλαγίους προφέρεσθαι Ζηνός Ζηνί Ζῆνα καὶ τῆς κύων κυνός

argument also it is possible to establish the fact that we do not require the Art of Grammar in order to speak good Greek. In familiar intercourse ordinary 191
people will either oppose us about certain phrases or will not oppose us. And if they oppose us, they will at once correct us, so that we have good Greek from those who live ordinary lives and not from the Grammarians. And if they are not vexed but concur in the 192
phrases we use as being clear and correct, we too shall abide by them.—Further, either all men, or most, or many converse in accordance with this "analogy" of theirs; but neither all, nor most, nor many do so; in fact hardly two or three are found to do so, and most men do not even know of it. Since, 193
then, we must necessarily follow the customary usage of the many and not that of the two, one must declare that the observing of the common usage is useful for speaking good Greek, and not analogy. Certainly, in the case of nearly all the things which are of use in ordinary life, the fact that one is not hindered in supplying one's needs is a sufficient criterion. Where- 194
fore, if "hellenism" has met with acceptance for two main reasons, its clear and agreeable presentation of the things described,[a]—and following on these qualities and externally joined to them is the employment of metaphor and emphasis and the other figures of speech,—we shall inquire by which of the two are 195
these qualities better secured, by common usage or by analogy, so that we may adhere to it. And we clearly see that they are better secured by common usage than by analogy. So then the former and not the latter must be employed. Now when the nominative case is Ζεύς the formations Ζηνός, Ζηνί, Ζῆνα

[a] *Cf.* § 176.

κυνί κύνα σαφές, ἀλλὰ καὶ ἀπρόσκοπον τοῖς πολλοῖς εἶναι φαίνεται· τοῦτο δέ ἐστι τὸ τῆς κοινῆς συνηθείας. τὸ δὲ ἀπὸ τῆς Ζεύς ὀρθῆς Ζεός λέγειν καὶ Ζεΐ καὶ Ζέα, καὶ ἀπὸ τῆς κύων σχηματίζειν κύωνος κύωνι κύωνα, ἢ ἀπὸ τῆς κυνός γενικῆς ἀξιοῦν τὴν ὀρθὴν κῦς ὑπάρχειν, καὶ ἐπὶ τῶν ῥηματικῶν φερήσω λέγειν καὶ βλεπήσω ὡς ποιήσω καὶ θελήσω, οὐ μόνον ἀσαφὲς ἀλλὰ καὶ γέλωτος ἔτι δὲ προσκοπῆς ἄξιον εἶναι
196 δοκεῖ· τοῦτο δὲ γίνεται ἀπὸ ἀναλογίας. τοίνυν, ὡς ἔφην, οὐ ταύτῃ χρηστέον ἀλλὰ τῇ συνηθείᾳ.

Μήποτε δὲ καὶ περιτρέπονται, καὶ ἐάν τε θελήσωσιν ἐάν τε καὶ μή, ἀναγκασθήσονται χρῆσθαι μὲν τῇ συνηθείᾳ παραπέμπειν δὲ τὴν ἀναλογίαν. σκοπῶμεν δ' ἐντεῦθεν τὸ λεγόμενον, τουτέστιν ἐκ
197 τῆς πρὸς αὐτοὺς ἀκολουθίας. ζητουμένου γὰρ τοῦ πῶς δεῖ λέγειν, χρῆσθαι ἢ χρᾶσθαι, φασὶν ὅτι χρᾶσθαι, καὶ ἀπαιτούμενοι τούτου τὴν πίστιν λέγουσιν ὅτι χρῆσις καὶ κτῆσις ἀνάλογά ἐστιν· ὡς οὖν κτᾶσθαι μὲν λέγεται κτῆσθαι δὲ οὐ λέγεται, οὕτω καὶ χρᾶσθαι μὲν ῥηθήσεται χρῆσθαι δὲ οὐ
198 πάντως. ἀλλ' εἰ ἐπακολουθῶν τις αὐτοῖς πύθοιτο "αὐτὸ δὲ τοῦτο τὸ κτᾶσθαι ὅτι ὀρθῶς εἴρηται, ἀφ' οὗ καὶ τὸ χρᾶσθαι ἀποδείκνυμεν, πόθεν ἴσμεν;" φήσουσιν ὅτι ἐν τῇ συνηθείᾳ λέγεται. τοῦτο δὲ λέγοντες δώσουσι τὸ τῇ συνηθείᾳ δεῖν ὡς κριτη-
199 ρίῳ προσέχειν, ἀλλὰ μὴ τῇ ἀναλογίᾳ. εἰ γὰρ ὅτι ἐν τῇ συνηθείᾳ λέγεται κτᾶσθαι, ῥητέον καὶ χρᾶσθαι, ὀφείλομεν παρέντες τὴν ἀναλογικὴν τέχνην ἐπὶ τὴν συνήθειαν ἀναδραμεῖν, ἀφ' ἧς κἀκείνη ἤρτηται.

Καὶ μὴν ἡ ἀναλογία ὁμοίων πολλῶν ὀνομάτων

as the oblique cases,—and κυνός, κυνί, κύνα, from the nominative κύων (" dog ")—are clear and also appear to ordinary men unobjectionable ; and these are the formations in common use. But to derive Ζεός, Ζεΐ, Ζέα from the nominative Ζεύς, and from κύων to form κύωνος, κύωνι, κύωνα, or from the genitive κυνός to claim that the nominative is κῦς, and in the case of the forms of verbs to speak of φερήσω and βλεπήσω, like ποιήσω and θελήσω, seems to be not only obscure but also deserving of ridicule and objectionable ; and these are formations due to analogy. So then, as I 196
have said, we should not employ this, but common usage.

Perhaps they may shift their ground ; but whether they wish it or not, they will be forced to employ customary usage and to dismiss analogy. Let us examine what they say by the method of pressing their own argument against themselves. When 197
it is asked whether one should say χρῆσθαι or χρᾶσθαι, they reply χρᾶσθαι, and on being asked for a proof of this, they assert that χρῆσις and κτῆσις are analogous ; as then we say κτᾶσθαι and do not say κτῆσθαι, so we shall say χρᾶσθαι and never χρῆσθαι. But if one should press them further and 198
ask how do we know that this word κτᾶσθαι itself, from which we deduce χρᾶσθαι, is a correct form, they will reply that it is the form customarily used. And in so saying they will be granting that one should adopt common usage as the criterion and not analogy. For if one ought to say χρᾶσθαι be- 199
cause κτᾶσθαι is the customary form, we ought to give up the art of analogy and go back to customary usage upon which the former depends.

Now analogy, in fact, is the comparison of many

ἐστὶ παράθεσις, τὰ δὲ ὀνόματα ταῦτα ἐκ τῆς
συνηθείας, ὥστε καὶ ἡ σύστασις τῆς ἀναλογίας
200 ἀπὸ τῆς συνηθείας πρόεισιν. τούτου δὲ οὕτως
ἔχοντος ἐρωτητέον τρόπῳ τῷδε. ἤτοι ἐγκρίνετε
τὴν συνήθειαν ὡς πιστὴν πρὸς διάγνωσιν ἑλλη-
νισμοῦ ἢ ἐκβάλλετε. εἰ μὲν ἐγκρίνετε, αὐτόθεν
συνῆκται τὸ προκείμενον, καὶ οὐ χρεία τῆς ἀνα-
λογίας· εἰ δὲ ἐκβάλλετε, ἐπεὶ καὶ ἡ ἀναλογία ἐκ
ταύτης συνίσταται, ἐκβάλλετε καὶ τὴν ἀναλογίαν.
καὶ πάλιν, ἄτοπον τὸ αὐτὸ καὶ ὡς πιστὸν προσ-
201 ίεσθαι καὶ ὡς ἄπιστον παραιτεῖσθαι. οἱ δὲ
γραμματικοὶ θέλοντες τὴν συνήθειαν ὡς ἄπιστον
ἐκβάλλειν καὶ πάλιν ταύτην ὡς πιστὴν παραλαμ-
βάνειν, τὸ αὐτὸ πιστὸν ἅμα καὶ ἄπιστον ποιήσουσιν.
ἵνα γὰρ δείξωσιν ὅτι οὐ διαλεκτέον κατὰ τὴν
συνήθειαν, εἰσάγουσι τὴν ἀναλογίαν· ἡ δὲ ἀναλογία
οὐκ ἰσχυροποιεῖται, εἰ μὴ συνήθειαν ἔχοι τὴν
202 βεβαιοῦσαν· τῇ ἄρα συνηθείᾳ ἐκβάλλοντες τὴν
συνήθειαν τὸ αὐτὸ πιστὸν ἅμα καὶ ἄπιστον ποιή-
σουσιν. ἐκτὸς εἰ μή τι φήσουσι μὴ τὴν αὐτὴν
συνήθειαν ἐκβάλλειν ἅμα καὶ προσίεσθαι, ἀλλ'
ἄλλην μὲν ἐκβάλλειν ἄλλην δὲ προσίεσθαι. ὅπερ
καὶ λέγουσιν οἱ ἀπὸ Πινδαρίωνος. ἀναλογία,
φασίν, ὁμολογουμένως ἐκ τῆς συνηθείας ὁρμᾶται·
203 ἔστι γὰρ ὁμοίου τε καὶ ἀνομοίου θεωρία, τὸ δὲ
ὅμοιον καὶ ἀνόμοιον ἐκ τῆς δεδοκιμασμένης λαμ-
βάνεται συνηθείας, δεδοκιμασμένη δὲ καὶ ἀρχαιο-
τάτη ἐστὶν ἡ Ὁμήρου ποίησις· ποίημα γὰρ οὐδὲν
πρεσβύτερον ἧκεν εἰς ἡμᾶς τῆς ἐκείνου ποιήσεως.
διαλεξόμεθα ἄρα τῇ Ὁμήρου κατακολουθοῦντες
204 συνηθείᾳ. ἀλλὰ πρῶτον μὲν οὐχ ὑπὸ πάντων
ὁμολογεῖται ποιητὴς ἀρχαιότατος εἶναι Ὅμηρος·

similar nouns, and these nouns are taken from customary speech, so that the substance of analogy is derived from customary speech. This being so, one 200
must question them in this way : Either you accept customary usage as reliable for the distinguishing of " hellenism " or you reject it. If you accept it, the problem before us is settled of itself, and there is no need of analogy ; but if you reject it, you reject analogy as well, since from it analogy derives its substance.—Again, it is absurd to admit a thing as trustworthy and to dismiss it as untrustworthy. But 201
the Grammarians, in their desire to reject common usage as untrustworthy and contrariwise to accept it as trustworthy, will cause the same thing to be at once trustworthy and untrustworthy. For in order to prove that we should not converse in accordance with common usage, they introduce analogy ; but analogy is not made secure unless it has common usage to support it ; so by rejecting common usage by means 202
of common usage, they will cause the same thing to be at once trustworthy and untrustworthy. Unless, indeed, they shall declare that they do not accept and reject simultaneously the same common usage, but reject one sort and accept another sort. And that is what is said by the School of Pindarion. Analogy, they say, confessedly sets out from common usage ; for it is the consideration of the like and the 203
unlike, and the like and the unlike are derived from approved common usage, and the poetry of Homer is that which is approved and most ancient ; for no poem older than his poetry has come down to us. Thus we shall converse following the usage of Homer. But, in the first place, it is not agreed by all that 204
Homer is the most ancient poet ; for some say that

ἔνιοι γὰρ Ἡσίοδον προήκειν τοῖς χρόνοις λέγουσιν, Λίνον τε καὶ Ὀρφέα καὶ Μουσαῖον καὶ ἄλλους παμπληθεῖς. οὐ μὴν ἀλλὰ καὶ πιθανόν ἐστι γεγονέναι μέν τινας πρὸ αὐτοῦ καὶ κατ' αὐτὸν ποιητάς, ἐπεὶ καὶ αὐτός πού φησι

τὴν γὰρ ἀοιδὴν μᾶλλον ἐπικλείουσ' ἄνθρωποι
ἥτις ἀκουόντεσσι νεωτάτη ἀμφιπέληται,

τούτους δὲ ὑπὸ τῆς περὶ αὐτὸν λαμπρότητος
205 ἐπεσκοτῆσθαι. καὶ εἰ ἀρχαιότατος δὲ ὁμολογοῖτο τυγχάνειν Ὅμηρος, οὐδὲν εἴρηται ὑπὸ τοῦ Πινδαρίωνος ἱκνούμενον. ὥσπερ γὰρ προηποροῦμεν περὶ τοῦ πότερόν ποτε τῇ συνηθείᾳ ἢ τῇ ἀναλογίᾳ χρηστέον, οὕτω καὶ νῦν διαφορήσομεν πότερον τῇ συνηθείᾳ ἢ τῇ ἀναλογίᾳ, καὶ εἰ τῇ συνηθείᾳ, ἆρα τῇ καθ' Ὅμηρον ἢ τῇ τῶν ἄλλων ἀνθρώπων·
206 πρὸς ὅπερ οὐδὲν εἴρηται. εἶτα κἀκείνην μάλιστα δεῖ τὴν συνήθειαν μεταδιώκειν ᾗ προσχρώμενοι οὐ γελασθησόμεθα· τῇ δὲ Ὁμηρικῇ κατακολουθοῦντες οὐ χωρὶς γέλωτος ἑλληνιοῦμεν, μάρτυροι λέγοντες καὶ "σπάρτα λέλυνται" καὶ ἄλλα τούτων ἀτοπώτερα. τοίνυν οὐδ' οὗτός ἐστιν ὁ λόγος ὑγιής, μετὰ τοῦ συγκεχωρῆσθαι τὸ κατασκευαζόμενον ὑφ' ἡμῶν, τουτέστι τὸ μὴ χρῆσθαι ἀνα-
207 λογίᾳ. τί γὰρ διήνεγκεν εἴτ' ἐπὶ τὴν τῶν πολλῶν εἴτ' ἐπὶ τὴν Ὁμήρου συνήθειαν ἐλθεῖν; ὡς γὰρ ἐπὶ τῆς τῶν πολλῶν τηρήσεώς ἐστι χρεία ἀλλ' οὐ τεχνικῆς ἀναλογίας, οὕτω καὶ ἐπὶ τῆς Ὁμήρου· τηρήσαντες γὰρ αὐτοὶ πῶς εἴωθε λέγειν, οὕτω καὶ
208 διαλεξόμεθα. τὸ δὲ ὅλον, ὡς αὐτὸς Ὅμηρος οὐκ ἀναλογίᾳ προσεχρήσατο ἀλλὰ τῇ τῶν κατ' αὐτὸν ἀνθρώπων συνηθείᾳ κατηκολούθησεν, οὕτω καὶ

Hesiod preceded him in time, Linos, too, and Orpheus and Musaeus and a host of others. And, in fact, it is probable that there were some poets before Homer and in his time (since he himself says somewhere [a]—

> Surely that song above all by men is most loudly applauded
> Which to their listening ears as the newest of songs resoundeth),

and that these poets were eclipsed by his own brilliance. And even if it should be agreed that Homer 205
is the most ancient, what Pindarion has asserted is not convincing. For just as we questioned before whether one should adopt common usage or analogy, so, too, now we shall be in doubt whether to adopt common usage or analogy; and if usage, is it to be that of Homer or that of all other men; and as to this Pindarion has said nothing. Further, we should 206
follow that usage above all the adoption of which will not bring ridicule upon us; but if we follow Homer's, our Greek speech will not escape ridicule, when we say *μάρτυροι* (for *μάρτυρες*) [b] and *σπάρτα λέλυνται* (for *σπάρτα λέλυται*),[c] and other things still more absurd. Neither, then, is this argument sound, besides the fact that our contention is conceded, namely, that analogy should not be used. For what's the odds 207
whether we have recourse to the usage of the many or to that of Homer? For just as there is need of observation, but not of technical analogy, in the case of the usage of the many, so there is also in the case of that of Homer; for when we have observed how he is wont to speak, we ourselves too will converse in the same way. To sum up, just as Homer himself made 208
no use of analogy but followed the usage of the men of

[a] *Cf.* Hom. *Od.* i. 351 f.
[b] *Cf.* Hom. *Il.* ii. 302.
[c] *Cf.* Hom. *Il.* ii. 135.

ἡμεῖς οὐκ ἀναλογίας πάντως ἑξόμεθα βεβαιωτὴν ἐχούσης Ὅμηρον, ἀλλὰ τὴν συνήθειαν τῶν καθ' αὑτοὺς ἀνθρώπων παραπλασόμεθα.

209 Ἄρτι μὲν οὖν ἐκ τῆς πρὸς τοὺς γραμματικοὺς ἀκολουθίας συνῆκται τὸ παρέλκειν μὲν τὴν ἀναλογίαν πρὸς ἑλληνισμόν, εὐχρηστεῖν δὲ τὴν τῆς συνηθείας παρατήρησιν· δῆλον δὲ ἴσως ἔσται ἐκ
210 τῶν ῥητῶν. ὁριζόμενοι γάρ τόν τε βαρβαρισμὸν καὶ τὸν σολοικισμὸν φασὶ "βαρβαρισμός ἐστι παράπτωσις ἐν ἁπλῇ λέξει παρὰ τὴν κοινὴν συνήθειαν" καὶ "σολοικισμός ἐστι παράπτωσις ἀσυνήθης κατὰ τὴν ὅλην σύνταξιν καὶ ἀνακόλουθος."
211 πρὸς ἃ δυνάμεθα λέγειν εὐθύς, ἀλλ' εἰ ὁ μὲν βαρβαρισμός ἐστιν ἐν ἁπλῇ λέξει ὁ δὲ σολοικισμὸς ἐν συνθέσει λέξεων, δέδεικται δὲ ἔμπροσθεν ὡς οὔτε ἁπλῆ ἔστι λέξις τις οὔτε σύνθεσις λέξεων,
212 οὐδέν ἐστι βαρβαρισμὸς ἢ σολοικισμός. πάλιν εἰ ἐν λέξει μιᾷ ὁ βαρβαρισμὸς νοεῖται καὶ ἐν συνθέσει λέξεων ὁ σολοικισμός, ἀλλ' οὐκ ἐν τοῖς ὑποκειμένοις πράγμασι, πῶς ἥμαρτον εἰπὼν "οὗτος," δείκνυμι δὲ γυναῖκα, ἢ "αὕτη," δείκνυμι δὲ νεανίαν; οὔτε γὰρ ἐσολοίκισα· οὐ γὰρ σύνθεσιν πολλῶν ἀκαταλλήλων λέξεων προηνεγκάμην, ἀλλ'
213 ἁπλῆν τὴν οὗτος λέξιν ἢ αὕτη· οὔτ' ἐβαρβάρισα· οὐδὲν γὰρ ἀσύνηθες εἶχεν ἡ οὗτος λέξις, ὡς ἡ παρὰ τοῖς Ἀλεξανδρεῦσιν ἐλήλυθαν καὶ ἀπελήλυθαν.

Πλὴν τοιαῦτα μὲν πολλὰ πρὸς τοὺς γραμμα-
214 τικοὺς ἐνδέχεται λέγειν· ἵνα δὲ μὴ δοκῶμεν ἐν πᾶσιν ἀπορητικοὶ τυγχάνειν, ἐπὶ τὴν ἐξ ἀρχῆς πρόθεσιν ἀναδραμόντες φήσομεν ὡς εἴπερ ὁ βαρ-

[a] *Cf.* § 231. [b] *Cf.* §§ 131, 165 ff.

[c] Instead of the regular 3 pers. plur. endings in *-θασι*.

his age, so too will we by no means cleave to an analogous form which has the authority of Homer, but will conform our speech to the usage of the men of our own age.

So it has just been concluded by the method of 209
pressing the Grammarians' arguments against themselves, that analogy is superfluous for "hellenism," whereas the observance of common usage is serviceable; and this will, no doubt, be plain from their
own assertions. Thus in defining "barbarism" and 210
"solecism" they say that "barbarism is a blunder in a single word contrary to common usage,"[a] and "solecism is a blunder contravening common usage in respect of the whole construction, and incoherent."
To this we can at once reply, that if the "barbarism" 211
is in a single word and the "solecism" in a combination of words, and it has been shown above[b] that no simple word nor any combination of words exists, then neither "barbarism" nor "solecism" is any-
thing. Again, if the "barbarism" is perceived in 212
one word and the "solecism" in a combination of words, but not in the underlying objects, how am I wrong in saying *οὗτος* when I point to a woman, or *αὕτη* when I point to a young man? For I have not committed a solecism, since I have not uttered a combination of many incongruous words, but only
the single word *οὗτος* or *αὕτη*; nor have I been 213
guilty of a barbarism, for the word *οὗτος* contains nothing contrary to common usage, as does the Alexandrians' *ἐλήλυθαν* and *ἀπελήλυθαν*.[c]

However, it is possible to bring many such argu-
ments against the Grammarians. But lest we should 214
seem to be sceptical about everything, returning to the original theme we shall assert that if a "bar-

βαρισμὸς παράπτωσίς ἐστι παρὰ τὴν κοινὴν συν-
ήθειαν ἐν μιᾷ λέξει θεωρούμενος, ὡσαύτως δὲ καὶ
ὁ σολοικισμὸς ἐν πολλαῖς λέξεσι τὴν ὑπόστασιν
λαμβάνων, καὶ ἔστι βάρβαρον μὲν τὸ τράπεσα
διὰ τὸ μὴ σύνηθες εἶναι τὸ ῥῆμα, σόλοικον δὲ τὸ
“ πολλὰ περιπατήσας κοπιᾷ μου τὰ σκέλη ” διὰ
τὸ μὴ λέγεσθαι τῇ κοινῇ συνηθείᾳ, ὡμολόγηται
ὅτι ἡ μὲν ἀναλογιστικὴ τέχνη ὄνομα κενόν ἐστι
πρὸς τὸ μὴ βαρβαρίζειν ἢ σολοικίζειν, δεῖ δὲ τὴν
συνήθειαν παρατηρεῖν καὶ ἀκολούθως αὐτῇ δια-
215 λέγεσθαι. εἰ μὲν γὰρ μετακαθίσαντες λέγοιεν
βαρβαρισμὸν ἁπλῶς παράπτωσιν ἐν ἁπλῇ λέξει,
δίχα τοῦ προσθεῖναι τὸ παρὰ τὴν κοινὴν συνή-
θειαν, καὶ σολοικισμὸν παράπτωσιν κατὰ τὴν ὅλην
σύνταξιν καὶ ἀνακόλουθον, χωρὶς τοῦ παραλαβεῖν τὸ
ἀσύνηθες, ἔτι[1] χεῖρόν τι κινήσουσιν ἑαυτοῖς πρᾶγμα.
τὰ γὰρ τοιαῦτα καθ' ὅλην τὴν σύνταξιν ἀνακολου-
θοῦντα[2] ἕξουσιν, “ Ἀθῆναι καλὴ πόλις, Ὀρέστης
καλὴ τραγῳδία, ἡ βουλὴ οἱ ἑξακόσιοι ”· ἃ δεήσει
σολοικισμοὺς λέγειν, οὐχὶ δέ γε σολοικισμοὶ
216 τυγχάνουσι διὰ τὸ σύνηθες. οὐκ ἄρα ψιλῇ τῇ
ἀκολουθίᾳ κριτέον τὸν σολοικισμόν, ἀλλὰ τῇ
συνηθείᾳ.

Εὖ δ' ἂν ἔχοι καὶ μετὰ τὴν ἐκ τῆς ἀκολουθίας
καὶ τῶν ῥητῶν ἔνστασιν ἔτι καὶ ἀπὸ τῆς κατὰ τὸ
217 ὅμοιον μεταβάσεως αὐτοὺς δυσωπεῖν. εἴπερ γὰρ
τοῦ ὁμοίου θεωρητικοὶ καθεστήκασιν, ἐπεὶ τῷ
εἰς ἀντικνήμιον τύπτεσθαι ἀνάλογόν ἐστι τὸ εἰς

[1] ἔτι scripsi : ἢ mss., Bekk. : ἤ Fabr.
[2] ἀνακολουθοῦντα Fabr. : ἀκολουθοῦντα mss., Bekk.

[a] The usual construction would have περιπατήσαντος, genitive agreeing with μου (“ My legs ache from walking much ”).

barism " is a blunder contrary to common usage and observed in a single word, and likewise a "solecism" finds its existence in a number of words, and the word τράπεσα (for τράπεζα, " a table ") is a barbarism because it is contrary to common usage, and the sentence πολλὰ περιπατήσας κοπιᾷ μου τὰ σκέλη [a] is a solecism because it is a construction not employed in common usage,—then, if so, it is conceded that the art of analogy is an empty name, useless for the avoidance of barbarisms or solecisms, and one ought to observe the rules of common speech, and converse in accordance therewith. For if they were to change 215
their ground and say simply that " barbarism " is a " blunder in a single word," without adding " contrary to common usage," and that " solecism " is " a blunder in respect of the whole construction and incoherent," without the addition of " contravening common usage," they will bring upon themselves a still worse consequence. For " in respect of the whole construction " they will have such " incoherent " sentences as these,—'Αθῆναι καλὴ πόλις, 'Ορέστης καλὴ τραγῳδία, ἡ βουλὴ οἱ ἑξακόσιοι,[b] which they will have to term "solecisms," whereas they are not solecisms because in common use. Hence, the sole- 216
cism is not to be judged by grammatical concord alone but by common usage.

After our objection based on pressing their arguments and on their rules [c] it would be well also to put them to shame in respect of transition based on analogy. For if they are observant of similarity, inas- 217
much as εἰς τὴν ῥῖνα τύπτεσθαι (" to be struck on the

[b] Termed " solecisms " because (1) 'Αθῆναι is plur. in agreement with the sing. καλὴ πόλις; (2) 'Ορέστης masc. with fem. καλὴ τραγῳδία; (3) ἡ βουλὴ fem. sing. with masc. plur. οἱ ἑξακόσιοι.

[c] Cf. § 209.

τὴν ῥῖνα τύπτεσθαι καὶ τὸ εἰς τὴν γαστέρα, λέγεται δὲ τὸ πρῶτον ἀντικνημιάζειν, ἀναλόγως καὶ τὸ γαστρίζειν ἢ μυκτηρίζειν. τὸ δὲ αὐτὸ καὶ ἐπὶ τοῦ ἱππάζεσθαι καὶ κατακρημνίζεσθαι καὶ ἡλιάζεσθαι ὑποδεικτέον. οὐ λέγομεν δὲ ταῦτα διὰ τὸ παρὰ τὴν κοινὴν εἶναι συνήθειαν· τοίνυν οὐδὲ τὸ κυήσω οὐδὲ τὸ φερήσω καὶ τὰ ἄλλα πάντα, ἅπερ ἀναλογίας ἐστὶν ὀφειλόμενα λέγεσθαι, διὰ τὸ μὴ
218 κατὰ τὴν συνήθειαν λέγεσθαι. οὐ μὴν ἀλλ' εἴπερ ἄριστα μὲν θρᾳκιστὶ διαλέγεσθαί φαμεν τὸν ὡς σύνηθές ἐστι Θρᾳξὶ διαλεγόμενον, καὶ κάλλιστα ῥωμαϊστὶ τὸν ὡς σύνηθες Ῥωμαίοις, ἀκολουθήσει καὶ τὸ ἑλληνιστὶ ὑγιῶς διαλέγεσθαι τὸν ὡς σύνηθες Ἕλλησι διαλεγόμενον, ἐὰν τῇ συνηθείᾳ ἀλλὰ μὴ τῇ διατάξει κατακολουθῶμεν. τῇ ἄρα συνηθείᾳ, οὐ τῇ ἀναλογίᾳ κατακολουθοῦντες ἑλληνιοῦμεν.
219 καθόλου τε ἤτοι σύμφωνός ἐστι τῇ συνηθείᾳ ἡ ἀναλογία ἢ διάφωνος. καὶ εἰ μὲν σύμφωνος, πρῶτον μὲν ὡς ἐκείνη οὐκ ἔστι τεχνική, οὕτως οὐδὲ αὕτη γενήσεται τέχνη· τὸ γὰρ ἀτεχνίᾳ συμφωνοῦν πάντως καὶ αὐτό ἐστιν ἄτεχνον. καὶ ἄλλως τὸ κατ' ἐκείνην ἑλληνικὸν καὶ κατὰ ταύτην[1] ἐκείνῃ συμφωνοῦσαν γενήσεται ἑλληνικόν, καὶ τὸ κατ'
220 ἐκείνην ἔσται τοιοῦτον. τούτου δ' οὕτως ἔχοντος οὐ δεησόμεθα τῆς ἀναλογίας πρὸς διάγνωσιν τοῦ ἑλληνισμοῦ, ἔχοντες εἰς τοῦτο τὴν συνήθειαν. εἰ δὲ διάφωνός ἐστιν αὕτη, πάντως ἑτέραν εἰσηγουμένη συνήθειαν παρ' ἐκείνην καὶ οἱονεὶ βάρβαρον ἀδόκιμος γενήσεται καὶ ὡς προσκοπὴν ἐμποιοῦσα τελέως ἄχρηστος.

[1] κατὰ ταύτην scripsi: κατ' αὐτὴν mss., Bekk. (? ⟨εἰ⟩ τὸ κατὰ ταύτην, deletis κατ' ἐκ. ἑλλ.).

nose ") and εἰς τὴν γαστέρα (" on the belly ") are analogous to εἰς ἀντικνήμιον τύπτεσθαι (" to be struck on the shin "), and this is expressed by ἀντικνημιάζειν, we must also say by analogy γαστρίζειν or μυκτηρίζειν. Other examples of the same rule are ἱππάζεσθαι (" to be driven ") and κατακρημνίζεσθαι (" to be thrown down headlong ") and ἡλιάζεσθαι (" to sun oneself "). But we do not employ these words since they are contrary to common usage. So, too, we do not use the forms κυήσω or φερήσω, or any of the other formations which by analogy ought to be used, because they do not conform to the common 218
usage. If, however, we declare that he talks the best Thracian who talks as is customary with the Thracians, and he the best Latin who talks as is customary with the Romans, then it will follow that he who talks as is customary with the Greeks is talking good Greek, if we are to follow customary usage and not an artificial system. Thus by following common usage and not 219
analogy we shall speak good Greek.—In general, too, analogy either agrees with common usage or disagrees. And if it agrees, then, firstly, since the latter is not a matter of art, so too the former will not be an art; for what agrees with the artless must certainly be artless itself.—Further, what is Greek according to common usage will also be Greek according to analogy which agrees therewith, and the Greek commonly used will be of that kind. This being so, we 220
shall not need analogy for distinguishing good Greek, since for this purpose we possess common usage. If, on the other hand, analogy disagrees, then, since it introduces another usage beside the common,—that of barbarians, as it were,—it will be disapproved and, as causing offence, it will be completely useless.

221 Ἐπιχειρητέον δὲ καὶ ἀπὸ τῆς συστάσεως τῆς τέχνης αὐτῶν. θέλουσι μὲν γὰρ καθολικά τινα θεωρήματα συστησάμενοι ἀπὸ τούτων πάντα τὰ κατὰ μέρος κρίνειν ὀνόματα, εἴτε ἑλληνικά ἐστιν εἴτε καὶ μή· οὐ δύνανται δὲ [καὶ] τοῦτο ποιεῖν διὰ τὸ μήτε τὸ καθολικὸν αὐτοῖς συγχωρεῖσθαι ὅτι καθολικόν ἐστι, μήτ᾽ ἄλλως ἀναπτυσσόμενον τοῦτο
222 τὴν τοῦ καθολικοῦ σώζειν φύσιν. λαμβανέσθω δὲ εἰς τοῦτο παράδειγμα ἀπ᾽ αὐτῶν τῶν γραμματικῶν. ζητήσεως γὰρ οὔσης ἐπί τινος τῶν κατὰ μέρος ὀνομάτων, οἷον ἐπὶ τοῦ εὐμενής, πότερον χωρὶς τοῦ σ προενεκτέον ἐστὶ τὴν πλάγιον πτῶσιν, εὐμενοῦ λέγοντας, ἢ σὺν τῷ σ, εὐμενοῦς, πάρεισιν οἱ γραμματικοὶ καθολικόν τι προφερόμενοι καὶ ἀπὸ τούτου τὸ ζητούμενον βεβαιοῦντες. φασὶ γὰρ "πᾶν ὄνομα ἁπλοῦν, εἰς ης λῆγον, ὀξύτονον, τουτὶ ἐξ ἀνάγκης σὺν τῷ σ κατὰ τὴν γενικὴν ἐξενεχθήσεται, οἷον εὐφυής εὐφυοῦς, εὐσεβής εὐσεβοῦς, εὐκλεής εὐκλεοῦς. τοίνυν καὶ τὸ εὐμενής ὀξυτόνως ἐκφερόμενον παραπλησίως τούτοις διὰ τοῦ σ ἐπὶ τῆς γενικῆς προενεκτέον, εὐμενοῦς
223 λέγοντας." οὐκ ᾔδεσαν δὲ οἱ θαυμάσιοι πρῶτον μὲν ὅτι ὁ εὐμενοῦ ἀξιῶν λέγειν οὐ δώσει αὐτοῖς καθολικὸν εἶναι τὸ παράπηγμα· τοῦτο γοῦν αὐτὸ τὸ εὐμενής ἁπλοῦν ὄνομα καθεστὼς καὶ ὀξύτονον οὐ φήσει σὺν τῷ σ ἐκφέρεσθαι, ἀλλὰ ἐκείνους τὸ
224 ζητούμενον ὡς ὁμολογούμενον συναρπάζειν. ἄλλως τε, εἰ καθολικόν ἐστι τὸ παράπηγμα, ἤτοι πάντα τὰ κατὰ μέρος ὀνόματα ἐπελθόντες καὶ τὴν ἐν αὐτοῖς ἀναλογίαν κατανοήσαντες συνέθεσαν αὐτό, ἢ οὐ πάντα. ἀλλὰ πάντα μὲν οὐκ ἐπεληλύθασιν· ἄπειρα γάρ ἐστι, τῶν δὲ ἀπείρων οὐκ ἔστι τις

One may argue also from the structure of their Art. 221
For after constructing certain universal rules they desire to judge by these all the particular words, as to whether they are good Greek or not ; but this they are unable to do since it is not conceded to them that their " universal " is universal, nor yet that it retains its nature as universal when applied in detail. To 222
make this clear let us take an example from the Grammarians themselves. When an inquiry is made respecting one of the particular words, such as εὐμενής, as to whether one should form the genitive case without the σ, and say εὐμενοῦ, or with the σ, εὐμενοῦς, the Grammarians come forward and propose a universal rule, and by this establish the point in question. For they assert that " every simple word ending in ης and oxytone must necessarily be declined in the genitive with σ, as for instance εὐφυής εὐφυοῦς, εὐσεβής εὐσεβοῦς, εὐκλεής εὐκλεοῦς. Hence, as εὐμενής, like these, is pronounced as oxytone, we must decline it in the genitive with σ, and say εὐμενοῦς." But our estimable friends failed to notice that, in 223
the first place, he who claims to say εὐμενοῦ will not grant them that their rule is universal : he will assert that this very word εὐμενής, which is a single word and oxytone, is not declined with a σ ; but they, by taking as settled the point in doubt, " beg the question."—Furthermore, if the rule is 224
universal, they have constructed it either after surveying all the particular words and noting the analogy they present, or else not all of them. But they have not surveyed all of them, for they are infinite in number, and there is no knowledge of

*γνῶσις. εἰ δὲ τινά, πόθεν ὅτι πᾶν ὄνομα τοιοῦτόν
ἐστιν; οὐ γὰρ ὅ τι τισὶ συμβέβηκεν ὀνόμασι, τοῦτο
225 καὶ πᾶσιν. ἀλλ' εἰσί τινες οἱ πρὸς τοῦτο γελοίως
ἀπαντῶντες καὶ λέγοντες ὅτι ἐκ πλειόνων ἐστὶ
τὸ καθολικὸν παράπηγμα. οὐχ ἑώρων γὰρ ὅτι
πρῶτον μὲν ἄλλο τί ἐστι τὸ καθολικὸν καὶ ἄλλο
τὸ ὡς ἐπὶ τὸ πολύ, καὶ τὸ μὲν καθολικὸν οὐδέ-
ποτε ἡμᾶς διαψεύδεται, τὸ δ' ὡς τὸ πολὺ κατὰ
226 τὸ σπάνιον· εἶθ' ὅτι καὶ εἰ ἐκ πολλῶν ἐστὶ τὸ
καθολικόν, οὐ πάντως τὸ τοῖς πολλοῖς ὀνόμασι
συμβεβηκός, τοῦτο ἐξ ἀνάγκης καὶ πᾶσι τοῖς
ὁμοειδέσι συμβέβηκεν, ἀλλ' ὃν τρόπον ἐν πολλοῖς
καὶ ἄλλοις φέρει τινὰ κατὰ μονοείδειαν ἡ φύσις,
οἷον ἐν ὄφεσι μὲν ἀπείροις οὖσι τὸν κεράστην
κερασφόρον, ἐν τετράποσι δὲ τὸν ἐλέφαντα προ-
βοσκίδι κεχρημένον, ἐν ἰχθύσι δὲ τὸν γαλεὸν
ζωοτόκον, ἐν λίθοις δὲ τὸν μάγνητα σιδηραγωγόν,
οὕτως εὔλογόν ἐστι καὶ ἐν πολλοῖς ὁμοιοπτώτοις
ὀνόμασιν εἶναί τι ὄνομα ὃ μὴ ὁμοίως τοῖς πολλοῖς
227 ὀνόμασι κλίνεται. ὅθεν παρέντες ζητεῖν εἰ ἀνά-
λογόν ἐστι τοῖς πολλοῖς, σκοπῶμεν πῶς αὐτῷ
χρῆται ἡ συνήθεια, πότερον ἀνάλογον ἐκείνοις ἢ
κατὰ ἴδιον τύπον· καὶ ὡς ἂν ᾖ χρωμένη, οὕτω καὶ
ἡμεῖς προοισόμεθα.*

*Περιδιωκόμενοι δὴ ποικίλως οἱ γραμματικοὶ
228 θέλουσιν ἀναστρέφειν τὴν ἀπορίαν. πολλαὶ γάρ,
φασίν, εἰσὶ συνήθειαι, καὶ ἄλλη μὲν Ἀθηναίων
ἄλλη δὲ Λακεδαιμονίων, καὶ πάλιν Ἀθηναίων
διαφέρουσα μὲν ἡ παλαιὰ ἐξηλλαγμένη δὲ ἡ νῦν,*

infinites. And if they have surveyed some, how do they know that every word is of a like kind? For that which is a property of some words is not a property of all. But to this there are some who make 225
an absurd reply, saying that the universal rule is based on the majority of cases. For they have failed to see that, firstly, what is "universal" is one thing and what holds good "for the most part" is another, and that which is universally true we never find false, but what is true "for the most part" is false occasionally; nor, secondly, have they seen that even if 226
the universal is composed of many, it is not always the case that the property of the many words is necessarily the property of all words similarly formed, but just as in many other things nature produces some with a unique form,—as, for instance, the horned "cerastes" among the infinite number of serpents, and amongst quadrupeds the elephant furnished with a proboscis, and amongst fish the viviparous shark, and amongst stones the magnet which attracts iron,—so also it is reasonable that amongst the many words of similar declension there should exist a certain word which is not declined like the many words. Hence, let us forgo any inquiry 227
as to its analogy with the many and consider how it is treated by common usage, whether as analogous to the many or as a peculiar type; and whichever way it is treated, in that way we too will pronounce it.

The Grammarians, then, being chased around by various means, desire to reverse the Sceptics' argument. The usages of speech, they say, are many,— 228
there is one of the Athenians, another of the Lacedaemonians; and of the Athenians, again, the old usage is of one sort and the present usage of a differ-

καὶ οὐχ ἡ αὐτὴ μὲν τῶν κατὰ τὴν ἀγροικίαν ἡ αὐτὴ δὲ τῶν ἐν ἄστει διατριβόντων, παρὸ καὶ ὁ κωμικὸς λέγει Ἀριστοφάνης

διάλεκτον ἔχοντα μέσην πόλεως,
οὔτ' ἀστείαν ὑποθηλυτέραν
οὔτ' ἀνελεύθερον ὑπαγροικοτέραν.

229 πολλῶν οὖν οὐσῶν συνηθειῶν, [ὡς] φασί, ποίᾳ χρησόμεθα; οὔτε γὰρ πάσαις κατακολουθεῖν δυνατὸν διὰ τὸ μάχεσθαι πολλάκις, οὔτε τινὶ ἐξ αὐτῶν, ἐὰν μή τις τεχνικῶς προκριθῇ. ἀλλὰ πρῶτον μέν, φήσομεν, τὸ ζητεῖν ποίᾳ χρηστέον συνηθείᾳ ἔστιν ἴσον τῷ εἶναί τινα τέχνην περὶ ἑλληνισμόν. αὕτη γάρ, φημὶ δ' ἡ ἀναλογία, ὁμοίου καὶ ἀνομοίου ἐστὶ θεωρία· τὸ δὲ ὅμοιον καὶ ἀνόμοιον λαμβάνετε ἀπὸ τῆς συνηθείας· κἂν μὲν ᾖ τετριμμένον, χρῆσθε αὐτῷ, εἰ δὲ μή, οὐκέτι.
230 πευσόμεθα οὖν καὶ ἡμεῖς, ἀπὸ ποίας συνηθείας λαμβάνετε τὸ ὅμοιον καὶ τὸ ἀνόμοιον; πολλαὶ γάρ εἰσι καὶ πολλάκις μαχόμεναι. ὅπερ δὲ ἀπολογούμενοι πρὸς τοῦτο ἐρεῖτε, τοῦτο καὶ παρ'
231 ἡμῶν ἀκούσεσθε. καὶ πάλιν ὅταν λέγητε τὸν βαρβαρισμὸν παράπτωσιν ἐν ἁπλῇ λέξει παρὰ τὴν συνήθειαν, ἀνταπορήσομεν λέγοντες ποίαν φατὲ πολλῶν οὐσῶν, καὶ ᾗ ἂν εἴπητε, ταύτῃ φήσομεν
232 καὶ ἡμεῖς ἀκολουθεῖν. κοινῆς οὖν οὔσης ἀπορίας οὐκ ἄπορος ἡ παρ' ἡμῶν ταύτης ἐστὶ λύσις. τῶν γὰρ συνηθειῶν αἱ μέν εἰσι κατὰ τὰς ἐπιστήμας αἱ δὲ κατὰ τὸν βίον. καὶ γὰρ ἐν φιλοσοφίᾳ ὀνομάτων τινῶν ἐστὶν ἀποδοχὴ καὶ ἐν ἰατρικῇ ἐξαιρέτως, καὶ ἤδη κατὰ μουσικὴν καὶ γεωμετρίαν. ἔστι δὲ καὶ βιωτική τις ἀφελὴς συνήθεια τῶν

ent sort; and that of country folk is not the same as that of townspeople, so that Aristophanes the comic poet says,[a]—

> Speaking like middle-class citizens all,
> Not with the fop's effeminate drawl,
> Nor with the rustics' vulgar bawl.

The usages, then, being many, which of them (they 229
ask) are we to adopt? For it is not possible to follow
them all, since they often conflict, nor yet some one of
them, unless some one be preferred on grounds of art.
But firstly, we shall reply, to inquire which usage
should be adopted is equivalent to asserting the exis-
tence of an art of "hellenism." And this art,—I mean
"analogy,"—is the consideration of the similar and
dissimilar[b]; and you take over the similar and dis-
similar from common usage; and if it be a form in cur-
rent use you employ it, otherwise you do not. So we 230
too shall inquire from which usage do you take over
the similar and dissimilar. For usages are many and
often conflicting. And the answer you give, in self-
defence, to this question is what you shall hear in turn
from us. And again, when you say that a "barbar- 231
ism" is a blunder in a single word contrary to common
usage, we shall retort with the question, "Which of
the many usages do you mean?" and whichever you
mention, that, we will say, is the one we too follow.
The doubt, then, being shared by us both, the solution 232
we give of it is not open to doubt. Of usages in speech
some are found in the sciences, some in ordinary con-
verse. Thus in philosophy and in medicine especially
certain terms are in favour, and so too in music and
geometry. And there is the plain untechnical usage of

[a] *Cf.* Arist. *Frag.* 552 (Dindorf).

[b] *Cf.* § 202.

233 ἰδιωτῶν, κατὰ πόλεις καὶ ἔθνη διαφέρουσα. ὅθεν ἐν φιλοσοφίᾳ μὲν τῇ τῶν φιλοσόφων στοιχήσομεν, ἐν ἰατρικῇ δὲ τῇ ἰατρικωτέρᾳ, ἐν δὲ τῷ βίῳ τῇ συνηθεστέρᾳ καὶ ἀπερίττῳ καὶ ἐπιχωριαζούσῃ.
234 παρὸ καὶ διχῶς τοῦ αὐτοῦ πράγματος λεγομένου πειρασόμεθα πρὸς τὰ παρόντα ἁρμοζόμενοι πρόσωπα τὸ μὴ γελώμενον προφέρεσθαι, ὁποῖόν ποτ' ἂν ᾖ κατὰ τὴν φύσιν. οἷον τὸ αὐτὸ ἀρτοφόριον καὶ πανάριον λέγεται, καὶ πάλιν τὸ αὐτὸ σταμνίον καὶ ἀμίδιον καὶ ἴγδις καὶ θυΐα. ἀλλὰ στοχαζόμενοι τοῦ καλῶς ἔχοντος καὶ σαφῶς καὶ τοῦ μὴ γελᾶσθαι ὑπὸ τῶν διακονούντων ἡμῖν παιδαρίων καὶ ἰδιωτῶν πανάριον ἐροῦμεν, καὶ εἰ βάρβαρόν ἐστιν, ἀλλ' οὐκ ἀρτοφορίδα, καὶ σταμνίον, ἀλλ'
235 οὐκ ἀμίδα, καὶ θυΐαν μᾶλλον ἢ ἴγδιν. καὶ πάλιν ἐν διαλέξει ἀποβλέποντες πρὸς τοὺς παρόντας τὰς μὲν ἰδιωτικὰς λέξεις παραπέμψομεν, τὴν δὲ ἀστειοτέραν καὶ φιλολόγον συνήθειαν μεταδιώξομεν· ὡς γὰρ ἡ φιλολόγος γελᾶται παρὰ τοῖς ἰδιώταις, οὕτως ἡ ἰδιωτικὴ παρὰ τοῖς φιλολόγοις. δεξιῶς οὖν ἑκάστῃ περιστάσει τὸ πρέπον ἀποδιδόντες δόξομεν ἀμέμπτως ἑλληνίζειν.

236 Ἄλλως τε, ἐπεὶ ἐγκαλοῦσιν ὡς ἀνωμάλῳ καὶ πολυειδεῖ τῇ συνηθείᾳ, καὶ ἡμεῖς ἀπὸ τῆς αὐτῆς ἀφορμῆς αὐτοῖς ἐγκαλέσομεν. εἰ γὰρ ἡ ἀναλογία ὁμοίου παράθεσίς ἐστι, τὸ δὲ ὅμοιον ἐκ τῆς συνηθείας, ἡ δὲ συνήθεια ἀνώμαλός τε καὶ ἄστατος, δεήσει καὶ τὴν ἀναλογίαν μὴ ἔχειν ἑστῶτα παρα-
237 πήγματα. καὶ τοῦτο πάρεστι διδάσκειν ἐπὶ τῶν ὀνομάτων καὶ τῶν ῥημάτων καὶ μετοχῶν καὶ

ordinary folk which differs from one State or Nation to another. Hence, in philosophy we shall fall in with the 233
usage of the philosophers, and in medicine with that proper to that science, and in ordinary intercourse with that which is more usual, free from affectation, and native to the district. Consequently, when the 234
same object is indicated by two names we shall try to suit ourselves to the persons present by employing the name which they do not laugh at, whatever the object's natural name may be. For instance, the same object is called *ἀρτοφόριον* ("bread-basket") and *πανάριον* or again *σταμνίον* and *ἀμίδιον* ("chamber-pot"), or *ἴγδις* and *θυΐα* ("mortar"). But, aiming at propriety and clearness and the avoidance of ridicule from our serving lads and ordinary folk, we shall use the terms *πανάριον* (even if it is barbarous [a]), not *ἀρτοφορίς*, and *σταμνίον*, not *ἀμίς*, and *θυΐα* rather than *ἴγδις*. And 235
again, in serious discussion, having regard to the company present, we shall put aside commonplace phrases and pursue after a more refined and cultured manner of speech; for just as the cultured manner is ridiculed by ordinary folk, so is the ordinary manner by men of culture. Thus, by discreetly adapting our style to suit each occasion, we shall be held to be speaking Greek faultlessly.

Moreover, when they charge common usage with 236
being inconsistent and multiform, we too will make our charge against them on the same grounds. For if "analogy" is "the juxtaposition of the similar," and the similar is taken from common usage, while common usage is inconsistent and variable, it must follow that analogy, too, has no invariable rules. And 237
one can show this in the case of nouns and verbs and

[a] *i.e.* Latin.

καθόλου τῶν ἄλλων ἁπάντων. οἷον ἐπὶ μὲν τῶν ὀνομάτων παρόσον τὰ κατὰ τὰς ὀρθὰς πτώσεις ἀνάλογα ὄντα καὶ ὅμοια, ταῦτα κατὰ τὰς πλαγίους ἀνομοίως τε καὶ οὐκ ἀναλόγως σχηματίζεται, οἷον Ἄρης Χάρης χάρτης—Ἄρεως Χάρητος χάρτου, καὶ Μέμνων Θέων λέων—Μέμνονος Θέωνος λέοντος, Σκόπας μέλας Ἄβας—Σκόπα μέλανος
238 Ἄβαντος. ἐπὶ δὲ τῶν ῥηματικῶν πολλὰ ὁμοίως κατὰ τὸν ἐνεστῶτα χρόνον λεγόμενα οὐκ ἀναλόγως ἐν τοῖς ἄλλοις χρόνοις σχηματίζεται, ἐνίων δὲ συζυγίαι τινὲς ἐκλελοίπασιν, οἷον αὐλεῖ ἀρέσκει —ηὔληκεν ἀρήρεκεν. καὶ κτείνεται μὲν λέγεται, ἔκτανκε δὲ οὐ λέγεται· ἀλήλιπται μὲν εἴποι τις ἄν, ἤλειπται δὲ οὐκέτι. ἐπὶ δὲ τῶν μετοχῶν βοῶν σαρῶν νοῶν—βοῶντος σαροῦντος νοοῦντος, καὶ ἐπὶ τῶν προσηγοριῶν ἄναξ ἄβαξ—ἄνακτος ἄβακος, γραῦς ναῦς—γραός νηός. ὡσαύτως δὲ
239 καὶ ἐπὶ τῶν τοιούτων. ἄρχων γὰρ λέγεται καὶ ὀνοματικῶς καὶ ὁ τὴν ἀρχὴν διέπων· ἀλλ' Ἄρχωνος μὲν γίνεται κατὰ πλάγιον πτῶσιν τὸ ὀνοματικόν, ἄρχοντος δὲ τὸ μετοχικόν. καὶ κατὰ ὅμοιον τρόπον μένων θέων νέων μετοχικὰ ὄντα καὶ ὀνοματικὰ διαφερούσας λαμβάνει τὰς κλίσεις· Μένωνος μὲν γὰρ γίνεται τὸ ὀνοματικόν, μένοντος δὲ ἡ μετοχή, καὶ Θέωνος μὲν τὸ ὀνοματικόν, θέοντος δὲ ἡ
240 μετοχή ἐστιν. πλὴν ἐκ τούτων συμφανὲς ὡς τῆς συνηθείας ἀνωμάλου καθεστώσης οὐχ ἕστηκε τὰ παραπήγματα τῆς ἀναλογίας, ἀλλ' ἀνάγκη ἀποστάντας αὐτῶν τοῖς κατὰ τὴν συνήθειαν σχηματισμοῖς προσέχειν, παρέντας τὸ ἀνάλογον.

participles and all the other forms without exception. For example, in the case of nouns, inasmuch as those which are analogous and similar in the nominative case are formed dissimilarly and not by analogy in the oblique cases,—for instance Ἄρης Χάρης χάρτης —Ἄρεως Χάρητος χάρτου, and Μέμνων Θέων λέων— Μέμνονος Θέωνος λέοντος, and Σκόπας μέλας Ἄβας— Σκόπα μέλανος Ἄβαντος. And in the case of verbs, 238
many which are of similar formation in the present tense are not formed by analogy in the other tenses, and the conjugations of some are partly defective,— for instance, αὐλεῖ ἀρέσκει—ηὔληκεν ἀρήρεκεν. And the form κτείνεται is used, but not ἔκτανκε, and one may say ἀλήλιπται, but not ἤλειπται. In the case of participles we find βοῶν σαρῶν νοῶν—βοῶντος σαροῦντος νοοῦντος: and in the case of substantives ἄναξ ἄβαξ—ἄνακτος ἄβακος, γραῦς ναῦς—γραός νηός. So too with similar cases. Thus, ἄρχων is used both as a 239
proper name and as meaning the holder of office, but the proper name becomes Ἄρχωνος in the genitive case, whereas the participial becomes ἄρχοντος. And in like manner μένων, θέων, νέων, which are participial and also proper nouns, take different declensions; for the proper noun becomes Μένωνος but the participial μένοντος, and the proper noun Θέωνος but the participial θέοντος. In short, it is evident from all this 240
that, as common usage is inconsistent, the rules of analogy are not fixed, but we must necessarily depart from them and hold by the forms used in common speech, forsaking analogy.

ΙΑ΄.—ΠΕΡΙ ΕΤΥΜΟΛΟΓΙΑΣ

241 Τὰ δὲ αὐτὰ λεκτέον πρὸς αὐτοὺς καὶ ὅταν δι' ἐτυμολογίας κρίνειν θέλωσι τὸν ἑλληνισμόν. πάλιν γὰρ ἤτοι σύμφωνός ἐστι τῇ συνηθείᾳ ἡ ἐτυμολογία ἢ διάφωνος· καὶ εἰ μὲν σύμφωνος, παρέλκει, εἰ δὲ διάφωνος, οὐ χρηστέον αὐτῇ ὡς προσκοπὴν ἐμποιούσῃ μᾶλλον τοῦ βαρβαρίζειν ἢ σολοικίζειν. καὶ καθόλου μετακτέον τὰς ὁμοίας ἀντιρρήσεις
242 ταῖς ἔμπροσθεν ἡμῖν ἀποδοθείσαις. ἰδιαίτερον δὲ ἐκεῖνο λεκτέον. τὸ ἐτυμολογίᾳ κρινόμενον ὄνομα ὅτι ἑλληνικόν ἐστιν, ἤτοι ἔτυμα πάντως ἔχειν ὀφείλει τὰ προηγούμενα αὐτοῦ ὀνόματα ἢ εἴς τινα τῶν φυσικῶς ἀναφωνηθέντων καταλήγειν. καὶ εἰ μὲν ἀπὸ ἐτύμων πάντων, κατὰ τοῦτο εἰς ἄπειρον τῆς ἐκπτώσεως γινομένης ἄναρχος ἔσται ἡ ἐτυμολογία, καὶ οὐκ εἰσόμεθα εἰ ἑλληνικόν ἐστι τὸ ἔσχατον λεγόμενον ὄνομα, ἀγνοοῦντες ποῖον ἦν
243 τὸ ἀφ' οὗ πρῶτον κατάγεται. οἷον εἰ ὁ λύχνος εἴρηται ἀπὸ τοῦ λύειν τὸ νύχος, ὀφείλομεν μαθεῖν εἰ καὶ τὸ νύχος ἀπό τινος ἑλληνικοῦ εἴρηται, καὶ τοῦτο πάλιν ἀπ' ἄλλου· καὶ οὕτως εἰς ἄπειρον γινομένης τῆς ἀνόδου, καὶ ἀνευρέτου καθεστῶτος τοῦ πρῶτον ἀναφωνηθέντος ὀνόματος, συνακαταληπτεῖται καὶ τὸ εἰ ἑλληνικῶς ὁ λύχνος εἴρηται.
244 εἰ δὲ ἐπί τινα τῶν ἀνετύμως κειμένων ὀνομάτων καταλήγοι τὸ ἐτυμολογούμενον ὄνομα, ὃν τρόπον ἐκεῖνα τὰ εἰς ἃ κατέληξεν οὐ διότι ἔστιν ἔτυμα παραδεξόμεθα, ἀλλὰ διότι τέτριπται κατὰ τὴν συνήθειαν, οὕτω καὶ τὸ δι' ἐτυμολογίας κρινόμενον

Chapter XI.—Concerning Etymology

We must also use the same arguments against them 241
when they propose to judge "hellenism" by "etymology." Once again, etymology either agrees or disagrees with common usage; and if it agrees it is superfluous, while if it disagrees one should not make use of it, as that would cause more offence than using barbarisms or solecisms. And in general one may transfer and apply counter-arguments similar to those already set forth. But this special objection should 242
be stated:—The word which is adjudged to be "hellenic" by etymology must either have the words which precede it as in all cases its *etyma* (or "true roots"), or be traced back to some word naturally pronounced. And if it is derived in all cases from *etyma*, since in this respect there is a regress *ad infinitum*, the etymology will be without a beginning, and we will not know whether the ultimate word is good Greek, seeing that we do not know the nature of the word from which it is first derived. Thus, if the 243
word λύχνος ("lamp") comes from λύειν τὸ νύχος ("dissolving the darkness"), we ought to find out whether νύχος comes from a Greek word, and this in turn from another; and as the regress thus goes on *ad infinitum* and the word first pronounced is indiscoverable therewith it is rendered impossible to ascertain whether λύχνος is a good Greek word. If, on the 244
other hand, the word of which the etymology is sought should be traced back to some words which are without *etyma* (or "roots"), just as we shall admit the words from which they are derived not because they are *etyma* but because they are current in common usage, so also we shall admit the word judged to be

παραδεξόμεθα οὐ διὰ τὴν ἐτυμολογίαν ἀλλὰ διὰ τὸ σύνηθες. οἷον προσκεφάλαιον ἀπὸ τοῦ τῇ κεφαλῇ προστίθεσθαι εἴρηται, ἡ δὲ κεφαλὴ καὶ τὸ πρός, ὅ ἐστι πρόθεσις, ἀνετύμως κέκληται.
245 τοίνυν ὡς ταῦτα χωρὶς ἐτυμολογίας πεπίστευται ὅτι ἔστιν ἑλληνικά, τῆς συνηθείας αὐτοῖς χρωμένης, οὕτω καὶ τὸ προσκεφάλαιον δίχα ἐτυμολογίας ἔσται πιστόν. ἄλλως τε ἐνίοτε τὸ αὐτὸ πρᾶγμα δυσὶν ὀνόμασι καλεῖται, τῷ μὲν ἐτυμολογίαν ἐπιδεχομένῳ τῷ δὲ ἀνετυμολογήτῳ, καὶ οὐ διὰ τοῦτο τὸ μὲν ἔτυμον λέγεται ἑλληνικὸν τὸ δὲ ἀνέτυμον βαρβαρικόν, ἀλλ' ὡς ἐκεῖνο ἑλληνικόν,
246 οὕτω καὶ τοῦτο. οἷον τὸ ὑφ' ἡμῶν καλούμενον ὑποπόδιον Ἀθηναῖοι καὶ Κῷοι χελωνίδα καλοῦσιν· ἀλλ' ἔστι τὸ μὲν ὑποπόδιον ἔτυμον, ἡ δὲ χελωνὶς ἀνέτυμον, καὶ οὐ διὰ τοῦτο οἱ μὲν Ἀθηναῖοι λέγονται βαρβαρίζειν ἡμεῖς δὲ ἑλληνίζειν, ἀλλ'
247 ἀμφότεροι ἑλληνίζειν. τοίνυν ὡς ἐκεῖνοι διὰ τὴν συνήθειαν καὶ οὐ διὰ τὴν τοῦ ὀνόματος ἐτυμότητα λέγονται ἑλληνίζειν, οὕτω καὶ ἡμεῖς διὰ τὸ ἐν τῇ αὐτῶν συνηθείᾳ τετριμμένον ἔχειν τὸ τοιοῦτον ὄνομα καὶ οὐ διὰ τὴν τῆς ἐτυμολογίας πίστιν ἑλληνιοῦμεν.

Ἀλλ' ὅτι μὲν τὸ τεχνικὸν μέρος τῆς γραμματικῆς ἀνυπόστατόν ἐστιν, αὐτάρκως ἐκ τῶν εἰρημένων δέδεικται· χωρῶμεν δὲ ἀκολούθως καὶ ἐπὶ τὸ ἱστορικόν.

ΙΒ'.—ΕΙ ΣΥΣΤΑΤΟΝ ΤΟ ΙΣΤΟΡΙΚΟΝ

248 Ὅτι μὲν οὖν ἀξιοῦται τοῦτο ὁλοσχερῶς εἶναι

Greek by etymology not because of its etymology but because it is commonly used. προσκεφάλαιον ("pillow"), for instance, is so named from being placed at the κεφαλή ("head"), but κεφαλή, and the πρός ("at") which precedes, are words without *etyma* ("roots"). So then, as these words are be- 245
lieved to be good Greek apart from etymology, since they are used in common speech, so too προσκεφάλαιον will be believed apart from etymology.—Again, the same object is sometimes called by two names, the one admitting of etymology, the other void of etymology, but neither is the *etymon* said to be good Greek for this reason nor the *non-etymon* to be barbaric, but as the former is good Greek so also is the latter. For example, that which is called by us 246
ὑποπόδιον ("foot-stool") the Athenians and Coans call χελωνίς: but ὑποπόδιον is an *etymon*, whereas χελωνίς is a *non-etymon*, yet the Athenians are not said on this account to be using a barbarism, and we to be speaking good Greek, but rather both are said to be speaking good Greek. Since, then, they are 247
said to be speaking good Greek not because the word they employ is an *etymon* but because it is in common use, so too we shall be speaking good Greek because the word we employ is current in our own customary speech and not because of our trust in etymology.

That the technical section of the Art of Grammar is without foundation has been proved sufficiently by what has now been said. Let us proceed in the next place to deal with the "historical" section.

Chapter XII.—Is the Historical Part consistent?

Now it is evident that "history" is regarded as 248

μέρος γραμματικῆς, συμφανές. Ταυρίσκος γοῦν
ὁ Κράτητος ἀκουστής, ὥσπερ οἱ ἄλλοι κριτικοί,
ὑποτάσσων τῇ κριτικῇ τὴν γραμματικήν, φησὶ
τῆς κριτικῆς εἶναι τὸ μέν τι λογικὸν τὸ δὲ τριβικὸν
249 τὸ δ' ἱστορικόν, λογικὸν μὲν τὸ στρεφόμενον περὶ
τὴν λέξιν καὶ τοὺς γραμματικοὺς τρόπους, τριβικὸν
δὲ τὸ περὶ τὰς διαλέκτους καὶ τὰς διαφορὰς τῶν
πλασμάτων καὶ χαρακτήρων, ἱστορικὸν δὲ τὸ περὶ
250 τὴν προχειρότητα τῆς ἀμεθόδου ὕλης. Διονύσιος
δὲ ὁ Θρᾷξ ἓξ μέρη γραμματικῆς εἶναι λέγων,
ἅπερ ἡμεῖς ἀνώτερον ὁλοσχερῶς τρία προση-
γορεύσαμεν, ἐν τούτοις καὶ τὸ ἱστορικὸν ἀποδί-
δωσιν· εἶναι γάρ φησι γραμματικῆς μέρη ἀνά-
γνωσιν ἐντριβῆ κατὰ προσῳδίαν, ἐξήγησιν κατὰ
τοὺς ἐνυπάρχοντας ποιητικοὺς τρόπους, λέξεων
καὶ ἱστοριῶν ἀπόδοσιν, ἐτυμολογίας εὕρεσιν, ἀνα-
λογίας ἐκλογισμόν, κρίσιν ποιημάτων, ἀτόπως
διαιρούμενος καὶ τάχα μὲν ἀποτελέσματά τινα
καὶ μόρια γραμματικῆς [οὐ] μέρη ταύτης ποιῶν,
251 ὁμολόγως δὲ τὴν μὲν ἐντριβῆ ἀνάγνωσιν καὶ τὴν
ἐξήγησιν καὶ τὴν κρίσιν τῶν ποιημάτων ἐκ τῆς
περὶ ποιητὰς καὶ συγγραφεῖς θεωρίας λαμβάνων,
τὴν δὲ ἐτυμολογίαν καὶ ἀναλογίαν ἐκ τοῦ τεχνικοῦ,
τοῖς δὲ τὸ ἱστορικὸν ἀντεκτιθείς, ἐν ἱστοριῶν
252 καὶ λέξεων ἀποδόσει κείμενον. Ἀσκληπιάδης
δὲ ἐν τῷ περὶ γραμματικῆς τρία φήσας εἶναι τὰ
πρῶτα τῆς γραμματικῆς μέρη, τεχνικὸν ἱστορικὸν
γραμματικόν, ὅπερ ἀμφοτέρων ἐφάπτεται, φημὶ
δὲ τοῦ ἱστορικοῦ καὶ τοῦ τεχνικοῦ, τριχῇ ὑποδι-
αιρεῖται τὸ ἱστορικόν· τῆς γὰρ ἱστορίας τὴν μέν
τινα ἀληθῆ εἶναί φησι τὴν δὲ ψευδῆ τὴν δὲ ὡς
ἀληθῆ, καὶ ἀληθῆ μὲν τὴν πρακτικήν, ψευδῆ δὲ

entirely a division of the Art of Grammar. Tauriscus, at least, the disciple of Crates, like the other critics, in subordinating grammar to criticism, declares that of the critical art one part is literary, another practical, and another historical; that which deals with 249
language and the grammatical tropes is literary, that concerned with dialects and the distinctions of formations and characters is practical, and that which treats of readiness in handling unarranged material is historical. And Dionysius the Thracian, in asserting 250
that there are six parts of grammar,—which we have described above [a] as three in all,—includes amongst them the historical; for he says that "the parts of grammar are skilled reading according to the scansion, explanation concerning the tropes which the poems contain, exposition of the phrases and histories, the discovery of etymologies, the reckoning of analogy, the judging of compositions,"—thus making absurd divisions and perhaps making out to be parts of grammar what are only certain results and particles of it, and avowedly taking away from the art which treats 251
of poets and composers the skilled reading and the explanation and the judgement of their compositions, and from the technical part etymology and analogy, in opposition to which he sets the historical part which consists of the exposition of phrases and histories. And Asclepiades, after stating in his treatise on 252
grammar that the primary parts of grammar are three, the technical, the historical and the grammatical (which has contact with both, I mean with the historical and the technical), subdivides the historical into three; for he says that "of history one division is true, one false, one as if true: the factual is true,

[a] *Cf.* § 91.

τὴν περὶ πλάσματα καὶ μύθους, ὡς ἀληθῆ δὲ οἷα
253 ἐστὶν ἡ κωμῳδία καὶ οἱ μῖμοι. τῆς δὲ ἀληθοῦς τρία πάλιν μέρη· ἡ μὲν γάρ ἐστι περὶ τὰ πρόσωπα θεῶν καὶ ἡρώων καὶ ἀνδρῶν ἐπιφανῶν, ἡ δὲ περὶ τοὺς τόπους καὶ χρόνους, ἡ δὲ περὶ τὰς πράξεις. τῆς δὲ ψευδοῦς, τουτέστι τῆς μυθικῆς, ἓν εἶδος μόνον ὑπάρχειν λέγει τὸ γενεαλογικόν. ὑποτάσσεσθαι δὲ τῷ ἱστορικῷ κοινῶς φησί, καθὼς καὶ Διονύσιος, τὸ περὶ τὰς γλώττας· ἱστορεῖ γὰρ ὅτι κρήγυον ἀληθές ἐστιν ἢ ἀγαθόν. ὡσαύτως δὲ καὶ τὸ περὶ παροιμιῶν καὶ ὅρων.

Ἀλλ᾽ ὅτι μὲν γραμματικῆς εἶναι μέρος βούλονται τὸ ἱστορικόν, ἐκ τούτων ἐστὶ συμφανές·
254 λοιπὸν δέ, ἐπεὶ οἱ πλείους ὡμολογήκασιν αὐτὸ ἄτεχνον εἶναι καὶ ἐκ τῆς ἀμεθόδου ὕλης τυγχάνειν, ἀπολελύκασι μὲν ἡμᾶς τῆς ἐπὶ πλεῖον πρὸς αὐτοὺς ἀντιρρήσεως, ὅμως δ᾽ οὖν ὑπὲρ τοῦ μὴ ἀνεπισήμαντον παρελθεῖν τὸν τόπον οὕτως ἐρωτητέον. ἤτοι τέχνη ἐστὶν ἡ γραμματικὴ ἢ οὐκ ἔστι τέχνη. καὶ εἰ μὲν οὐκ ἔστιν, αὐτόθεν συμβεβίβασται τὸ προκείμενον· εἰ δὲ τέχνη ἐστίν, ἐπεὶ τὰ τῆς τέχνης μέρη πάντως ἐστὶ τεχνικά, τὸ δ᾽ ἱστορικὸν ἀμέθοδον ὡμολόγηται τυγχάνειν, οὐκ ἂν εἴη τῆς
255 γραμματικῆς μέρος τὸ ἱστορικόν. καὶ ὅτι τῷ ὄντι τοιοῦτόν ἐστιν, αὐτόθεν σχεδὸν ὑπέπιπτεν. οὐ γὰρ ὥσπερ ἀπὸ καθολικῆς τινὸς μεθόδου καὶ τεχνικῆς δυνάμεως λέγει ὁ μὲν ἰατρὸς ὅτι τόδε τὸ ἐπὶ μέρους ὑγιεινόν ἐστι καὶ τόδε νοσερόν, ὁ δὲ μουσικὸς ὅτι τόδε ἡρμοσμένον καὶ τόδε ἀνάρμοστον, καὶ ἡρμοσμένον μὲν κατὰ τήνδε τὴν συμφωνίαν ἀλλ᾽ οὐχὶ κατὰ τήνδε, οὕτω καὶ ὁ γραμματικὸς

that of fictions and legends is false, and as if true are such forms as comedy and mimes." And of true 253
history, again, there are three parts : one sort is that about the persons of gods and heroes and notable men, another about places and times, the third about actions. And of false history (that is, the legendary) there is, he says, one kind only, the genealogical. And he says, like Dionysius, that the section dealing with "*glosses*" (rare or obsolete words) commonly comes under the head of the historical part ; for it reports that κρήγυον [a] means "true" or "good." And so too the section about proverbs and definitions.

From all this it is plain that they wish the "historical" to be a part of the Art of Grammar. Moreover, 254
since the majority of them have agreed that it is non-technical and consists of unmethodical matter, they have relieved us from having to argue further against them ; notwithstanding, in order not to pass over the topic and leave it unnoticed, we must argue with them in this wise :—Grammar either is or is not an art. And if it is not an art, the question before us is settled of itself. But if it is an art, since the parts of an art must certainly be technical, and it is agreed that the historical part is without method, the historical will not be a part of the Art of Grammar. And it is almost self-evident that this 255
is really so. For whereas, arguing from a general method and a technical faculty, the physician pronounces that this particular thing is healthy, this other diseased, and the musician that this is in harmony and that out of harmony,—in harmony because of this note, out of harmony because of that note,—it is not so with the Grammarian : he cannot

[a] *Cf.* Hom. *Il.* i. 106.

δύναται ἀπὸ ἐπιστημονικῆς τινὸς καὶ καθολικῆς θεωρίας ἀπαγγέλλειν ὅτι ὁ μὲν Πέλοπος ὦμος ἐλεφάντινος ἦν ὑπὸ τοῦ Ἄρεως ἢ ὑπὸ Δήμητρος βρωθείς, ἡ δὲ τοῦ Ἡρακλέους κεφαλὴ ἐψέδνωτο ῥυεισῶν αὐτοῦ τῶν τριχῶν ὅτε ὑπὸ τοῦ ἐφορμῶντος
256 τῇ Ἡσιόνῃ κήτους κατεπόθη, ἀλλ᾽ ἵνα τούτων ποιήσηται τὴν ἔκθεσιν, ὀφείλει πᾶσι τοῖς κατὰ μέρος ἱστοροῦσι περὶ αὐτῶν ἐντυχεῖν. τὸ δὲ πάντων τῶν κατὰ μέρος ποιεῖσθαι τὴν ἀνάληψιν αὐτοῖς ἐντυγχάνοντα τοῖς κατὰ μέρος οὐκ ἔστι τεχνικόν. οὐκ ἄρα ἔκ τέχνης τινὸς μεθοδεύεται
257 τοῖς γραμματικοῖς τὸ ἱστορικόν. καὶ μὴν ἐπεὶ τῆς ἱστορίας ἡ μέν τίς ἐστι τοπικὴ ἡ δὲ χρονικὴ ἡ δὲ περὶ τὰ πρόσωπα ἡ δὲ περὶ τὰς πράξεις, δῆλον ὡς εἰ μὴ τεχνική ἐστιν ἡ τῶν τόπων καὶ ἡ τῶν χρόνων ἀπόδοσις, οὔθ᾽ ἡ τῶν προσώπων οὔθ᾽ ἡ τῶν πράξεων τεχνικὴ γενήσεται· τί γὰρ διήνεγκε τούτων ἢ ἐκείνων ἀποκρατεῖν; ἀλλὰ μὴν οὐδὲν ἔχει τεχνικὸν τὸ ἀποδιδόναι τοπικὴν ἱστορίαν, λέγοντας ὅτι, εἰ τύχοι, Βριλησὸς μὲν καὶ Ἀράκυνθος τῆς Ἀττικῆς ἐστὶν ὄρος, Ἀκάμας δὲ τῆς Κυπρίας ἀκρωτήριον, ἢ χρονικὴν ἐκτίθεσθαι, καθάπερ ὅτι Ξενοφάνης Κολοφώνιος ἐγένετο περὶ τὴν τεσσαρακοστὴν ὀλυμπιάδα. τοῦτο γὰρ καὶ ὁ μὴ ὢν γραμματικὸς ἄλλως δὲ περίεργος δυνή-
258 σεται ποιεῖν. τοίνυν οὐδὲ τὸ περὶ προσώπων καὶ πράξεων ἀπαγγέλλειν τεχνικὸν γενήσεται, οἷον ὅτι Πλάτων μὲν ὁ φιλόσοφος Ἀριστοκλῆς πρότερον ἐκαλεῖτο καὶ ἐτέτρητο τὸ οὖς ἐλλόβιον

declare, on the ground of any scientific and general consideration, that the shoulder of Pelops, after it was devoured by Ares or by Demeter, was of ivory and that Heracles' head became bald as his hair fell off when he was swallowed by the sea-monster which was attacking Hesionê,[a] but in order that he may 256
record these events he must have met with all the particular historians of these events. But to repeat all the particular events by meeting the particular historians is not a technical method. Therefore the historical part is not methodically derived by the Grammarians from any art. Moreover, since one 257
kind of history deals with places, another with times, another with persons and another with actions, it is plain that if the exposition of places and times is not technical, neither will that of persons and actions be technical; for what difference is there to make you prefer these rather than those? But, in fact, there is nothing technical about recounting a "history" of a place, as when we assert (shall we say?) that Brilesus, and Aracynthus too, is a mountain of Attica, and Acamas a headland of Cyprus; or about a statement of time, as that Xenophanes of Colophon was born in the fortieth Olympiad.[b] For even the man who is no Grammarian but otherwise occupied will be able to do this. Neither, then, will the making announce- 258
ments about persons and actions be technical, as for instance that Plato the philosopher was first called Aristocles and that, when a youth, he had an ear

[a] The story is that Heracles, in rescuing Hesionê, spent three days (like Jonah) in the monster's belly, hacking it to pieces.

[b] *i.e.* 620 B.C., but as this date is too early, we ought perhaps to read πεντηκοστὴν, "fiftieth" (*i.e.* N′ for M′); *cf.* Diog. Laert. ix. 20.

φορήσας ὅτ' ἦν μειρακίσκος, Πυθιὰς δὲ ἡ Ἀριστοτέλους θυγάτηρ τρισὶν ἀνδράσιν ἐγαμήθη, πρῶτον μὲν Νικάνορι τῷ Σταγειρίτῃ, οἰκείῳ ὄντι Ἀριστοτέλους, δευτέρῳ δὲ Προκλεῖ Δημαράτου τοῦ Λακεδαιμονίων βασιλέως ἀπογόνῳ, ὃς καὶ δύο ἐξ αὐτῆς τεκνοῦται παῖδας, Προκλέα τε καὶ Δημάρατον τοὺς παρὰ Θεοφράστῳ φιλοσοφήσαντας, τρίτῳ δὲ Μητροδώρῳ ἰατρῷ, Χρυσίππου μὲν τοῦ Κνιδίου μαθητῇ Ἐρασιστράτου δὲ ὑφηγητῇ, ᾧ
259 γίνεται παῖς Ἀριστοτέλης. ταῦτα γὰρ καὶ τὰ τούτοις ὅμοια πρὸς τῷ τελέως ἀχρηστεῖν ἔτι οὐδεμίαν ἐμφαίνει τεχνικὴν δύναμιν, ὥστε οὐδὲ ἡ τῶν ἱστοριῶν[1] ἀπόδοσίς ἐστιν ἔντεχνος. ἄλλως τε καθὼς[2] ἀνώτερον ὑπεδείξαμεν, οὔτε τῶν ἀπείρων οὔτε τῶν ἄλλοτε ἄλλως γινομένων ἔστι
260 τις τεχνικὴ γνῶσις. αἵ δέ γε κατὰ μέρος ἱστορίαι ἄπειροί τε διὰ τὸ πλῆθός εἰσι, καὶ οὐχ ἑστῶσαι διὰ τὸ μὴ τὰ αὐτὰ περὶ τοῦ αὐτοῦ παρὰ πᾶσιν ἱστορεῖσθαι. οἷον (οὐκ ἄτοπον γὰρ ἵνα συμφυέσι τε καὶ οἰκείοις χρησώμεθα τῶν πραγμάτων παραδείγμασιν) ὑπόθεσιν γὰρ ἑαυτοῖς ψευδῆ λαμβάνοντες οἱ ἱστορικοὶ τὸν ἀρχηγὸν ἡμῶν τῆς ἐπιστήμης Ἀσκληπιὸν κεκεραυνῶσθαι λέγουσιν, οὐκ ἀρκούμενοι τῷ ψεύσματι, ἐν ᾧ καὶ ποικίλως αὐτὸ
261 μεταπλάττουσι, Στησίχορος μὲν ἐν Ἐριφύλῃ εἰπὼν ὅτι τινὰς τῶν ἐπὶ Θήβαις πεσόντων ἀνιστᾷ, Πολύανθος δὲ ὁ Κυρηναῖος ἐν τῷ περὶ τῆς Ἀσκληπιαδῶν γενέσεως ὅτι τὰς Προίτου θυγατέρας

[1] ἱστοριῶν cj. Bekk. : ἱστορικῶν MSS., Bekk.
[2] καθὼς cj. Bekk. : καὶ ὡς MSS., Bekk.

pierced and wore an earring, and that Pythias, the daughter of Aristotle, was married to three men, first to Nicanor of Stageira, a connexion of Aristotle, secondly to Procleus, a descendant of Damaratus the Lacedaemonians' king (who had by her two sons, the Procleus and Damaratus who studied philosophy with Theophrastus), and thirdly to Metrodorus the physician, a disciple of Chrysippus [a] of Cnidos and teacher of Erasistratus, whose son was Aristotle.[b] For these 259
stories and those like them, besides being perfectly useless, evince no technical faculty, so that the recital of histories too is void of art.—Furthermore, as we have shown above, there is no technical knowledge either of things infinite or of things which vary from hour to hour. But particular histories are both in- 260
finite, because of their great number, and without fixity, because the same facts are not recorded by all respecting the same person. For instance (for it is not out of place to use familiar and appropriate examples of the facts), the historians adopting a false assumption say that Asclepius, the founder of our science, was struck by lightning,[c] and not content with this falsehood they invent many variations of it,—Stesichorus [d] saying in *Eriphylê* that it was be- 261
cause he had raised up some of the men who had fallen at Thebes,—Polyanthus of Cyrenê, in his work on the origin of the Asclepiades, that it was because he had cured the daughters of Proetus who had

[a] A physician, *circ.* 350 B.C.,—not the famous Stoic philosopher.

[b] *i.e.* Aristotle junior, called after his grandfather.

[c] *Cf.* Virgil, *Aen.* vii. 770 ff. Sextus, as a medical man, calls Asclepius "the founder of *our* Science."

[d] Stesichorus was a famous lyric poet of Sicily, *circ.* 580 B.C.

κατὰ χόλον Ἥρας ἐμμανεῖς γενομένας ἰάσατο, Πανύασις δὲ διὰ τὸ νεκρὸν Τυνδάρεω ἀναστῆσαι, Στάφυλος δὲ ἐν τῷ περὶ Ἀρκάδων ὅτι Ἱππόλυτον ἐθεράπευσε φεύγοντα ἐκ Τροιζῆνος κατὰ τὰς παραδεδομένας κατ᾽ αὐτοῦ ἐν τοῖς τραγῳδουμένοις
262 φήμας, Φύλαρχος δὲ ἐν τῇ ἐννάτῃ διὰ τὸ τοὺς Φινέως υἱοὺς τυφλωθέντας ἀποκαταστῆσαι, χαριζόμενον αὐτῶν τῇ μητρὶ Κλεοπάτρᾳ τῇ Ἐρεχθέως, Τελέσαρχος δὲ ἐν τῷ Ἀργολικῷ ὅτι τὸν Ὠρίωνα ἐπεβάλετο ἀναστῆσαι. οὐ τοίνυν τῆς οὕτως ἀπὸ ψευδοῦς ὑποθέσεως ἀρχομένης καὶ ἀδιεξιτήτου κατὰ πλῆθος καὶ πρὸς τὴν ἑκάστου προαίρεσιν μεταπλαττομένης γένοιτ᾽ ἄν τις τεχνικὴ θεωρία.

263 Πρὸς τούτοις ἐπεὶ τῶν ἱστορουμένων τὸ μέν ἐστιν ἱστορία τὸ δὲ μῦθος τὸ δὲ πλάσμα, ὧν ἡ μὲν ἱστορία ἀληθῶν τινῶν ἐστὶ καὶ γεγονότων ἔκθεσις, ὡς ὅτι Ἀλέξανδρος ἐν Βαβυλῶνι δι᾽ ἐπιβούλων φαρμακευθεὶς ἐτελεύτα, πλάσμα δὲ πραγμάτων μὴ γενομένων μὲν ὁμοίως δὲ τοῖς γενομένοις λεγομένων, ὡς αἱ κωμικαὶ ὑποθέσεις καὶ οἱ μῖμοι,
264 μῦθος δὲ πραγμάτων ἀγενήτων καὶ ψευδῶν ἔκθεσις, ὡς ὅτι τὸ μὲν τῶν φαλαγγίων καὶ ὄφεων γένος Τιτήνων ἐνέπουσιν ἀφ᾽ αἵματος ἐζωγονῆσθαι, τὸν δὲ Πήγασον λαιμοτομηθείσης τῆς Γοργόνος ἀπὸ τῆς κεφαλῆς ἐκθορεῖν, καὶ οἱ μὲν Διομήδους ἑταῖροι εἰς θαλασσίους μετέβαλον ὄρνις, ὁ δὲ
265 Ὀδυσσεὺς εἰς ἵππον, ἡ δὲ Ἑκάβη εἰς κύνα,—τοιαύτης δὲ οὔσης τῶν ἱστοριῶν διαφορᾶς, ἐπεὶ οὐκ ἔστι τέχνη τις περὶ τὰ ψευδῆ καὶ ἀνύπαρκτα,

[a] An epic poet, of Halicarnassus, *circ.* 480 B.C.

[b] A Boeotian hunter, slain by Artemis; and placed, after death, among the stars.

become mad owing to the wrath of Hera,—Panyasis,[a] that it was owing to his raising up the dead body of Tyndareôs,—Staphylus, in his book about the Arcadians, that it was because he had healed Hippolytus when he was fleeing from Troezen, according to the reports handed down about him in the tragedies,
—Phylarchus, in his ninth book, that it was because 262
he restored their sight to the blinded sons of Phineus, as a favour to their mother Cleopatra, the daughter of Erechtheus,—Telesarchus in his *Argolicum*, that it was because he set himself to raise up Orion.[b] Thus, of an assumption which begins with a falsehood and is so multiform that it cannot be checked, and changes its shape at each man's fancy, there can be no technical treatment.

Moreover, since of the subjects of history one part 263
is history, another legend, another fiction,—and of these history is the recording of certain things which are true and have happened, as that Alexander died at Babylon through having been poisoned by plotters; and fiction is the narrating things which are not real events but are similar to real events in the telling, such as the hypothetical situations in comedies and
mimes; and legend is the narrating of events which 264
have never happened and are false, like the story that the species of venomous spiders and snakes were born alive from the blood of the Titans, and that Pegasus sprang from the head of the Gorgon when her throat was cut,[c] and that the companions of Diomede were changed into sea-birds, Odysseus into
a horse, and Hecuba into a dog,—such then being 265
the variety in histories, since there exists no art which deals with things false and unreal, and the

[c] *Cf.* Hesiod, *Theog.* 280.

ψευδῆ δέ ἐστι καὶ ἀνύπαρκτα τὰ περὶ τοὺς μύθους καὶ τὰ πλάσματα, περὶ ἃ μάλιστα τοῦ ἱστορικοῦ μέρους ἡ γραμματικὴ καταγίνεται, οὐκ ἂν εἴη τις τέχνη περὶ τὸ ἱστορικὸν μέρος τῆς γραμμα-
266 τικῆς. ὅθεν καταγελᾶν ἄξιον τῶν λεγόντων ὅτι εἰ καὶ ἡ ὕλη τῆς ἱστορίας ἐστὶν ἀμέθοδος, ἡ μέντοι κρίσις ταύτης γενήσεται τεχνική, δι' ἧς γινώσκομεν
267 τί τε ψευδῶς ἱστόρηται καὶ τί ἀληθῶς. πρῶτον μὲν γὰρ οὐ παραδεδώκασιν ἡμῖν οἱ γραμματικοὶ τῆς ἀληθοῦς ἱστορίας κριτήριον, ἵνα καὶ ἐξετάζωμεν πότε ἀληθής ἐστιν αὕτη καὶ πότε ψευδής. εἶτα καὶ μηδεμιᾶς οὔσης ἀληθοῦς ἱστορίας παρὰ τοῖς γραμματικοῖς οὐδὲ τὸ τοῦ ἀληθοῦς κριτήριον ὑποστατόν ἐστιν, ἐπεὶ πῶς οὐκ ἔργον, τινὸς μὲν λέγοντος ὅτι Ὀδυσσεὺς ὑπὸ Τηλεγόνου παιδὸς κατὰ ἄγνοιαν ἀνῄρηται, τινὸς δὲ ὅτι λάρου κέντρον θαλασσίας τρυγόνος ἀφέντος αὐτοῦ τῇ κεφαλῇ διεφώνησεν, ἄλλου δὲ ὅτι εἰς ἵππον μετέβαλε τὴν μορφήν, θέλειν ἐν οὕτως ἀπερρωγόσι πράγμασιν εὑρεῖν τἀληθές; πρότερον γὰρ δεῖ ὑποστῆναι ἐν τοῖς διαφωνοῦσι τὸν ἀληθεύοντα, καὶ τότε ζητεῖν
268 τί ἐστιν· πάντων δὲ ἀπίθανα καὶ ψευδῆ λεγόντων οὐδὲ τεχνικῷ τινὶ κριτηρίῳ δίδοται πάροδος.

Καὶ μὴν οὐδὲ δι' ὧν ἂν ἱστορία καλῶς γραφείη διδάσκουσιν οἱ γραμματικοί, ἵνα κατ' ἀναφορὰν τὴν ὡς ἐπὶ τὰ τοιαῦτα θεωρήματα λέγωμεν τεχνικόν τι μέρος ὑπάρχειν παρ' αὐτοῖς τὸ ἱστο-
269 ρικόν· τοῦτο γὰρ ῥητορικῶν ἐστὶ τὸ ἔργον. ὥστε εἰ καὶ αὐτοὶ ἐξομολογοῦνται ἀμέθοδόν τι εἶναι παράπηγμα τὴν ἱστορίαν καὶ ἡμεῖς ἐπελογισάμεθα, καὶ ἄλλως οὐδὲν τεχνικὸν θεώρημα πρὸς τὴν γνῶσιν αὐτῆς ἢ σύστασιν παραδεδώκασι, ῥητέον

legends and fictions, which form the main subjects of the historical part with which grammar is concerned, are false and unreal, it will follow that there exists no art which deals with the historical part of grammar. Hence they deserve to be laughed at who assert that 266 even if the subject-matter of history lacks method, yet the judging of it will be a matter of art, by means of which we ascertain what is falsely related and what truly. For, firstly, the Grammarians have not 267 furnished us with a criterion of true history, so that we might determine when it is true and when false. In the next place, as the Grammarians have no history that is true, the criterion of truth is also non-existent; for when one man says that Odysseus was killed in ignorance by his son Telegonus,[a] and another that he breathed his last when a sea-gull[b] dropped on his head the spike of a roach, and yet another that he was transformed into a horse, surely it is a hard task to try to discover the truth in such incoherent accounts. For we must establish first which of these dissentient narrators is telling the truth, and then inquire as to the facts; but when all relate what is improb- 268 able and false no opening is given for a technical criterion.

Nor yet do the Grammarians instruct us as to how history should rightly be written, so that by a reference to such rules we might declare that with them the historical part is a technical one; for this is the task of the Rhetoricians. So that, if even they them- 269 selves confess that history is a chronicle without method and we have confirmed this by logic, and if, moreover, they have provided no technical rule for understanding or constructing it, we must declare

[a] His son by Circê.

[b] *Cf.* Hom. *Od.* xi. 134.

καὶ κατὰ τὸ ἱστορικὸν μέρος ἀσύστατον εἶναι τὴν γραμματικήν.

ΙΓ'.—ΤΟ ΠΕΡΙ ΤΟΥΣ ΠΟΙΗΤΑΣ ΜΕΡΟΣ

270 Ἤδη μὲν δυνάμει καὶ τὸ περὶ ποιητὰς καὶ συγγραφεῖς μέρος τῆς γραμματικῆς ἡμῖν ἀνῄρηται, δείξασι τὸ ἀδύνατον τοῦ κατὰ τὰς τεχνολογίας καὶ τοῦ ἱστορικοῦ[1]· χωρὶς γὰρ τούτων οὐκ ἀπευθύνεταί τις ποιήσεως ἐξήγησις. ὅμως δ' οὖν καὶ τὰ ἐν τούτῳ τῷ μέρει πειρασόμεθα κοινότερον δυνάμενα λέγεσθαι σκοπεῖν, καὶ μάλιστα ὅτι οὕτως ἐπιτεθαρρήκασιν αὐτῷ οἱ γραμματικοὶ ὡς καὶ τὸ βιωφελὲς τῆς γραμματικῆς καὶ πρὸς εὐδαιμονίαν ἀναγκαῖον ἐξ αὐτοῦ τολμᾶν πιστοῦσθαι. φασὶ γοῦν ὡς ἡ ποιητικὴ πολλὰς δίδωσιν ἀφορμὰς πρὸς σοφίαν καὶ εὐδαίμονα βίον, ἄνευ δὲ τοῦ ἀπὸ γραμματικῆς φωτὸς οὐχ οἷόν τε τὰ παρὰ τοῖς ποιηταῖς διορᾶν ὁποῖά ποτε ἐστίν· χρειώδης ἄρα ἡ γραμ-
271 ματική. τὸ δ' ὅτι συχνὰς δίδωσιν ἡ ποιητικὴ ἀφορμὰς πρὸς εὐδαιμονίαν δῆλον ἐκ τοῦ τὴν ὄντως κρατίστην καὶ ἠθοποιὸν φιλοσοφίαν ἀπὸ τῆς παρὰ τοῖς ποιηταῖς γνωμολογίας τὴν ἀρχὴν ἐρριζῶσθαι, καὶ διὰ τοῦτο τοὺς φιλοσόφους, εἴ ποτε παραινετικῶς τι λέγοιεν, ταῖς ποιητικαῖς φωναῖς ὡσπερεὶ σφραγίζεσθαι τὸ ὑπ' αὐτῶν λεγόμενον. καὶ ὁ μὲν ἐπ' ἀρετὴν παρακαλῶν φησὶν

> ἀρετὴ δὲ κἂν θάνῃ τις οὐκ ἀπόλλυται·

ὁ δὲ φιλαργυρίαν φεύγειν ἐγκελευόμενος προφέρεται τὸ

> μὴ Πλοῦτον εἴπῃς· οὐχὶ θαυμάζω θεὸν
> ὃν χὠ κάκιστος ῥᾳδίως ἐκτήσατο·

[1] τοῦ ἱστορικοῦ scripsi: τὸ ἱστορικόν MSS., Bekk.

that as regards its historical part also the Art of Grammar is without foundation.

Chapter XIII.—The Part dealing with Writers

The part of Grammar which deals with poets and prose-writers we have virtually abolished already by showing the impossibility of the technological part and of the historical; for without these no exposition of composition can proceed rightly. Nevertheless, we shall endeavour to examine also what can be said in a general way of this part, especially because the Grammarians are so confident about it that they dare to prove from it the practical usefulness of grammar and its necessity for happiness. They assert, at least, that poetry furnishes many aids to wisdom and a happy life, but without the light of grammar it is not possible to discern clearly what the poets' sayings really mean; therefore grammar is useful. And that poetry furnishes many aids to happiness is plain from the fact that the best and character-forming philosophy had its original roots in the gnomic sayings of the poets, and on this account the philosophers, when giving exhortations, always stamped, as it were, their injunctions with phrases from the poets. Thus one of them, in exhorting to virtue, says—

270 (beside the line beginning "The part of Grammar"); 271 (beside the line ending "And that")

> Yet virtue doth not perish though one die.[a]

And another, enjoining us to flee covetousness, proclaims—

> Speak not of Wealth: I honour not a god
> Whom e'en the basest makes his own with ease.[b]

[a] *Cf.* Eurip. *Temen.* (*Frag.* 734 Nauck).
[b] *Cf.* Eurip. *Aeolus* (*Frag.* 20 Nauck).

ὁ δὲ αὐτάρκειαν ὑπαγορεύων συμπιστοῦται τὸ δόγμα ἐκ τοῦ τὸν Εὐριπίδην λέγειν

τί γὰρ δέει βροτοῖσι πλὴν δυοῖν μόνον,
Δήμητρος ἀκτῆς πώματός θ᾽ ὑδρηχόου,
ἃ δὴ πάρεστι καὶ πέφυχ᾽ ἡμᾶς τρέφειν;

272 καὶ τὸ μὲν τοὺς ἄλλους φιλοσόφους τοῦτο ποιεῖν οὐ παράδοξον, αὐτοὺς δὲ εὑρήσομεν τοὺς τῆς γραμματικῆς κατηγόρους, Πύρρωνά τε καὶ Ἐπίκουρον, ἐξομολογουμένους τὸ ἀναγκαῖον αὐτῆς· ὧν ὁ μὲν Πύρρων ἱστορεῖται τὴν Ὁμηρικὴν διὰ παντὸς ποίησιν ἀναγινώσκων, μὴ ἄν ποτε τοῦτο ποιήσας εἴπερ μὴ ἐγίνωσκεν αὐτὴν χρησίμην καὶ
273 διὰ τοῦτο τὴν γραμματικὴν ἀναγκαίαν, ὁ δὲ Ἐπίκουρος φωρᾶται τὰ κράτιστα τῶν δογμάτων παρὰ ποιητῶν ἀνηρπακώς. τόν τε γὰρ ὅρον τοῦ μεγέθους τῶν ἡδονῶν, ὅτι ἡ παντός ἐστι τοῦ ἀλγοῦντος ὑπεξαίρεσις, ἐξ ἑνὸς στίχου δέδεικται λαβών,

αὐτὰρ ἐπεὶ πόσιος καὶ ἐδητύος ἐξ ἔρον ἕντο·

τόν τε θάνατον, ὅτι οὐδέν ἐστι πρὸς ἡμᾶς, Ἐπίχαρμος αὐτῷ προμεμήνυκεν, εἰπὼν

ἀποθανεῖν ἢ τεθνάναι οὔ μοι διαφέρει.

ὡσαύτως δὲ καὶ τὰ νεκρὰ τῶν σωμάτων ἀναισθητεῖν παρ᾽ Ὁμήρου κέκλοφε, γράφοντος

κωφὴν γὰρ δὴ γαῖαν ἀεικίζει μενεαίνων.

274 καὶ μὴν οὐ ταῦτα μόνον τοῖς ποιηταῖς δεξιῶς εἰρῆσθαι φαίνεται ἀλλὰ καὶ τὰ περὶ θεῶν, οἷόν ἐστι καὶ

And another, in recommending self-sufficiency, supports his doctrine by a quotation from Euripides[a]—

> What do men need save only these two things,
> Earth's corn and draughts of water from the spring?
> These are to hand, our natural nutriment.

That the rest of the philosophers do this is not 272
paradoxical, but we shall find even those accusers of grammar, Pyrrho and Epicurus, acknowledging its necessity. Of these it is recorded that Pyrrho was constantly reading Homer's poetry, which he would never have done unless he knew that it was useful,
and that grammar therefore was necessary; and 273
Epicurus has been detected as guilty of having filched the best of his dogmas from the poets. For he has been shown to have taken his definition of the intensity of pleasures,—that it is "the removal of everything painful,"—from this one verse—

> When they had now put aside all longing for drinking and eating.[b]

And as to death, that "it is nothing to us," Epicharmus had already pointed this out to him when he said,—

> To die or to be dead concerns me not.

So, too, he stole the notion that dead bodies have no feeling from Homer, where he writes,[c]—

> 'Tis dumb clay that he beats with abuse in his violent fury.

Moreover, it is evident that is it not only these things
which have been happily expressed by the poets, but 274
also notions about the gods, such as that expressed by

[a] *Cf.* Eurip. *Frag.* 884 (Nauck).
[b] *Cf.* Hom. *Il.* i. 469.
[c] *Cf.* Hom. *Il.* xxiv. 54.

τὸ παρὰ τῷ Εὐριπίδῃ λεχθὲν ἐν Φρίξῳ,

> ὅστις δὲ θνητῶν οἴεται καθ' ἡμέραν
> κακόν τι πράσσων τοὺς θεοὺς λεληθέναι,
> δοκεῖ πονηρά, καὶ δοκῶν ἁλίσκεται
> ὅταν σχολὴν ἄγουσα τυγχάνῃ δίκη.

ἀλλ' εἴπερ ταῦτα καὶ τὰ τούτοις ἐοικότα χρειώδη ἐστί, λαμβάνεται δ' οὐ χωρὶς γραμματικῆς, ἔσται
275 καὶ ἡ γραμματικὴ τῶν βιωφελῶν. ἔχοι δ' ἄν τινα, φασίν, ἐξαιρέτως καὶ ταῖς τῶν μανθανόντων αὐτὴν πατρίσιν ἀναγκαῖα. Λεβεδίων γοῦν διαφερομένων πρὸς τοὺς ἀστυγείτονας περὶ Καμανδωδοῦ ὁ γραμματικὸς τὸ Ἱππωνάκτειον παραθέμενος ἐνίκα,

> μηδὲ . . . μοι μῦ
> λαλεῖν Λεβεδίην ἰσχάδ' ἐκ Καμανδωδοῦ.

ὁμιλητικούς τε παρεχομένη τοὺς προσέχοντας αὐτῇ εὐθὺς καὶ ταύτῃ [καὶ] τοῖς πέλας κατὰ πολλὰς
276 περιστάσεις ὀνησιφόρος γίνεται. πάρεστι δὲ τὸ λεγόμενον σκοπεῖν ἐξ αὐτῶν τῶν ἀποτελεσμάτων. Σώστρατος γάρ, ὥς φασίν, ἀποσταλεὶς ὑπὸ Πτολεμαίου πρὸς τὸν Ἀντίγονον βασιλικῆς τινὸς ἕνεκα χρείας, κἀκείνου εἰκαιότερον ἀποκρινομένου, ἐπέτυχεν εἰπὼν

> οὕτω δὴ κέλεαι, γαιήοχε κυανοχαῖτα;
> τόνδε φέρω Διὶ μῦθον ἀπηνέα τε κρατερόν τε;
> ἦ τι μεταστρέψεις; στρεπταὶ μέν τε φρένες ἐσθλῶν.

ταῦτα γὰρ ἀκούσας Ἀντίγονος μετεβάλετο.

---

[a] *Cf.* Eurip. *Phrixus* (*Frag.* 832 Nauck).

Euripides in his *Phrixus*,[a]—

> Whoe'er of mortals, sinning day by day,
> Deemeth the gods are blind to his misdeeds,
> Thinks wrongly, and in thinking thus is caught
> When Justice, haply, has some leisure time.

But if these and similar sayings are useful, and are not apprehended without the Art of Grammar, then grammar too will be profitable for life. And it will 275
also contain, they say, some things which are especially necessary for the countries of those who learn it. Thus, when the Lebedians [b] were quarrelling with their neighbours about Camandodus, the Grammarian won the victory by quoting from Hipponax [c]—

> Mutter not to me
> Of figs Lebedian from Camandodus.

Also, by rendering those who attend to it sociable, in this way too it obviously becomes helpful to their neighbours in many circumstances. The truth of this 276
statement one can judge from the actual results. For when Sostratus,[d] as they say, was sent to Antigonus by Ptolemy to make some request for the king, and the former gave a rather hasty reply, Sostratus gained his point by saying,—

> Biddest thou thus, Earth-shaker, whose locks are the blue of the sea-waves?
> Shall I bear unto Zeus this harsh and stubborn answer?
> Wilt thou not alter at all? Yet the minds of the good may be altered.[e]

For on hearing this Antigonus changed his mind.

[b] *i.e.* inhabitants of the (Ionian) Greek city Lebedos, on the coast of Lydia.

[c] Hipponax of Ephesus wrote satirical verse in iambics, *circ.* 530 B.C.

[d] Famous architect under Ptolemy I, King of Egypt, who was opposed by Antigonus ("King of Asia" Minor), *circ.* 320-310 B.C.

[e] *Cf.* Hom. *Il.* xv. 201 ff.

277 Πολλῶν δὴ τοιούτων λεγομένων εἰς τὸ χρησιμώτατον εἶναι τὸ τῆς γραμματικῆς μέρος τὸ περὶ ποιητὰς καὶ συγγραφεῖς καταγιγνόμενον, δείγματος χάριν τοῖς ἐκκειμένοις ἀρκεσθέντες λέγωμεν πρὸς ἕκαστον αὐτῶν. τὸ τοίνυν βιωφελῆ εἶναι τὴν ποιητικὴν γνωμολογίαν καὶ φιλοσοφίας ἀρχήν, ταύτης δὲ ἀποδοτικὴν ὑπάρχειν τὴν γραμματικήν,
278 ὄντως γραμματικόν ἐστιν. πρῶτον μὲν γάρ, ἵνα συνδράμωμεν αὐτοῖς μηδὲν ποιητικῆς κατειπόντες, ἀλλ' οὖν γε ἐκεῖνο πρόδηλόν ἐστιν ὅτι ὁπόσα μὲν βιωφελῆ καὶ ἀναγκαῖα εὑρίσκεται παρὰ ποιηταῖς, οἷά ἐστι τὰ γνωμικὰ καὶ παραινετικά, ταῦτα σαφῶς αὐτοῖς πέφρασται καὶ οὐ δεῖται γραμματικῆς, ⟨ὁπόσα δὲ δεῖται γραμματικῆς⟩[1] καθάπερ τὰ ἐν ξέναις ἱστορίαις κείμενα ἢ αἰνιγματωδῶς ἐκφερόμενα, ταῦτ' ἔστιν ἄχρηστα, ὥστε καὶ τῇ ἀπ' ἐκείνων ὠφελείᾳ μὴ συνεισέρχεσθαι αὐτῶν τὸ χρειῶδες τῆς γραμματικῆς καὶ τῇ τούτων μαται-
279 ότητι συμπεριφέρεσθαι. εἶτα φάσις μόνον ἐστὶν ἡ γνώμη, καθάπερ τὸ τοιοῦτο,

σοφὸν γὰρ ἓν βούλευμα τὰς πολλὰς χέρας
νικᾷ, σὺν ὄχλῳ δ' ἀμαθία πλεῖστον κακόν·

φάσει δὲ οὐ πείθεται ὁ νοῦς περὶ τοῦ καλῶς εἰρῆσθαι ἢ μὴ [εἰρῆσθαι], ἀλλ' ἀποδείξεων δεῖται. αἱ δὲ ἀποδείξεις τῶν καθηκόντως λεγομένων ἢ μὴ οὐ γραμματικῆς εἰσὶν ἀλλὰ φιλοσοφίας· τοίνυν καὶ ταύτῃ περισσὴν καὶ ματαίαν συμβέβηκεν εἶναι τὴν γραμματικήν. καὶ μὴν εἴπερ διὰ τὸ πολλὰ καλῶς εἰρῆσθαι τοῖς ποιηταῖς καὶ βιωφελῶς

[1] ⟨ὁπόσα δὲ δεῖται γραμματικῆς⟩ addidi: ⟨ὁπόσα δ' ἀσαφῆ⟩ add. Shorey.

Seeing that many such arguments are used to 277
prove that the part of grammar which deals with poets and prose-writers is most useful, let us content ourselves with the samples we have now set forth and reply to each one of them. Thus, the statement that the gnomic sayings of the poets are useful for life and the origin of philosophy, and that grammar is expository of them, is truly worthy of the Grammarians. For, in the first place,—to concur with them in 278
making no charge against poetry,—it is certainly plain that all the sayings in the poets which are found useful for life and necessary,—such as are those of a gnomic and hortatory character,—are expressed by them clearly and have no need of grammar ; while all those which have need of it,—such as those which consist of foreign stories or are enigmatically expressed,—are useless ; so that the usefulness of grammar both fails to accompany the benefit of the former sayings and spins round with the idle talk of the latter. Secondly, the gnomic sentence is only 279
an assertion, such as this,[a]—

> For one wise head excels a hundred hands,
> But crowd-clapp'd folly is a monstrous ill.

But reason does not put trust in an assertion, as to whether it is rightly stated or not, but demands proofs. And proofs of fair statements or unfair are matters for philosophy and not for grammar ; so that it turns out that in this respect also grammar is superfluous and idle.—Furthermore, if it is because of the many fine and profitable sayings of the poets that

[a] *Cf.* Eurip. *Antiopê* (*Frag.* 220 Nauck).

χρησίμη ἐστὶν ἡ προφῆτις γραμματικὴ αὐτῶν, ἐπεὶ πολλαπλασίονα τούτων διαστρόφως καὶ ἐπὶ λύμῃ τοῦ βίου παρ' αὐτοῖς ἐξενήνεκται, ἄχρηστος γενήσεται. καθὰ γὰρ ἔστι τις ὁ εἰπὼν

μὴ Πλοῦτον εἴπῃς· οὐχὶ θαυμάζω θεὸν
ὃν χὡ κάκιστος ῥᾳδίως ἐκτήσατο,

οὕτως ἔστι καὶ ὁ τοὐναντίον ἀποφαινόμενος,

ὦ χρυσὲ δεξίωμα κάλλιστον βροτοῖς,
ὡς οὔτε μήτηρ ἡδονὰς τοίας ἔχει,
οὐ παῖδες ἀνθρώποισιν, οὐ φίλος πατήρ,
οἵας σὺ χοἱ σὲ δώμασιν κεκτημένοι.

καὶ πάλιν

εὖ πρᾶσσε· τὰ φίλων δ' οὐδὲν ἢν τις δυστυχῇ.

καὶ

κάλλιστα μουσῶν φθέγγεται πλουτῶν ἀνήρ.

280 ἀναποδείκτως μὲν οὖν λεγομένων τῶν οὕτως ἐναντίων ἐπιρρεπέστερον ἔχουσιν ἄνθρωποι πρὸς τὴν τοῦ χείρονος ἐκλογήν, καὶ διὰ τοῦτο βλαπτικὴ ἀναφαίνεται ἡ ποιητική· διακρινομένων δὲ αὐτῶν, καὶ τῶν μὲν ἀθετουμένων τῶν δὲ προκρινομένων, χρειώδης γίνεται οὐχ ἡ γραμματικὴ ἀλλ' ἡ διακρίνειν δυναμένη φιλοσοφία. ποιητικοῖς τε μαρτυρίοις χρῶνται οὐχ οἱ γνησίως φιλοσοφοῦντες (τούτων γὰρ ὁ λόγος αὐτάρκης ἐστὶ πρὸς πειθώ) ἀλλ' οἱ τὸν πολὺν καὶ ἀγοραῖον φενακίζοντες
281 ὄχλον· οὐ γὰρ δυσχερὲς ποιητὰς μαχομένους καὶ εἰς ὅ τι ἂν θέλωσιν ᾄδοντας δεῖξαι, ὅτε καὶ οἱ προηγουμένως φιλοσοφοῦντες πολλὰ μαχομένως λέγουσιν. τῶν δὲ γραμματικῆς κατηγόρων ὁ μὲν Πύρρων παρ' ἕκαστα τὴν Ὁμηρικὴν διετύλισσε

grammar, their interpreter, is useful, it will be useless when they utter sentiments, much more numerous than the former, of a perverse character and ruinous for practical life. For just as there is one who declares,[a]—

> Speak not of Wealth; I honour not a god
> Whom e'en the basest makes his own with ease;

so there is another who asserts the opposite,

> Hail, gold, thou boon most fair to mortal men!
> For neither mother doth such joys provide
> To men, nor children, no, nor father dear,
> As thou, and whoso hold thee in their homes.[b]

And again,—

> Fare well! The luckless man finds friendship naught.[c]

And,—

> Sweetest of music is the rich man's voice.

However, as statements so contrary as these are 280
made without proof, men are more prone to choose that which is worse, and for this reason poetry is openly seen to be injurious; and if they are being distinguished, and the one sort set aside and the other preferred, what is useful is not grammar but that which is capable of making the distinction, namely philosophy. Further, it is not the genuine philosophers who make use of testimonies from the poets (for with them the argument is sufficient of itself to carry conviction), but those who humbug the vulgar crowd; for there is no difficulty in showing 281
that the poets are at odds and sing to whatever tune they please, when even the leading philosophers make many conflicting statements. Of the accusers of grammar, Pyrrho used to unroll the poems of

[a] *Cf.* § 271. [b] *Cf.* Eurip. *Danaë* (*Frag.* 326 Nauck).
[c] *Cf.* Eurip. *Phoeniss.* 403.

ποίησιν οὐ πάντως διὰ τὴν εἰρημένην αἰτίαν, ἀλλὰ τάχα μὲν ψυχαγωγίας χάριν καὶ ὡς εἰ κωμῳδῶν ἠκροᾶτο, τάχα δὲ καὶ τοὺς ποιητικοὺς παρατηρῶν
282 τρόπους καὶ χαρακτῆρας· λέγεται γὰρ αὐτὸν καὶ ποίησιν εἰς τὸν Μακεδόνα Ἀλέξανδρον γράψαντα μυρίοις χρυσοῖς τετιμῆσθαι. οὐκ ἀπέοικε δὲ καὶ ἄλλας αἰτίας ὑπάρχειν, περὶ ὧν ἐν τοῖς Πυρρωνείοις
283 διεξήλθομεν. ὁ δὲ Ἐπίκουρος οὐκ ἐκ τῶν Ὁμηρικῶν εἴληφε τὸν ὅρον τοῦ μεγέθους τῶν ἡδονῶν. μακρῷ γὰρ διαφέρει τὸ λέγειν ὅτι ἐπαύσαντό τινες πίνοντες καὶ ἐσθίοντες καὶ τὴν αὐτῶν ἐπιθυμίαν πληροῦντες (τοῦτο γάρ ἐστι τὸ

αὐτὰρ ἐπεὶ πόσιος καὶ ἐδητύος ἐξ ἔρον ἕντο)

τοῦ φάναι ὅρον εἶναι τῶν περὶ τὰς ἡδονὰς μεγεθῶν τὴν τοῦ ἀλγοῦντος ὑπεξαίρεσιν· τοῦτο γὰρ οὐ πάντως κρέασι καὶ οἴνῳ ἀλλὰ καὶ τοῖς λιτοτάτοις
284 πέφυκε γίνεσθαι. ἄλλως τε ὁ μὲν ποιητὴς ἐπὶ προσφερομένων μόνων ἐποιήσατο τὴν ἀπόφασιν, Ἐπίκουρος δὲ ἐπὶ πάντων τῶν ἀπολαυστῶν, ἐν οἷς ἐστὶ καὶ ἡ ἀφροδίσιος μῖξις, περὶ ἧς πάντες ἴσασιν οἵαν ἔσχε γνώμην Ὅμηρος. τό τε τὸν θάνατον [μὲν] μηδὲν εἶναι πρὸς ἡμᾶς εἴρηται μὲν ἴσως τῷ Σώφρονι, ἀποδέδεικται δὲ Ἐπικούρῳ, καὶ ἔστιν οὐ τὸ εἰπεῖν ἀλλὰ τὸ ἀποδεῖξαι θαυμαστόν.
285 εἶτα οὐδὲ κατὰ τοῦτο ἔφησεν ὁ Ἐπίκουρος τὸν θάνατον μηδὲν εἶναι πρὸς ἡμᾶς, καθὸ ἀδιάφορόν ἐστιν ἢ ζῆν ἢ μή· πολλῷ γὰρ αἱρετώτερον τὸ ζῆν διὰ τὸ αἰσθανομένων εἶναι τὸ ἀγαθόν· ἀλλ' ἐν ἀναισθησίᾳ οὔτε κακόν τι εἶναι οὔτε ἀγαθόν. τὸ

---

[a] *Cf.* § 272. [b] *Cf.* § 273.

[c] Sophron of Syracuse, writer of Mimes, *circ.* 440 B.C.: for

Homer one by one, and that not altogether for the reason stated,[a] but possibly because of the pleasure they gave him, as if he were listening to a comedy, and perhaps with an eye to his poetic tropes and characters; for Pyrrho himself, it is said, wrote a 282
poem for Alexander of Macedon and was rewarded with thousands of gold pieces. And it is not unlikely that there were also other reasons which we have discussed in our *Pyrrhonea*. And Epicurus did not take 283
his definition of the intensity of pleasures from the Homeric poems.[b] For there is a vast difference between saying that certain persons ceased from drinking and eating and satisfying their appetite (for that is the meaning of

> When they had now put aside all longing for drinking and eating)

and saying that the definition of the intensities of pleasures is "the removal of the painful"; for this is the natural effect not of meats and wine only but also of the simplest foods. Moreover, the poet made his 284
statement about viands only, but Epicurus about all enjoyable things, amongst which is sexual intercourse, and all know what opinion Homer held about it. That "death is nothing to us" was said, no doubt, by Sophron,[c] but was proved by Epicurus, and it is not the saying but the proving that is admirable. Moreover, Epicurus did not 285
assert that "death is nothing to us" in the sense that to be alive or not is a matter of indifference; for it is far preferable to be alive because the good belongs to the sentient; his meaning was that where sense is absent there is nothing evil or good.

the sentiment *cf.* "nil igitur mors est ad nos neque pertinet hilum" (Lucret. iii. 830), and *P.H.* iii. 229.

μὲν γὰρ ἀναισθητεῖν τὰ νεκρὰ τῶν σωμάτων οὐχ ὁ ποιητὴς μόνος οἶδεν ἀλλὰ καὶ ὁ σύμπας βίος. μήτηρ γοῦν πολλάκις υἱὸν θρηνοῦσα φησὶν "ἀλλὰ σὺ μὲν τούτων οὐκ ἐπαισθάνῃ, ἐγὼ δὲ ταλαιπωρῶ"· καὶ ἐνατενίζουσα ἐπιφθέγγεται "τίς δέ
286 ἐστιν ἔτι σοι τούτων ὄνησις;" οὐ μὴν ἀλλ' ἐὰν ἐξετάζῃ τις, τὴν ἐναντίαν ἔχοντα δόξαν εὑρήσει τὸν ποιητήν. αἱ μὲν γὰρ ψυχαὶ κοινῶς διψῶσιν αἵματος

(ἀλλ' ἀποχάζεο βόθρου, ἄπισχε δὲ φάσγανον ὀξὺ
αἵματος, ὄφρα πίω καί τοι νημερτέα εἴπω),[a]

ὁ δὲ Τιτυὸς ὑπὸ γυπῶν διὰ τὴν ἐπιθυμίαν ἡπατοφαγεῖται, ὁ δὲ Τάνταλος ἕστηκεν ἐν λίμνῃ,

ἡ δὲ προσέκλυζε γενείῳ·
στεῦτο δὲ διψάων, πιέειν δ' οὐκ εἶχεν ἑλέσθαι.[b]

287 καὶ μὴν ὅσον ἐπὶ τῷ ὑπ' Εὐριπίδου λεχθέντι περὶ θεῶν, τὴν αὐτὴν καὶ οἱ ἰδιῶται δόξαν ἔχουσιν. ἴσον γάρ ἐστι τῷ

ὅστις δὲ θνητῶν οἴεται τοὐφήμερον
κακόν τι πράσσων τοὺς θεοὺς λεληθέναι,
δοκεῖ πονηρά, καὶ δοκῶν ἁλίσκεται
ὅταν σχολὴν ἄγουσα τυγχάνῃ δίκη

καὶ τὸ οὕτω παρὰ τοῖς πολλοῖς λεγόμενον,

ὀψὲ θεῶν ἀλέουσι μύλοι, ἀλέουσι δὲ λεπτά·

288 μόνῳ δὲ διενήνοχε τῷ μέτρῳ. ἂν δὲ καὶ ἐξετάσῃ τις, πολλῷ χείρονα τῆς τῶν ἰδιωτῶν ὑπολήψεως εὑρήσει τὰ τῶν ποιητῶν καὶ ὁ μὲν σκηνικὸς

---

[a] *Cf.* Hom. *Od.* xi. 95 ff.

[b] *Cf.* Hom. *Od.* xi. 583 f.; *Adv. Phys.* i. 69.

And that dead bodies are devoid of sense is a fact known not only to the poet but also to all mankind. Thus, a mother in her frequent laments over her son cries—" But thou perceivest not these things, it is I that am in distress " ; and as she gazes at him fixedly she cries—" What benefit from these things hast thou still ? " Nevertheless, if one searches he will find the 286
poet holding the opposite opinion. For the souls commonly are thirsty for blood,—

> Nay, keep away from the trench, from its blood averting thy sharp sword,
> So shall I drink my draught and announce to thee tidings unerring.[a]

And Tityos, because of his lust, has his liver eaten by vultures, and Tantalus stands in a lake,—

> Against his chin was it dashing :
> Hard strove he in his thirst, but failed to catch any for drinking.[b]

Furthermore, as regards what is said by Euripides 287
about the gods, ordinary folk too hold the same opinion. For the sentence [c]—

> Whoe'er of mortals, sinning day by day,
> Deemeth the gods are blind to his misdeeds,
> Thinks evil thoughts and thinking thus is caught
> When Justice, haply, has some leisure time—

is matched by the sentence commonly quoted—

> The mills of God grind slowly, but they grind exceeding small[d] ;

for the difference is only in the metre. And if one 288
investigates one will find that the sentiments of the poets are much worse than the notions of ordinary folk. Yet he who has been proclaimed the philo-

[a] *Cf.* § 274.

[d] *Cf.* Plutarch, p. 549 D.

ἀναγορευθεὶς φιλόσοφος ἔτι μετριώτερος φαίνεται, λέγων μὴ εἰδέναι ᾧ προσεύχεται.

ὦ γῆς ὄχημα κἀπὶ γῆς ἔχων ἕδραν,
ὅστις ποτ' εἶ σύ, δυστόπαστος εἰσιδεῖν,
Ζεύς, εἴτ' ἀνάγκη φύσεος εἴτε νοῦς βροτῶν,
προσευξάμην σέ.

289 Ὅμηρος δὲ καὶ Ἡσίοδος κατὰ τὸν Κολοφώνιον Ξενοφάνη

ὡς πλεῖστ' ἐφθέγξαντο θεῶν ἀθεμίστια ἔργα,
κλέπτειν μοιχεύειν τε καὶ ἀλλήλους ἀπατεύειν.

Κρόνος μὲν γάρ, ἐφ' οὗ τὸν εὐδαίμονα βίον γεγονέναι λέγουσι, τὸν πατέρα ἠνδροτόμησε καὶ τὰ τέκνα κατέπιεν, Ζεύς τε ὁ τούτου παῖς ἀφελόμενος αὐτὸν τῆς ἡγεμονίας

γαίης νέρθε καθεῖσε καὶ ἀτρυγέτοιο θαλάσσης,
τῆλε μάλ' ἧχι βάθιστον ὑπὸ χθονός ἐστι βέρεθρον.

290 τῷ δὲ Διὶ ἐπιβουλεύουσιν οἱ συγγενεῖς, παρὸ καὶ ὑπὸ Θέτιδος βοηθεῖται,

ὁππότε μιν ξυνδῆσαι Ὀλύμπιοι ἤθελον ἄλλοι,
Ἥρη τ' ἠδὲ Ποσειδάων καὶ Παλλὰς Ἀθήνη·

ὠμότατος γάρ ἐστι, καὶ τὴν μὲν ἀδελφὴν καὶ γυναῖκα ἱεροσύλου τρόπον κρεμάσας οὐκ ἀρκεῖται, ἀλλὰ καὶ ὀνειδίζει λέγων

ἦ οὐ μέμνῃ ὅτε τ' ἐκρέμω ὑψόθεν, ἐκ δὲ ποδοῖιν
ἄκμονας ἧκα δύω, περὶ χερσὶ δὲ δεσμὸν ἴηλα
χρύσεον ἄρρηκτον, σὺ δ' ἐν αἰθέρι καὶ νεφέλησιν
ἐκρέμω, ἠλάστεον δὲ θεοὶ κατὰ μακρὸν Ὄλυμπον;

---

[a] *Cf.* Eurip. *Troad.* 884.
[b] *Cf. Adv. Phys.* i. 193. [c] *Cf. P.H.* iii. 210.
[d] *Cf.* Hom. *Il.* xiv. 204, and viii. 14.

sopher of the Stage seems still rather moderate when he says that he does not know to whom he is praying,[a]—

> Thou stay of earth, who hast on earth thy throne,
> To see and know thee, who thou art, O Zeus,
> Doth baffle wit! Art thou Necessity
> Of Nature? Or mankind's Intelligence?
> Howbeit, I invoke thee.

But Homer and Hesiod, according to Xenophanes of 289
Colophon,[b]—

> Tell full many a tale of the lawless deeds of immortals,
> Thieving and wenching and dealing deceitfully one with another.

For Cronos,[c] in whose time, they say, was the life of blessedness, castrated his father and swallowed his children; while Zeus, his son, after robbing him of his dominion,—

> Under the earth he flung him and under the barren ocean,
> Far away, under the earth where the pit of destruction is deepest.[d]

But Zeus' own brethren conspire against Zeus, and 290
so he is helped by Thetis,—

> Whenas the other Olympians all were wishful to bind him,
> Hera along with Poseidon, with Pallas Athenê to aid them.[e]

For he is most savage, and when he had hung up his sister and wife like a temple-robber, not satisfied with that he also reviles her, saying[f]—

> Hast thou already forgot how from heav'n thou didst hang, with two boulders
> Tied to thy feet, and about thy hands a chain did I fasten
> Gold-wrought, not to be broken, and thou wast suspended in heaven
> High in the clouds; and the gods were enraged throughout lofty Olympus?

[e] *Cf.* Hom. *Il.* i. 399 f.

[f] *Cf. Il.* xv. 18 ff. (addressing Hera).

291 τὸν δὲ Ἥφαιστον ὀργισθεὶς ῥίπτει ἀπὸ τοῦ οὐρανοῦ, ὁ δὲ

> κάππεσεν ἐν Λήμνῳ, ὀλίγος δ' ἔτι θυμὸς ἐνῆεν.

τὸν δὲ ἀδελφὸν ὑπερορᾷ

> οἰκί' ἔχοντα
> σμερδαλέ' εὐρώεντα, τά τε στυγέουσι θεοί περ.

πρόσεστι δὲ αὐτῷ πρὸς τῇ ἀποτομίᾳ καὶ ἀκρασία, ὃς θεασάμενος τὴν Ἥραν ἐπὶ τῆς Ἴδης κεκοσμημένην οὐ καρτερεῖ μέχρι τῶν ἀποδεδειγμένων αὐτοῖς θαλάμων ἐλθεῖν, ἀλλ' ἐπὶ τοῦ ὄρους χαμαὶ ῥίψας ἑαυτὸν συγκυλίνδεται τῇ γυναικί,

> τοῖσι δ' ὑπὸ χθὼν δῖα φύεν νεοθηλέα ποίην,
> λωτόν θ' ἑρσήεντα ἰδὲ κρόκον ἠδ' ὑάκινθον.

292 ποικίλης οὖν πεφωραμένης τῆς ποιήσεως ἀνωφελὴς ἡ γραμματικὴ μὴ δυναμένη ἀποδεῖξαι τίσι πιστευτέον ἐστὶν ὡς ἀληθέσι καὶ τίσιν ἀπιστητέον ὡς μυθικοῖς ψεύσμασιν.

293 Ἀλλὰ πόλει φασὶ χρησίμην εἶναι τὴν γραμματικήν, ἐπεὶ καὶ Λεβεδίοις νίκης αἴτιον ἐγένετο ἐκ ποιητικῆς μαρτύριον. ἕνεκα δὲ τούτου καὶ τὴν ὀρχηστικὴν ἀναγκαίαν λέγομεν εἶναι, ἐπεὶ Σώστρατος ὁ Ἀντιόχου ὀρχηστής, λαβόντος ὑποχείριον τὴν Πριήνην τοῦ βασιλέως πατρίδα οὖσαν αὐτοῦ, καὶ παρὰ τὸ συμπόσιον τὴν ἐλευθερίαν ἀναγκαζόμενος ὀρχεῖσθαι, οὐ καλὸν ἔφη τῆς πατρίδος αὐτοῦ δουλευούσης αὐτὸν ἐλευθερίαν ὀρχεῖσθαι· καὶ διὰ τοῦτο ἐλευθερωθῆναι τὴν πόλιν.

---

[a] *Cf. Il.* i. 593. [b] *Cf. Il.* xx. 65, said of Pluto.
[c] *Cf. Il.* xiv. 347 f. [d] § 275.
[e] A Greek city in Caria. Antiochus was king of Syria.

And in his rage he hurls Hephaestus out of heaven, 291
and he—

> Crashed in Lemnos, and little of life was remaining within him.[a]

And he treats his brother with contempt, as—

> Having for dwelling
> Grim halls, dismal and dank, detested e'en by immortals.[b]

And in addition to his ruthlessness he is given to incontinence, for on beholding Hera finely decked out on Ida he cannot bear waiting to retire to their appointed chamber, but flings himself to the ground on the mountain and rolls over with his wife,—

> Grass, fresh grass, for their couch did the earth shoot forth ever kindly,
> Hyacinths, too, with the crocus, and lotus besprinkled with dewdrops.[c]

The varied nature of poetry being thus exposed, 292
grammar becomes useless since it cannot show us which parts of it we should believe as being true, and which we should disbelieve as mythical falsehoods.

Still, they declare that the Art of Grammar is useful 293
for the State, seeing that even for the Lebedians[d] a testimony from poetry was the cause of victory. But we reply that for the same reason the art of dancing is necessary since Sostratus, the dancer of Antiochus, —when the king had subjugated Prienê,[e] which was his homeland, and he was being forced at the banquet to dance "the liberty dance,"[f]—declared that it was not right for him to dance a "liberty-dance" while his country was enslaved; and because of this

[f] Danced at the Feast of "Eleutheria" ("Liberty") held at Plataea in memory of the victory over the Persians.

294 εἶτα ἄλλο μέν ἐστι τὸ πόλει χρήσιμον, ἄλλο δὲ τὸ ἡμῖν αὐτοῖς. σκυτοτομικὴ γοῦν καὶ χαλκευτικὴ πόλει μέν ἐστιν ἀναγκαῖον, ἡμῖν δὲ χαλκεῦσι γενέσθαι καὶ σκυτοτόμοις πρὸς εὐδαιμονίαν οὐκ ἀναγκαῖον. διόπερ καὶ ἡ γραμματικὴ οὐκ ἐπεὶ πόλει χρησίμη καθέστηκεν, ἐξ ἀνάγκης καὶ ἡμῖν ἐστὶν [ἡ] τοιαύτη. ἡ μὲν γὰρ ὁμιλητικὴ οὐκ ἀπὸ
295 γραμματικῆς περιγίνεσθαι πέφυκεν ἀλλ' ἀπὸ κοινῆς τινὸς ἐντρεχείας, εἰ μή τι καὶ Δημάδης ὁ ῥήτωρ γραμματικὸς ἦν,[1] πολλοῖς τῶν Ἀθηναίων μετὰ τὴν ἐν Χαιρωνείᾳ ἧτταν συναιχμαλωτισθείς, καὶ εἰπὼν πρὸς τὸν Φίλιππον ἀναγκάζοντα εὐωχεῖσθαι

τίς γάρ κεν ἀνήρ, ὃς ἐναίσιμος εἴη,
πρὶν τλαίη πάσασθαι ἐδητύος ἠδὲ ποτῆτος,
πρὶν λῦσαί θ' ἑτάρους καὶ ἐν ὀφθαλμοῖσιν ἰδέσθαι;

296 Ταῦτα μὲν οὖν πρὸς τὰς τῶν γραμματικῶν ἐπιχειρήσεις λεγέσθω· προηγουμένως δὲ ῥητέον ὡς εἰ μὲν μόνοι ἦσαν οἱ ποιηταὶ βιωφελεῖς, τάχα ἂν ἡ γραμματικὴ βιωφελὴς ἐγίνετο περὶ τούτους πονουμένη, νῦν δὲ ἐπεὶ οὗτοι μὲν ἢ ἀνωφελεῖς ε σὶν ἢ ὀλιγωφελεῖς, φιλόσοφοι δὲ καὶ οἱ λοιποὶ συγγραφεῖς διδάσκουσι τὰ ὠφέλιμα τῶν πραγμά-
297 των, οὐ δεόμεθα γραμματικῆς. καὶ ὅτι οἱ συγγραφεῖς μᾶλλον ἢ οἱ ποιηταὶ τὰ χρήσιμα τῷ βίῳ δηλοῦσιν, εὐεπιλόγιστον. οἱ μὲν γὰρ τοῦ ἀληθοῦς στοχάζονται, οἱ δὲ ἐκ παντὸς ψυχαγωγεῖν ἐθέλουσιν, ψυχαγωγεῖ δὲ μᾶλλον τὸ ψεῦδος ἢ τἀληθές. τοίνυν ἐκείνοις ἢ ὅτι τούτοις προσεκτέον τοῖς

[1] ἦν cj. Bekk. : ὢν MSS., Bekk.

his city was set free. Moreover, usefulness for the 294
State is one thing, usefulness for ourselves another thing. Thus the arts of the cobbler and the coppersmith are necessary to the State, but it is not necessary for our happiness that we should become coppersmiths and cobblers. Hence, the Art of Grammar is not necessarily useful to us because it is useful to the State. For the art of conversation is not usually gained from grammar but from a general quickness of
wit,—unless, indeed, Demades,[a] the orator, was a 295
Grammarian, who (when he was captured along with many Athenians after the defeat at Chaeronea) said to Philip who was forcing him to join in a feast,—

> Was there ever a man with a feeling for justice
> Who in his heart could endure to share in eating and drinking
> Ere he had freed his companions and seen them standing before him?[b]

Let thus much, then, be said in opposition to the 296
arguments of the Grammarians; but before all it should be stated that if the poets only were useful for life, then perhaps grammar, as occupied with them, had been useful for life, but as it is, seeing that the poets are either useless or of little use, whereas the philosophers and the other prose-writers instruct us in things that are useful, we do not need grammar.
And the fact that the prose-writers rather than the 297
poets show what is useful for life is easy to demonstrate. For the former aim at the truth, but the latter seek by every means to attract the soul, and the false attracts more than the true. They, therefore, who expressly pursue what is false must listen to

[a] Demades, the Athenian orator, opponent of Demosthenes, captured at Chaeronea in 338 B.C.
[b] *Cf.* Hom. *Od.* x. 383 ff.

298 ἐπίτηδες τὸ ψεῦδος μεταδιώκουσιν. καθόλου τε, ὅσον ἐπὶ τοῖς ποιηταῖς, οὐχ οἷον ἀνωφελὴς τῷ βίῳ ⟨ἡ γραμματικὴ⟩[1] ἀλλὰ καὶ βλαβερωτάτη. ἐπιτείχισμα γὰρ ἀνθρωπίνων παθῶν ἡ ποιητικὴ καθέστηκεν· καὶ ὡς

γέρων γέροντι γλῶσσαν ἡδίστην ἔχει,

οὕτως οἱ μὲν ἐρωτομανεῖς καὶ μέθυσοι τὰς Ἀλκαίου καὶ Ἀνακρέοντος ποιήσεις ἀναγνόντες προσεκκαίονται, οἱ δὲ ὀργίλοι Ἱππώνακτα καὶ Ἀρχίλοχον ἀλείπτας ἔχουσι τῆς περὶ αὐτοὺς κακίας.

299 Τὰ μὲν οὖν ὑπὸ τῶν ἄλλων λεγόμενα κατὰ τὸν τόπον, καὶ μάλιστα τῶν Ἐπικουρείων, ἐστὶ τοιαῦτα· ἡμεῖς δὲ μηδὲν κατειπόντες τῆς ποιητικῆς ἄλλως ποιώμεθα τὰς ἀντιρρήσεις πρὸς τοὺς ἀξιοῦντας γραμματικὴν ἔχειν τέχνην τῶν παρὰ ποιηταῖς καὶ
300 συγγραφεῦσι λεγομένων διαγνωστικήν. ἐπεὶ τοίνυν πᾶν σύγγραμμα καὶ πᾶσα ποίησις ἐκ λέξεων τῶν δηλουσῶν καὶ πραγμάτων τῶν δηλουμένων συνέστηκε, δεήσει τὸν γραμματικόν, εἴπερ ἔχει τέχνην διαρθρωτικὴν τῶν παρὰ συγγραφεῦσι καὶ ποιηταῖς λεγομένων, ἤτοι τὰς λέξεις μόνον ἢ τὰ ὑποκείμενα πράγματα γινώσκειν ἢ τὸ συναμφότερον. ἀλλὰ τὰ μὲν πράγματα, κἂν ἡμεῖς μὴ λέγωμεν, φαίνεται μὴ γινώσκειν. τούτων γὰρ τὰ μέν ἐστι φυσικὰ τὰ δὲ μαθηματικὰ τὰ δὲ ἰατρικὰ τὰ δὲ μουσικά, καὶ δεῖ τὸν μὲν φυσικοῖς ἐπιβάλλοντα πράγμασιν εὐθὺς φυσικὸν εἶναι καὶ τὸν μουσικοῖς μουσικὸν εἶναι καὶ τὸν μαθηματικοῖς εὐθὺς εἶναι μαθηματικόν, καὶ ἐπὶ τῶν ἄλλων ὁμοίως. ὁ μέντοι γραμματικὸς ὅτι οὐκ ἔστιν ἐν τῷ αὐτῷ πάνσοφος καὶ

[1] ⟨ἡ γραμματικὴ⟩ add. cj. Bekk.

these rather than to those.—Further, to generalize, 298
so far as regards the poets, grammar is not merely useless for life but even most injurious. For poetry is a stronghold of human passions, and as

> To ancient men an ancient's is the tongue most sweet,[a]

so too the love-maddened and the wine-bibbers are further inflamed by reading the poems of Alcaeus and Anacreon, while the irascible have for *aliptae* (or " trainers ") in their vice Hipponax and Archilochus.

Such, then, are the things said by the rest, and 299
especially by the Epicureans, regarding this topic; but let us, without decrying poetry at all, set forth our counter-arguments against those who claim that there is an Art of Grammar capable of making distinctions in the statements of poets and prose-writers. Now since every composition and every 300
poem consists of indicating words and things indicated, it will be necessary for the Grammarian, if he possesses an art capable of discriminating among the things said by poets and prose-writers, to understand either the words only or the objects behind them or both. But that they do not understand the objects, even if we do not say so, is evident. For some of these are physical, some mathematical, some medical, some musical, and he who deals with the physical must of course be a physicist, he who deals with the musical a musician, and of course he who deals with the mathematical a mathematician, and similarly with the other sciences. That the Grammarian, however, is not simultaneously all-wise and skilled in every

[a] *Cf.* Meineke, *Com. Gr.* 5, p. 668.

πάσης ἐπιστήμης δαήμων, σὺν τῷ καὶ αὐτόθεν προσπίπτειν, ἔτι κἀκ τῶν ἀποτελεσμάτων ἐλέγ-
301 χεται. ποῦ γάρ τις δύναται τῶν ὠφρυωμένων γραμματικῶν Ἡράκλειτον συνεῖναι καὶ Πλάτωνι παρακολουθῆσαι λέγοντι "τῆς ἀμερίστου καὶ ἀεὶ κατὰ ταὐτὰ ἐχούσης οὐσίας καὶ τῆς περὶ τὰ σώματα μεριστῆς τρίτον ἐξ ἀμφοῖν συνεκεράσατο οὐσίας εἶδος, τῆς τε ταὐτοῦ φύσεως καὶ τῆς θατέρου" καὶ ἤδη τὰ ἑξῆς περὶ τὴν λέξιν, ἃ πάντες οἱ Πλάτωνος ἐξηγηταὶ ἐσίγησαν. ἢ ποῦ τοῖς Χρυσίππου διαλεκτικοῖς θεωρήμασιν ἢ Ἀρχιμήδους τε καὶ Εὐδόξου μαθηματικοῖς ἐπιβάλλειν ἰσχύσει;
302 καὶ μὴν ὡς ἐν τούτοις ἐστὶ τυφλός, οὕτω κἂν τοῖς περὶ αὐτῶν γραφεῖσι ποιήμασιν, οἷον Ἐμπεδοκλέους λέγοντος

χαίρετ', ἐγὼ δ' ὑμῖν θεὸς ἄμβροτος, οὐκέτι θνητός,
πωλεῦμαι μετὰ πᾶσι τετιμένος,

καὶ πάλιν

ἀλλὰ τί τοῖσδ' ἐπίκειμ' ὡσεὶ μέγα χρῆμά τι πράσσων
εἰ θνητῶν περίειμι πολυφθερέων ἀνθρώπων;

303 ὁ μὲν γὰρ γραμματικὸς καὶ ὁ ἰδιώτης ὑπολήψονται κατ' ἀλαζονείαν καὶ τὴν πρὸς τοὺς ἄλλους ἀνθρώπους ὑπεροψίαν ταῦτ' ἀνεφθέγχθαι τὸν φιλόσοφον, ὅπερ ἀλλότριόν ἐστι τοῦ κἂν μετρίαν ἕξιν ἐν φιλοσοφίᾳ ἔχοντος, οὐχ ὅτι γε τοῦ τοσούτου ἀνδρός· ὁ δὲ ἀπὸ φυσικῆς ὁρμώμενος θεωρίας, σαφῶς γινώσκων ὅτι ἀρχαῖον ὅλως τὸ δόγμα ἐστί, τοῖς ὁμοίοις τὰ ὅμοια γινώσκεσθαι, ὅπερ ἀπὸ Πυθα-

science is proved by the results of experience, besides being obvious of itself. For how can any of the grave 301
Grammarians comprehend Heracleitus or follow the meaning of Plato where he says,[a]—"Between the Being which is indivisible and remains always the same and the Being which is divisible in bodies, He blended a third form of Being compounded of the twain, that is to say, out of the Same and the Other," and the rest of the context, about which all the interpreters of Plato keep silence? Or how will he be able to deal with the dialectical investigations of Chrysippus or the mathematical of Archimedes and Eudoxus? Moreover, as he is blind in respect of 302
these things, so he is also in respect of the poems written about them, as when Empedocles says,—

> Hail ye, but I as a deathless god, no longer a mortal,
> Walk in your midst and am honoured by all.

And again,—

> Why do I urge these things, as though it were some great matter
> If I do far surpass all mortals doomed to destruction?

For the Grammarian and the ordinary man will 303
suppose that the philosopher gave utterance to these sayings out of boastfulness and contempt for the rest of mankind,—a thing alien to one who is even moderately versed in philosophy, not to speak of a man of such eminence. But the man who sets out from physical investigation knows clearly that the dogma "like is known by like" is nothing but an old one[b] which is thought to have come down from

[a] *Cf. Timaeus* 35 A.
[b] *Cf. Adv. Log.* i. 92, 121.

γόρου δοκοῦν κατεληλυθέναι κεῖται μὲν καὶ παρὰ Πλάτωνι ἐν τῷ Τιμαίῳ, εἴρηται δὲ πολὺ πρότερον ὑπ' αὐτοῦ 'Εμπεδοκλέους,

γαίῃ μὲν γὰρ γαῖαν ὀπώπαμεν, ὕδατι δ' ὕδωρ,
ἠέρι δ' ἠέρα δῖον, ἀτὰρ πυρὶ πῦρ ἀίδηλον,
στοργὴν δὲ στοργῇ, νεῖκος δέ τε νείκεϊ λυγρῷ,

συνήσει ὅτι ὁ 'Εμπεδοκλῆς θεὸν ἑαυτὸν προσηγόρευσεν, ἐπεὶ μόνος καθαρὸν ἀπὸ κακίας τηρήσας τὸν νοῦν καὶ ἀνεπιθόλωτον τῷ ἐν ἑαυτῷ θεῷ τὸν
304 ἐκτὸς κατείληφεν. 'Αράτου τε μὴν γράφοντος

ὅσσον ἀπ' ὀφθαλμοῖο βολῆς ἀπολάμπεται αὐγή,
ἑξάκις ἂν τόσση μιν ὑποδράμοι· αὐτὰρ ἑκάστη
ἴση μετρηθεῖσα δύω περιτέλλεται ἄστρα

οὐ γραμματικοῦ τοῦτο νοῆσαι, ὅτι ἡλίκη ἐστὶν ἡ ἀπὸ τῆς ἡμῶν ὄψεως πρὸς τὴν ἀνατολὴν ἐκβαλλομένη εὐθεῖα, ἑξάκις αὕτη ληφθεῖσα τὸν ζωδιακὸν καταμετρήσει κύκλον ὥστε δύο αὐτὴν ἀποτέμνεσθαι ζώδια, ἀλλὰ μαθηματικοῦ, γραμμικῶς αὐτὸ ἀποδεικνύντος, ὅτι τὸ ἕκτον τοῦ ζωδιακοῦ κύκλου μέρος ἀπὸ τῆς μέχρι τῆς ἀνατολῆς ἐκβαλλομένης
305 εὐθείας καθέστηκεν. Τίμωνός τε τοῦ Φλιασίου τὸν Πύρρωνα ἡλίῳ ἀπεικάζοντος ἐν οἷς φησὶ

μοῦνος δ' ἀνθρώποισι θεοῦ τρόπον ἡγεμονεύεις,
ὃς περὶ πᾶσαν ἑλῶν[1] γαῖαν ἀναστρέφεται,
δεικνὺς εὐτόρνου σφαίρας πυρικαύτορα κύκλον,

δόξει μὲν τοῖς γραμματικοῖς κατὰ τιμὴν αὐτὸ λέγειν καὶ διὰ τὴν περὶ τὸν φιλόσοφον ἐπιφάνειαν· ἄλλος δὲ ἐπιστήσει μήποτε καὶ μάχεται [τὰ

[1] ἑλῶν cj. Bekk. : ἐλὼν mss., Bekk.

Pythagoras and is found also in Plato's *Timaeus*; and it was stated much earlier by Empedocles himself,—

> Verily earth by earth we behold, and water by water,
> Aether divine by aether, and fire the destructive by fire,
> Love, moreover, by love, and hate by dolorous hatred.

Such a man will understand that Empedocles called himself a god because he alone had kept his mind free from evil and unmuddied and by means of the god within him apprehended the god without. Again, 304
when Aratus writes,[a]—

> Far as the gleam of the ray from the eye doth reach in its shining,
> Full six times such a space would he intercept; of the spaces
> Each being equal in measure contains two signs of the star-world,—

it is not the part of the Grammarian to conceive that when the length of the straight line extending from our eye to the sun-rise is taken, six times this length will measure out the circle of the zodiac, so that it cuts off two of its Signs; rather it is the part of the mathematician, who proves by geometry that the sixth part of the circle of the zodiac is formed by the straight line which extends to the sun-rise. Also, when Timon of
Phlius compares Pyrrho to the sun,[b] where he says,— 305

> Thou, thou alone, art the guide of mankind, most like to the Sun-god,
> Who the whole earth surveys as he revolves in his course,
> Showing his globe well-turn'd, as a circle all-fiery and flaming,—

he will seem to the Grammarians to be saying it by way of commendation and on account of the philosopher's brilliance. But another man will ponder

[a] *Cf.* Aratus, *Phaenom.* 541.
[b] *Cf.* Lucretius (iii. 1043 f.), of Epicurus,—"omnes Praestinxit stellas, exortus ut aetherius Sol."

παραδείγματα][1] τῷ σκεπτικῷ βουλήματι τὰ ὑπὸ τοῦ Φλιασίου εἰς τὸν Πύρρωνα λεχθέντα, εἴγε ὁ μὲν ἥλιος τὰ πρότερον μὴ βλεπόμενα τῷ φωτὶ καταυγάζων δείκνυσιν, ὁ δὲ Πύρρων καὶ τὰ προδήλως ἡμῖν ληφθέντα τῶν πραγμάτων εἰς ἀδηλό-
306 τητα περιιστάναι βιάζεται. τὸ δὲ οὐχ οὕτως ἔχειν φαίνεται τῷ φιλοσοφώτερον ἐπιβάλλοντι, ἀλλ' ἡλίου τρόπον ἐπέχειν φησὶ τὸν Πύρρωνα καθόσον ὡς ὁ θεὸς τὰς τῶν ἀκριβῶς εἰς αὐτὸν ἀτενιζόντων ὄψεις ἀμαυροῖ, οὕτω καὶ ὁ σκεπτικὸς λόγος τὸ τῆς διανοίας ὄμμα τῶν ἐπιμελέστερον αὐτῷ προσεχόντων συγχεῖ, ὥστε ἀκαταληπτεῖν περὶ ἑκάστου τῶν κατὰ δογματικὴν θρασύτητα τιθεμένων.
307 εἰ δὲ δεῖ περὶ ἰατρικῆς διεξέρχεσθαι θεωρίας, καὶ παριστᾶν ⟨ἔστιν⟩[2] ὡς καὶ ἐπίθετον πολλάκις προσριφὲν ὑπὸ ποιητοῦ βαθὺν ἐμφαίνει καὶ, ἐπιστημονικὸν νοῦν, οἷόν ἐστι τὸ "βαθύσχοινον λεχεποίην" παρ' Ὁμήρῳ. σημαίνει γὰρ ὃ μὴ δύναται νοῆσαι γραμματικός, ⟨ὅτι⟩[3] παραστατικὸν πρὸς συνουσίαν ἐστὶ τὸ τῆς σχοίνου σπέρμα, λέχος καλοῦντος
308 τοῦ ποιητοῦ τὴν μῖξιν. ἢ τὸ παρὰ τῷ Εὐριπίδῃ ἐπὶ τῇ Λυκομήδους θυγατρὶ Δηιδαμείᾳ λεγόμενον,

ἡ παῖς νοσεῖ σου κἀπικινδύνως ἔχει.
πρὸς τοῦ; τίς αὐτὴν πημονὴ δαμάζεται;
μῶν κρυμὸς αὐτῆς πλευρὰ γυμνάζει χολῇ;[4]

πυνθάνεται γὰρ μή τι πλευριτικὴ γέγονε διὰ τὸ τοὺς πλευριτικοὺς βήσσοντας ὑπόχολον ἀνάγειν. ὧν οὐδὲν οἶδεν ὁ γραμματικός.

[1] [τὰ παραδείγματα] del. Heintz.
[2] ⟨ἔστιν⟩ addidi (*adduci potest* Herv.).
[3] ⟨ὅτι⟩ add. cj. Bekk.
χολῇ Valckenaer : χολῆς MSS., Bekk.

whether the statements made by the man of Phlius about Pyrrho do not conflict with the Sceptics' doctrine, since the sun exposes things not previously seen by illuminating them with its light, whereas Pyrrho forces the things we have already perceived clearly to revert into obscurity. Yet it is plain to 306
him who considers it more philosophically that this is not the case, but that he is declaring that Pyrrho is suspending judgement like the sun, in so far as that god dims the vision of those who gaze at him intently, and the Sceptics' argument likewise confuses the eye of the intellect in those who carefully attend to it, so that they fail to apprehend every one of the things postulated by the rash Dogmatists. And if one has 307
to discuss medical doctrine, one can show how an epithet thrown out by a poet often conveys a deep and scientific meaning, as for instance Homer's *βαθύσχοινον λεχεποίην* ("deep in rushes, with bed of grass").[a] For it signifies what the Grammarian is unable to perceive,—namely that the seed of the rush (σχοῖνος) is conducive to venery, as the poet gives the name of "bed" (λέχος) to sexual intercourse. Another example is the saying in Euripides[b] 308
about Deïdameia, the daughter of Lycomedes,—

> Thy daughter's ill and her condition's dangerous.
> Because of what? What sort of plague doth hold her down?
> Is it a chill which thus torments her side with bile?

For he is inquiring whether she is suffering from pleurisy, since sufferers from pleurisy bring up bile when they cough. But none of these things is known to the Grammarian.

[a] *Cf. Il.* iv. 383. [b] *Cf.* Eurip. *Frag.* 683 (Nauck).

309 Καίτοι περιττὸν ἴσως ἐστὶν ἀπὸ τῶν ἀρχαιοτέρων καὶ τάχα ἐπιστημονικῶν δυσωπεῖν τοὺς ἀπὸ τῆς γραμματικῆς, ὅτε καὶ τὸ τυχὸν ἐπιγραμμάτιον οὐχ οἷοί τέ εἰσι νοῆσαι, καθάπερ καὶ τὸ ὑπὸ τοῦ Καλλιμάχου εἰς Διόδωρον τὸν Κρόνον συγγραφέν,

> ἠνίδε κου κόρακες τεγέων ἔπι κοῖα συνῆπται
> κρώζουσιν, καὶ κῶς αὖθι γενησόμεθα.

310 ὅτι γὰρ διαλεκτικώτατος ἦν ὁ Κρόνος καὶ ἐδίδασκε πῶς κριτέον ἐστὶ τὸ ὑγιὲς συνημμένον, ὥστε διὰ τὸ ἐπικρατεῖν ἤδη τὴν διδασκαλίαν καὶ τοὺς ἐπὶ τῶν δωμάτων κόρακας ἐκ πολλῆς τῆς κατηχήσεως κράζειν τὴν κατ' αὐτὸν τοῦ συνημμένου κρίσιν, εἴποι ἂν ὁ γραμματικός, καὶ μέχρι τούτου συνήσει
311 τὸ καὶ παιδίοις γνώριμον· ἐλθὼν δὲ καὶ ἐπὶ τὸ "καὶ κῶς αὖθι γενησόμεθα" ἡσυχάσει, μὴ εὑρίσκων τὸ δηλούμενον πρᾶγμα. φιλοσόφου γὰρ ἦν εἰπεῖν ὅτι ἀρέσκει τῷ Διοδώρῳ μηδὲν κινεῖσθαι. τὸ γὰρ κινούμενον ἤτοι ἐν ᾧ ἔστι τόπῳ κινεῖται ἢ ἐν ᾧ μὴ ἔστιν· οὔτε δὲ τὸ πρῶτον οὔτε τὸ δεύτερον· οὐκ ἄρα κινεῖταί τι. τῷ δὲ μηδὲν κι-
312 νεῖσθαι τὸ μηδὲν φθείρεσθαι ἀκολουθεῖ. ὡς γὰρ διὰ τὸ μήτε ἐν ᾧ ἔστι τόπῳ κινεῖσθαί τι μήτε ἐν ᾧ μὴ ἔστιν οὐδὲν κινεῖται, οὕτως ἐπεὶ τὸ ζῶον οὔτε ἐν ᾧ ζῇ χρόνῳ ἀποθνήσκει οὔτε ἐν ᾧ μὴ ζῇ, οὐδέποτε ἄρα ἀποθνήσκει. εἰ δὲ τοῦτο, ἀεὶ ζῶντες κατ' αὐτὸν καὶ αὖθις γενησόμεθα.

---

[a] See vol. i. p. xvi.

[b] *Cf. P.H.* ii. 242, *Adv. Phys.* ii. 48. For Diodorus on the syllogism *cf. P.H.* ii. 110.

However, it is perhaps superfluous to put the Grammarians to shame in respect of antiquated, or perhaps of scientific, diction when they are incapable of understanding even any chance epigram, such as that composed by Callimachus referring to Diodorus Cronos,[a]— 309

> Lo and behold how the daws on the roof-tops tell us by croaking
> What things are conjoined, also how we shall live on.

For the fact that Cronos was a great dialectician and 310
taught us how the logical validity of a hypothetical syllogism is to be judged, so that because of the influence of his doctrine even the daws on the houses, from hearing it constantly repeated, croak out his criterion of the valid "syllogism" (or "conjunction"), —all this the Grammarian might relate, and so far he will understand what is common knowledge even to children. But when he comes to "also how we shall 311
live on," he will be dumbfounded, through not detecting what is indicated. For it belongs to the philosopher to explain that it is a tenet of Diodorus that nothing moves.[b] For that which moves, moves either in the place where it is or in the place where it is not; but neither the first nor the second (is true); therefore nothing moves. And that nothing perishes 312
follows from the fact that nothing moves. For just as nothing moves because a thing moves either in the place where it is or in the place where it is not, so too, since the living creature does not die either during the time in which it lives or during that in which it does not live, consequently it never dies. And if this is so, since we are ever-living, according to him, "we shall live on."

313 Οὐκοῦν τὰ μὲν πράγματα οὐ νοοῦσιν οἱ γραμματικοί. λείπεται τοίνυν τὰ ὀνόματα νοεῖν αὐτούς. ὃ πάλιν ἐστὶ ληρῶδες. πρῶτον μὲν γὰρ οὐδὲν ἔχουσι τεχνικὸν εἰς τὸ λέξιν γινώσκειν. οὔτε γὰρ ἐκ τέχνης τινὸς μεμαθήκασιν ὅτι οἱ παρὰ τῷ Σοφοκλεῖ ποιμένες "ἰὼ βαλλήν" λέγοντες ἰὼ βασιλεῦ λέγουσι φρυγιστί, ἀλλὰ παρ' ἄλλων ἀκούσαντες. διήνεγκε δὲ οὐδὲν ἢ βαρβάρου λέξεως ἑρμηνευτὰς γίνεσθαι ἢ τῆς κατὰ γλῶσσαν προ-
314 ενεχθείσης, ὁμοίως οὔσης ἀσυνήθους ἡμῖν. εἶτα καὶ τοῦτ' ἀδύνατόν ἐστιν ἀπείρων οὐσῶν λέξεων καὶ ἄλλως παρ' ἄλλοις ὀνοματοποιηθεισῶν ἢ ἐπὶ πράγμασιν οἷς ἡμεῖς οὐκ ἴσμεν τεθεισῶν. οἷόν ἐστι τὸ ⟨ἐβαρβάριζε τὸ ὅλον, ἕλκη ἔχον ἐν τῇ χειρί, τοῦ μὲν⟩[1] ἐβαρβάριζεν ἀντὶ τοῦ ἐσύριζε κειμένου, βάρβαροι γὰρ οἱ Σύροι, τοῦ δὲ ὅλου ἀντὶ τοῦ παντός, ὅλον γὰρ καὶ πᾶν συνώνυμον, τοῦ δὲ ἕλκους ἀντὶ τῆς σύριγγος, εἶδος γὰρ ἕλκους ἡ σύριγξ. ὥστε τὸ ὅλον γίνεσθαι τοιοῦτον "ἐσύ-
315 ριζεν ὁ Πάν, σύριγγας ἔχων ἐν τῇ χειρί." ἄλλως τε καὶ ποῦ ἴσασιν ἐνίας τῶν ἐπιστημονικῶν λέξεων οἱ γραμματικοί, καθάπερ τὴν παρὰ Ἀριστοτέλει ἐντελέχειαν ἢ τὸ τί ἦν εἶναι; ἢ ποῦ συνήσουσι τίνα δύναμιν ἔχει παρὰ σκεπτικοῖς ἡ "οὐδὲν μᾶλλον" φωνή, πότερον πυσματική ἐστιν ἢ ἀξιωματική, καὶ ἐπὶ τίνος τάσσεται, ἆρά γε τοῦ
316 ἐκτὸς ὑποκειμένου ἢ τοῦ περὶ ἡμᾶς πάθους; τί δὲ καὶ ἐροῦσιν ἐκ λέξεών τινων συντεθέντος τινὸς ποιήματος;

[1] ⟨ἐβαρβάριζε . . . μὲν⟩ add. cj. Fabr.

[a] *Cf.* Soph. *Frag.* 144, Aesch. *Pers.* 658.

So then, the Grammarians do not understand the 313
objects (behind the words). It only remains, then, that they understand the words. And that again is nonsense. For, firstly, they have no technical means for getting to know terms. For it is not by any art, but by hearing it from others, that they have learnt that when the shepherds in Sophocles[a] say "ἰὼ βαλλήν" what they say is the Phrygian for ἰὼ βασιλεῦ ("O king"). And there is no difference between interpreting a barbarous word and one which is obsolete, as they sound equally strange to us. And, 314
secondly, even this is impossible, as words are infinite in number and are constructed differently by different people, or applied to objects of which we have no knowledge. For instance, ἐβαρβάριζε τὸ ὅλον, ἕλκη ἔχον ἐν τῇ χειρί, where ἐβαρβάριζε stands for ἐσύριζε ("piped"), for the Syrians[b] are barbarians, and ὅλον stands for πᾶν ("all"), πᾶν and ὅλον being synonyms, and ἕλκος ("sore") for σύριγξ ("pipe" or "fistula"), for a fistula is a kind of sore. So the whole phrase amounts to this—ἐσύριζεν ὁ Πάν, σύριγγας ἔχων ἐν τῇ χειρί ("Pan played the pipe, holding the pipes in his hand"). Further, how do the Gram- 315
marians know some of the scientific terms, such as Aristotle's ἐντελέχεια ("entelechy") or τὸ τί ἦν εἶναι ("essence")? Or how will they comprehend the meaning of the Sceptics' formula οὐδὲν μᾶλλον ("no more"),[c] whether it is interrogative or declaratory, and to what it refers,—whether to the external object or to our subjective feeling? Or what will they say 316
when a poem is composed of certain obscure expressions?

[b] Note the word-play on Σύροι-ἐ-σύριζε.
[c] *Cf. P.H.* i. 191.

ἦ γάρ σοι δισσοῖσιν ὑπ᾽ οὔρεσι δισσὸς ἐραστὴς
ἔφθιτο, καὶ νεάτην μοῖρ᾽ ἀνέθηκε φύσιν.
ἄρθρῳ ἐν ἀσπιδόεντι βεβηκότα γυῖα καθ᾽ ὅλμου
βλαισά[1] τροχαντήρων ἄχρι περιστρέφεται,
σμερδαλέαι δ᾽ ὑπένερθεν ἀλώπεκες, ἄχρι δοχαίης
αἰῶνος χαλαρᾷ σύνδραμον ἁρμονίῃ.[2]

317 τοὺς γὰρ ἐραστὰς οἵτινές εἰσι καὶ τὰ ὄρη καὶ τὸ ἀσπιδόεν ἄρθρον καὶ τοὺς τροχαντῆρας, ἔτι δὲ καὶ τὸν ὅλμον καὶ τὰς ἀλώπεκας δοχαίην τε καὶ αἰῶνα καὶ ἁρμονίαν, μήτε τροπικῶς μήτε κατὰ ἱστορίαν ἀλλὰ κυρίως ἐξενεχθέντα ὀνόματα, κἂν μυριάκις ἐπιστήσωσιν, οὐ συνήσουσιν.

318 Εἰ οὖν μήτε τὰ πράγματα μήτε τὰς λέξεις ἴσασιν, παρὰ δὲ ταῦτα οὐδέν ἐστιν ἡ ποίησις ἢ τὸ σύγγραμμα, οὐκ ἂν ἔχοιεν τέχνην ἐξηγητικὴν τῶν παρὰ ποιηταῖς καὶ συγγραφεῦσι λεγομένων. ἄλλως τε καὶ εἰ χρῄζομεν γραμματικῆς, ἐπὶ τῶν ἀρίστων ποιημάτων χρῄζομεν ἀλλ᾽ οὐ τῶν μοχθηρῶν. ἄριστον δὲ ποίημά ἐστι κατ᾽ αὐτοὺς τὸ
319 σαφές· ἀρετὴ γὰρ ποιήματος ἡ σαφήνεια, καὶ μοχθηρὸν τὸ ἀσαφὲς παρὰ γραμματικῇ. οὔτε οὖν ἐπὶ ἀρίστου ἐστὶ ποιήματος χρειώδης διὰ τὸ μὴ δεῖσθαι ἐξηγήσεως σαφὲς ὄν, οὔτε ἐπὶ τοῦ
320 μοχθηροῦ διὰ τὸ αὐτόθεν εἶναι μοχθηρόν. τό τε ἀνεπικρίτως διαφωνούμενον ἀκατάληπτόν ἐστιν,

[1] βλαισὰ scripsi : βᾶσα MSS., Bekk. (βαιὰ Hermann).
[2] σμερδαλέαι . . . ἀλώπεκες . . . χαλαρᾷ σύνδραμον ἁρμονίῃ Hermann : σμερδαλέα . . . ἀλώπεκος . . . χαλαρὰν σύνδρομον ἁρμονίης MSS., Bekk.

Under a twofold mountain a twofold lover did perish :
Destiny, nevertheless gave them a strange new form.
Set in joints like a serpent's the crooked limbs of the body
Turn'd and twisted around down to the joints of the hips ;
Dread to behold were the foxes beneath, lax Harmony's race-mates,
Down to the secret hold storing the ages to come.[a]

For who the ἐρασταί (" lovers ") are, and the ὄρη 317
(" mountains "), and ἀσπιδόεν ἄρθρον (" serpent-like joint "), and τροχαντῆρες (" hip-joints "), and also ὅλμος (" trunk " of the body), and ἀλώπεκες (" foxes " = " loins "), and δοχαίη (" receptacle," " storehouse "), and αἰών (" age," " posterity "), and ἁρμονία (" harmony "),—words used neither metaphorically nor as technical terms but in their proper sense,—these the Grammarians will not understand though they ponder them ten thousand times.

If, then, they know neither the objects nor the 318
words, and the poem or the treatise is nothing beyond these, they will not possess an art capable of explaining what is said by the poets and prose-writers.—And further, if we have need of the Art of Grammar, we need it to deal with the best poems and not with the bad ones. But, according to them, the clear
poem is the best ; for the excellence of a poem is its 319
clearness, and, for the Art of Grammar, want of clearness is bad. Thus grammar is not needed either in the case of the best poem, since it, being clear, requires no explanation, nor in the case of the
bad poem because it is obviously bad.—Also, that 320
which is disputed without a decision is non-appre-

[a] " The secret hold " etc. means " partes genitales, quibus spes posteritatis continetur " (Fabr.) ; and " Harmony " means " venerei amores " (Fabr.). The lovers were Cadmus and his wife Harmonia, who were both turned into serpents.

ἀνεπικρίτως δ' ἔτι διαφωνοῦσιν ἐν ταῖς ἐξηγήσεσιν οἱ γραμματικοὶ περὶ τῆς τοῦ συγγραφέως διανοίας· ἀκατάληπτος ἄρα ἐστὶν ἡ τοῦ συγγραφέως διάνοια, καὶ διὰ τοῦτο ἄχρηστος ἡ γραμματική.

Ἀλλὰ γὰρ πρὸς μὲν τοὺς ἀπὸ τούτου τοῦ μαθήματος ἀναγομένους ἐπὶ τοσοῦτον εἰρήσθω· ἀπ' ἄλλης δὲ ἀρχῆς σκεψώμεθα καὶ πρὸς τοὺς ῥήτορας ἃ δεῖ λέγειν.

hensible ; but the Grammarians, in their explanation regarding the writer's meaning, still dispute without coming to a decision ; the writer's meaning, therefore, is non-apprehensible, and on this account grammar is useless.

Well then, let thus much be said in reply to those whose studies are based on this branch of learning ; and, making a fresh start, let us also consider what we ought to say in reply to the Rhetoricians.

B

ΠΡΟΣ ΡΗΤΟΡΑΣ

1 Τοῖς περὶ γραμματικῆς διεξοδευθεῖσιν ἡμῖν ἀκόλουθον ἂν εἴη καὶ περὶ ῥητορικῆς λέγειν, ἀνδρικωτέρας ἤδη καθεστώσης καὶ τὸ πλέον ἐπ' ἀγορᾶς καὶ βημάτων ἐξεταζομένης. ἀλλ' ἐπεὶ κοινὸν ὑπάρξεώς τε καὶ ἀνυπαρξίας ἐστὶν ἡ ἔννοια, καὶ οὐδὲν τούτων ἕτερον οἷόν τέ ἐστι ζητεῖν μὴ προλαβόντας ὅ ἐστι τὸ ζητούμενον, φέρε πρῶτον σκεψώμεθα τί ἂν εἴη ῥητορική, τὰς ἐπιφανεστάτας εἰς τοῦτο τῶν φιλοσόφων ἀποδόσεις παρατιθέμενοι.
2 Πλάτων μὲν οὖν ἐν τῷ Γοργίᾳ κατὰ διοριστικὴν ἔφοδον τοιοῦτον ἔοικεν ἐξ ἐπισυνθέσεως ὅρον τῆς ῥητορικῆς ἀποδιδόναι "ῥητορική ἐστι πειθοῦς δημιουργὸς διὰ λόγων, ἐν αὐτοῖς τοῖς λόγοις τὸ κῦρος ἔχουσα, πειστική, οὐ διδασκαλική," τὸ μὲν "διὰ λόγων" προστιθεὶς τάχα παρόσον πολλά ἐστι τὰ πειθὼ τοῖς ἀνθρώποις ἐνεργαζόμενα χωρὶς λόγου, καθάπερ πλοῦτος καὶ δόξα καὶ ἡδονὴ καὶ
3 κάλλος. οἱ γοῦν παρὰ τῷ ποιητῇ δημογέροντες, καίπερ ἐκπεπολεμωμένοι καὶ τελέως ἀπηλλοτριωμένοι πρὸς τὴν Ἑλένην ὡς κακῶν αἰτίαν γενομένην αὐτοῖς, ὅμως ὑπὸ τοῦ περὶ αὐτὴν κάλλους πείθονται,

BOOK II

AGAINST THE RHETORICIANS

Following on our discussion of the Art of Grammar 1
we shall speak of the art of Rhetoric,—in itself a more manly art, and one which is for the most part tested in the Assembly and the Law-courts. But since the notion of a thing holds of it equally whether it be existent or non-existent, and it is impossible to investigate either of these states without having grasped beforehand what the object of investigation is, come and let us consider first what Rhetoric is by comparing the most notable accounts of it given by the philosophers.

Now Plato in the *Gorgias*,[a] using his method of de- 2
finition, seems to set forth a composite definition of Rhetoric such as this,—" Rhetoric is the creator of persuasion by means of words, having its efficacy in the words themselves, and being persuasive, not instructive " ; and he added the phrase " by means of words " probably because there are many things which effect persuasion in men without speech, such
as wealth and glory and pleasure and beauty. Thus 3
the elders of the folk (in the poet Homer), although exhausted by war and wholly estranged from Helen as the cause of their woes, are yet persuaded by her

[a] *Cf. Gorg.* 453 A.

καὶ προσιούσης τοιαῦτά τινα πρὸς ἀλλήλους διεξίασιν,

οὐ νέμεσις Τρῶας καὶ ἐυκνήμιδας Ἀχαιοὺς
τοιῇδ᾽ ἀμφὶ γυναικὶ πολὺν χρόνον ἄλγεα πάσχειν.

4 Φρύνη τε, ὥς φασίν, ἐπεὶ συνηγοροῦντος αὐτῇ Ὑπερίδου ἔμελλε καταδικάζεσθαι, καταρρηξαμένη τοὺς χιτωνίσκους καὶ γυμνοῖς στήθεσι προκυλινδουμένη τῶν δικαστῶν πλεῖον ἴσχυσε διὰ τὸ κάλλος τοὺς δικαστὰς πεῖσαι τῆς τοῦ συνηγοροῦντος ῥητορείας. τὸ δὲ αὐτὸ καὶ ἐπὶ χρημάτων ἐστὶν ἡδονῆς τε καὶ δόξης· τούτων γὰρ ἕκαστον εὑρήσομεν οὕτω πεῖθον ὡς πολλάκις τινὰ τῶν καθηκόντων ὑπερβαίνειν. οὐ τοίνυν ἀσκόπως ὁ Πλάτων ἀποβλέπων εἰς τὴν δι᾽ αὐτῶν γινομένην πειθὼ ἔλεξεν ὅτι ῥητορική ἐστι πειθοῦς δημιουργὸς οὐχ
5 ὁπωσοῦν ἀλλὰ διὰ λόγων. καὶ μὴν οὐκ ἐπεὶ λόγοις πείθει, πάντως ἐστὶ ῥητορική (καὶ γὰρ ἡ ἰατρικὴ καὶ αἱ ὁμοειδεῖς ταύτῃ τέχναι διὰ λόγου πείθουσιν), ἀλλ᾽ εἴ τις ἐν αὐτοῖς προηγουμένως τοῖς λόγοις ὑποκειμένην ἔχει τὴν ἰσχύν, καὶ οὐ κοινῶς, ἐπείπερ καὶ ἡ γεωμετρία καὶ ἀριθμητικὴ καὶ πᾶσα ἡ τῷ γένει θεωρητικὴ τέχνη ἐν λόγοις προηγουμένως ἔχει τὸ κῦρος, ἀλλ᾽ ὅταν σὺν τούτοις μὴ διδασκαλικήν, ὥσπερ γεωμετρία, ἀλλὰ πειστικὴν ποιῆται τὴν πειθώ· ὅπερ ἦν ἴδιον ῥητορικῆς.

6 Ξενοκράτης δὲ ὁ Πλάτωνος ἀκουστὴς καὶ οἱ ἀπὸ τῆς στοᾶς φιλόσοφοι ἔλεγον ῥητορικὴν ὑπάρχειν ἐπιστήμην τοῦ εὖ λέγειν, ἄλλως μὲν Ξενοκράτους τὴν ἐπιστήμην λαμβάνοντος καὶ ἀρχαϊκῷ

beauty, and at her approach they address one another in this wise,[a]—

Cause for anger is none that the Trojans and well-greav'd Achaeans
Suffer for many a year sad woes for a woman so beauteous.

Phrynê too, as they say, when Hypereides was pleading for her and she was on the point of being condemned, tore asunder her garments and with her breasts bare flung herself at the feet of the judges, and because of her beauty had more power to persuade her judges than the rhetoric of her advocate. The same, too, holds good of money and pleasure and glory; for we shall find each one of these so persuasive as often to make men overstep some of the moral duties. It was not without reason, then, that Plato, having regard to the persuasion they induce, declared that rhetoric is "the creator of persuasion," not by any means you choose but "by means of words." It is not, however, always rhetoric when there is persuasion by words (for the art of medicine also and arts similar to it persuade by speech), but only if it is an art which has its efficacy dependent mainly on words alone, and that not like the rest, —since geometry, too, and arithmetic and every art classed as theoretical depends for its efficacy mainly on words,—but whenever, in addition, it employs persuasion not in order to instruct, like geometry, but to induce belief; for that is the special mark of rhetoric. 4 5

But Xenocrates, the disciple of Plato, and the Stoic philosophers asserted that rhetoric is "the science of good speech"; though Xenocrates took the word "science" in one sense,—*i.e.* in the old way, 6

[a] Hom. *Il.* iii. 156 f.

νόμῳ, ἀντὶ τῆς τέχνης, ἄλλως δὲ τῶν στωικῶν, ἀντὶ τοῦ βεβαίας ἔχειν καταλήψεις, ἐν σοφῷ μόνῳ φυομένην. τὸ δὲ λέγειν ἀμφότεροι παραλαμβάνουσιν ὡς διαφέρον τοῦ διαλέγεσθαι, ἐπειδήπερ τὸ μὲν ἐν συντομίᾳ κείμενον κἀν τῷ λαμβάνειν
7 καὶ διδόναι λόγον διαλεκτικῆς ἐστὶν ἔργον, τὸ δὲ λέγειν ἐν μήκει καὶ διεξόδῳ θεωρούμενον ῥητορικῆς ἐτύγχανεν ἴδιον. ἔνθεν γοῦν καὶ Ζήνων ὁ Κιτιεὺς ἐρωτηθεὶς ὅτῳ διαφέρει διαλεκτικὴ ῥητορικῆς, συστρέψας τὴν χεῖρα καὶ πάλιν ἐξαπλώσας ἔφη "τούτῳ," κατὰ μὲν τὴν συστροφὴν τὸ στρογγύλον καὶ βραχὺ τῆς διαλεκτικῆς τάττων ἰδίωμα, διὰ δὲ τῆς ἐξαπλώσεως καὶ ἐκτάσεως τῶν δακτύλων τὸ πλατὺ τῆς ῥητορικῆς δυνάμεως αἰνιττόμενος.

8 Ἀριστοτέλης δὲ ἐν τῷ πρώτῳ τῶν ῥητορικῶν τεχνῶν ἁπλούστερον παραδίδωσι τὴν ῥητορικὴν τέχνην λόγων. καὶ ζητουμένου πρὸς αὐτὸν ὅτι καὶ ἡ ἰατρικὴ τέχνη ἐστὶ λόγων ἰατρικῶν, ἀπολογούμενοί τινές φασιν ὅτι ἡ ἰατρικὴ τοὺς λόγους ἐφ' ἕτερόν τι ἀναφέρει τέλος, καθάπερ τὴν ὑγείαν,
9 ἡ δὲ ῥητορικὴ ἄντικρύς ἐστι λόγων τέχνη. καὶ ἄλλους δὲ ἐκτίθεται ὁ ἀνὴρ οὗτος ὅρους, περὶ ὧν οὐκ ἀναγκαῖόν ἐστιν ἡμῖν λέγειν τοῖς μὴ προηγουμένως τὸν περὶ ῥητορικῆς διεξοδεύουσι λόγον, ἀλλ' ὅσον τοῦ τὴν ἰδιότητα ταύτης περινοῆσαι πρὸς τὴν χειρισθησομένην ἡμῖν ἀντίρρησιν. ἧς ἀρχὴ γένοιτ' ἂν εὐθέως ἀπὸ τῆς ἐκκειμένης νοήσεως. ἐπεὶ γὰρ τέχνην ἢ ἐπιστήμην λόγων ἢ τοῦ λέγειν καὶ πειθοῦς περιποιητικὴν βούλονται τυγχάνειν τὴν ῥητορικὴν οἱ τὴν ἔννοιαν αὐτῆς ἀποδιδόντες, πει-

as a synonym for "art,"—whereas the Stoics took it in another, as meaning "the holding firm apprehensions," which is inherent only in the Sage. But both parties assume that "speech" differs in sense from dialectical discussion, since speaking which is concise and consists in giving and receiving an account is the task of dialectic, but to speak on a 7
given subject at length and with detailed exposition is the special characteristic of rhetoric. Hence, Zeno of Citium, when asked what is the difference between dialectic and rhetoric, clenched his fist and then opened it out and said, "This,"—comparing the compact and short character of dialectic to the clenching, and suggesting the breadth of the rhetorical style by the opening and extension of his fingers.[a]

In the first of his books on *The Art of Rhetoric* 8
Aristotle describes rhetoric more simply as "the art of speech." And when it is objected against him that the art of Medicine also is the art of medical speech, some say by way of defence that the medical art directs its speech to another end, such as health, whereas rhetoric is precisely the art of words alone. And Aristotle puts forward other definitions also, 9
which it is unnecessary for us to mention as we are not primarily discussing the description of rhetoric, but only in so far as it enables us to perceive its special character with a view to the refutation we propose to take in hand. And a beginning of this may be made at once from the concepts already stated. For since those who offer a concept of rhetoric assert that it is an art, or a science, of speech, or productive of speech and persuasion, holding fast

[a] *Cf.* Cicero, *de Fin.* ii. 6, *Orat.* 32.

ρασόμεθα καὶ ἡμεῖς τῶν τριῶν τούτων ἐχόμενοι διδάσκειν τὸ ἀνυπόστατον αὐτῆς.

10 Πᾶσα τοίνυν τέχνη σύστημά ἐστιν ἐκ καταλήψεων συγγεγυμνασμένων καὶ ἐπὶ τέλος εὔχρηστον τῷ βίῳ λαμβανουσῶν[1] τὴν ἀναφοράν· ἡ δὲ ῥητορικὴ οὐκ ἔστι σύστημα ἐκ καταλήψεων, ὡς παρα-
11 στήσομεν· οὐκ ἄρα ἔστιν ἡ ῥητορική. τῶν γὰρ ψευδῶν οὐκ εἰσὶ καταλήψεις, ψευδῆ δέ ἐστι τὰ λεγόμενα τῆς ῥητορικῆς εἶναι θεωρήματα, τοιαῦτα ὄντα "οὕτω παραπειστέον τοὺς δικαστὰς καὶ ὀργὴν κινητέον ἢ ἔλεον καὶ μοιχῷ συνηγορητέον ἢ ἱεροσύλῳ"· ⟨οἷ'⟩[2] ἐμφαίνει τὸ καθήκειν οὕτω τοὺς δικαστὰς παραπείθειν καὶ ὀργὴν ἢ ἔλεον κινεῖν· ἅπερ οὐκ ἔστιν ἀληθῆ, καὶ διὰ τοῦτο ἀκατάληπτα. οὐ τοίνυν αὐτῶν εἰσὶν αἱ καταλήψεις· ᾧ συνεισέρχεται τὸ μηδὲ τὴν ῥητορικὴν ὑπάρχειν.
12 καθάπερ οὖν οὐκ ἂν εἴποιμεν τὴν τοιχωρυχικὴν εἶναί τινα τέχνην παραινοῦσαν τὸ "οὕτω δεῖ τοῖχον διορύττειν," καὶ τὴν κλεπτικὴν τὸ "οὕτω καθήκει κλέπτειν καὶ βαλαντιοτομεῖν" (ψευδῆ γάρ ἐστι ταῦτα, καὶ οὔτε καθήκοντα οὔτε θεωρήματα), οὕτως οὐδὲ τὴν ῥητορικὴν ὑποληπτέον ἔχειν τεχνικὴν ὑπόστασιν, ἐπὶ τοιούτοις παραγγέλμασι σαλεύουσαν. ἀμέλει γέ τοι καὶ οἱ περὶ Κριτόλαον τὸν περιπατητικόν, καὶ πολὺ πρότερον οἱ περὶ Πλάτωνα, εἰς τοῦτο ἀπιδόντες ἐκάκισαν αὐτὴν ὡς
13 κακοτεχνίαν μᾶλλον ἢ τέχνην καθεστηκυῖαν. καὶ μὴν ἐπεὶ πᾶσα τέχνη ἤτοι ἑστηκὸς ἔχει τὸ τέλος καὶ πάγιον, ὡς φιλοσοφία καὶ γραμματική, ἢ τοῦ ὡς τὸ πολὺ ἐχόμενον, καθάπερ ἰατρική τε καὶ κυβερνητική, δεήσει καὶ τὴν ῥητορικήν, εἴπερ

to these three descriptions we shall endeavour to show its unreality.

Now every art is "a system composed of co- 10
exercised apprehensions directed to an end useful for life"[a]; but, as we shall establish, rhetoric is not a system of apprehensions; therefore rhetoric does
not exist. For of things false there are no appre- 11
hensions, but what are said to be the rules of rhetoric are false, being such as these,—"The judges must be misled by persuasion in this way," "One must excite anger or pity," "One must plead the cause of the adulterer or temple-robber,"—rules which declare the duty of thus misleading the judges and exciting anger or pity; but these are not true and consequently are not apprehensible. So there are no apprehensions of them; whence it follows that
neither does rhetoric exist.—As, then, we would not 12
say that burgling is an art which advises—"This is the way one ought to burgle a house," or thieving an art which instructs us that "This is the right way to steal and to cut purses" (for these things are false, and neither duties nor rules), so we must not suppose that rhetoric has any technical foundation when it is based on such shaky injunctions. And in fact Critolaüs the Peripatetic, and Plato long before him, in consideration of this, condemned rhetoric as being
a base artifice rather than an art.—Moreover, since 13
every art has an end which is either stable and fixed, like philosophy and grammar, or holds good for the most part, like medicine and piloting, rhetoric too,

[a] *Cf. P.H.* iii. 188 for this Stoic definition.

---

[1] *λαμβανουσῶν* scripsi: *λαμβανόντων* MSS., Bekk.
[2] ⟨*οἷ'*⟩ addidi (*ἐμφαίνοντα, κ.* cj. Bekk.).

14 ἔστι τέχνη, τὸ ἕτερον τούτων ἐπαγγέλλεσθαι. οὔτε δὲ ἑστηκὸς ἔχει πᾶν τέλος[1] (οὐδὲ γὰρ ἀεὶ περιγίνεται [περὶ] τῆς τῶν ἀντιδίκων νίκης, ἀλλ' ἔσθ' ὅτε ἕτερον μὲν προτίθεται ὁ ῥήτωρ ἕτερον δὲ
15 ἐξακολουθοῦν ἔχει τέλος) οὔτε τοῦ ὡς τὸ πολὺ ἐφιέμενον, ἐπεὶ πᾶς ῥήτωρ ἑαυτῷ συγκρινόμενος πολλάκις ἐλείφθη μᾶλλον ἢ ἐνίκησεν, ἅτε διὰ παντὸς ἑτέρου τὰς ἐπιχειρήσεις αὐτοῦ διαλύοντος.
16 οὐκ ἄρα τέχνη ἐστὶν ἡ ῥητορική. εἴπερ τε ἐνδέχεται γενέσθαι ῥήτορα μὴ μετασχόντα τῆς ῥητορικῆς τέχνης, οὐκ ἂν εἴη τις τέχνη ῥητορική. ἐνδέχεται δέ γε ἱκανῶς καὶ κατὰ τρόπον ῥητορεύειν μὴ μετασχόντα ῥητορικῆς, ὡς καὶ περὶ Δημάδου παρειλήφαμεν· κωπηλάτης γὰρ ὢν ὁμολογεῖται ἄριστος γεγονέναι ῥήτωρ, καὶ σὺν τούτῳ ἄλλοι παμπληθεῖς. τοίνυν οὐκ ἔστι τέχνη ἡ ῥητορική.
17 ἄλλως τε καὶ ἐπεὶ τούτοις ἀπιστοῦμεν ὡς τοιούτοις γεγονόσι καὶ ἐν ἕξει κἀκ[2] τοιαύτης τινὸς τριβῆς ἐπὶ τὸ ῥητορεύειν παρεληλυθόσιν, ἀλλ' οὖν γε ἐν τῷ καθ' ἡμᾶς βίῳ πολλοὺς πάρεστιν ὁρᾶν λέγοντας μὲν εὐφυῶς ἐπὶ δικαστηρίων καὶ ἐν ἐκκλησίαις, τὰ δὲ τεχνικὰ τῆς ῥητορικῆς παραγγέλματα μὴ
18 γινώσκοντας. καὶ ἀντιστρόφως, εἰ οἱ ἐξηκριβωκότες [ἐπὶ πλεῖον] καὶ ἐπὶ πλεῖον ἐκπονήσαντες τὸν τεχνικὸν τῆς ῥητορικῆς λόγον ἀδυνατοῦσι ῥητορεύειν ἐπὶ δικαστηρίων καὶ ἀγορᾶς, οὐ ῥητέον τεχνικὴν μέθοδον εἶναι τὴν ῥητορικήν. ἀλλὰ μὴν ὡς ὁ σύμπας οἶδε βίος, οἱ σοφιστεύοντες ἐπ' ἄκρον μὲν τὴν ῥητορικὴν ἐξήσκησαν τεχνολογίαν, ἰχθύων δὲ ἀφωνότεροι ἐπὶ τῆς ὑπαίθρου θεωροῦνται.

[1] an παν⟨τελῶς τὸ⟩ τέλος? (καὶ πάγιον τὸ τ. Herv.).
[2] κἀκ cj. Bekk.: καὶ mss., Bekk.

if it is an art, will have to profess one or other of these. But it has not an end which is always stable 14
(for it does not always avoid being worsted by the opponents, and sometimes the rhetor proposes one end but finds the resultant end to be different) ; nor 15
does it attain its end for the most part, since every rhetor, when his own experiences are compared, turns out to have been often defeated rather than victorious, because the opponent constantly rebuts his arguments. Rhetoric, therefore, is not an art.—Also, if 16
it is possible to become an orator without being acquainted with the art of rhetoric, there will be no art of rhetoric. But it is possible to make a speech quite successfully and well without having studied rhetoric, as we have been told in the case of Demades [a] ; for though he was a boatman, it is agreed that he became a very fine orator; and besides him there are numerous other instances. Hence, rhetoric is not an art.—Moreover, if we do not believe that these men 17
were such as we have described and that they attained to oratory by experience and by practice, yet still we may see many in our daily life who speak admirably in the law-courts and assemblies though they have no knowledge of the technical rules of rhetoric. And conversely, if those who have studied 18
closely and worked hard at the technique of rhetorical speech are incapable of making a speech at the law-courts and assembly, one must deny that rhetoric is a technical method. But in fact, as all the world knows, our lecturers, though they have studied the technology of rhetoric to the utmost point, are seen to be more mute than fishes in open court. Thus none 19

[a] *Cf. Adv. Gram.* 295.

19 τοίνυν οὐ κατὰ τέχνην εἰσί τινες ῥήτορες. ὅθεν καὶ γελᾶν ἔστιν ἐπ' αὐτούς, ὅταν πρὸς τοῦτον ἀπολογούμενοι τὸν ἔλεγχον φάσκωσιν ὅτι, ὥσπερ αἱ ἀκόναι τέμνειν μὲν οὐ πεφύκασιν ὀξύνουσαι δὲ τὴν μάχαιραν τέμνειν παρασκευάζουσιν, οὕτω καὶ αὐτοὶ ἀδυνατοῦσι μὲν εἰπεῖν ὑπὸ τῆς ἀηθείας, ἄλλους δὲ διὰ τέχνης προάγοντες λέγειν ποιοῦσιν. οὐκ ᾔδεσαν γὰρ οἱ θαυμάσιοι τὴν ἀνομοιότητα ταύτης τῆς εἰκόνος, εἴ γε ἡ μὲν ἀκόνη οὐχ ἣν εἶχε δύναμιν, ταύτην τῷ σιδήρῳ ἐμποιεῖν πέφυκεν, αὐτοὶ δὲ ἐπαγγέλλονται ὡς προηγούμενον ἔργον, ἣν ἔχουσι τέχνην, ταύτην τῷ πέλας περιποιήσειν.

20 Οἱ δὲ περὶ τὸν Κριτόλαον καὶ οἱ ἀπὸ τῆς Ἀκαδημίας, ἐν οἷς ἐστὶ Κλειτόμαχος καὶ Χαρμίδας, εἰώθασι καὶ οὗτοι τοιαῦτά τινα λέγειν, ὅτι τὰς μὲν τέχνας οὐκ ἐκβάλλουσιν αἱ πόλεις πάνυ τι βιωφελεῖς οὔσας ἐπιστάμεναι, ὡς οὐδὲ τοὺς μὲν οἰκονομικοὺς τῶν οἴκων ἐκβάλλομεν τοὺς δὲ βουκόλους ἐκ τῆς ἀγέλης, τὴν μέντοι ῥητορικὴν πάντες πανταχόθεν ὡς πολεμιωτάτην ἐδίωξαν, ὥσπερ ὁ μὲν Κρητικὸς νομοθέτης εἴρξας ἐπιβαίνειν τῆς νήσου τοὺς ἐν λόγοις ἀλαζονευσαμένους,
21 ὁ δὲ Σπαρτιάτης Λυκοῦργος, ὡς ἂν ζηλωτὴς Θάλητος τοῦ Κρητὸς γενόμενος, τὸν αὐτὸν τοῖς Σπαρτιάταις νόμον εἰσηγήσατο· παρ' ἣν αἰτίαν πολλοῖς ὕστερον χρόνοις τὸν ἐπὶ ξένης ῥητορικὴν ἐκπονήσαντα νεανίαν ἐπανελθόντα ἐκόλασαν οἱ ἔφοροι, τὴν αἰτίαν προσθέντες τῆς καταδίκης ὡς δολεροὺς λόγους ἐπὶ παρακρούσει τᾶς Σπάρτας ἐμελέτησεν. καὶ αὐτοὶ δὲ διέμειναν ῥητορικὴν
22 μισοῦντες, ἀφελεῖ δὲ βραχυλογίᾳ χρώμενοι. ὅθεν καὶ ὁ ἀντιχειροτονηθεὶς Ἀθηναίοις ὑπ' αὐτῶν

are orators by reason of art. Hence one may well laugh at them when, in defending themselves against this confutation, they assert that just as whetstones are not made to cut but by sharpening the knife enable it to cut, so also they themselves are incapable of making speeches through want of practice, yet they enable other men to do so by helping them to improve by means of art. For our fine gentlemen failed to notice the lack of similarity in this comparison, in that the whetstone is not of a nature to implant its own quality in the iron, whereas they profess, as their primary task, to convey to their neighbour the art which they themselves possess.

And Critolaüs and the men of the Academy, 20
including Cleitomachus and Charmidas,[a] are wont to argue like this,—that the cities do not expel the arts knowing them to be extremely useful for life, just as we do not expel skilled domestics from our houses or cowmen from our herds, but all men in every quarter have hunted down rhetoric as most inimical; for example, the Cretan lawgiver forbade those who prided themselves on their oratory to land on his
island, and the Spartan Lycurgus, having become an 21
admirer of Thales the Cretan, introduced the same law for the Spartans; and for this reason, many years later, the Ephors punished on his return a young man who had studied rhetoric abroad, alleging as the cause of his condemnation that he practised a deceitful mode of speaking in order to lead Sparta astray. And the Spartans themselves continued to loathe rhetoric and to employ speech which is simple and
short. Hence, too, the man whom they had elected 22

[a] Both disciples of Carneades; *cf.* vol. i. p. xxxiii.

πρὸς Τισσαφέρνην πρεσβευτής, τῶν Ἀθηναίων μακρὰς καὶ ποικίλας ῥήσεις διεξιόντων, δύο τῇ βακτηρίᾳ γραμμὰς κατὰ τοῦ ἐδάφους χαράξας, τὴν μὲν εὐθεῖαν καὶ μικρὰν τὴν δὲ ἐπιμήκη καὶ σκολιάν, "τούτων" εἶπεν, "ὦ βασιλεῦ, ὁποτέραν θέλεις ἑλοῦ," αἰνιττόμενος διὰ μὲν τῆς ἐπιμήκους καὶ σκολιᾶς γραμμῆς τὴν τερθρείαν τὴν ῥητορικήν, διὰ δὲ τῆς βραχείας ἅμα καὶ εὐθείας τὴν ἀφελῆ
23 καὶ σύντομον εὐθυρρημοσύνην, δι' ἣν οὐκ ἐν οἰκείοις μόνοις ἀλλὰ καὶ ξένοις τὴν ἀπεριττότητα τοῦ λόγου μεταδιώκουσιν. ἐκ μέσων μέντοι γε καὶ τὸν Χίων πρεσβευτὴν περὶ ἐξαγωγῆς πυροῦ δεόμενον, ἐπεὶ μακρῶς ἡρμήνευσε τὴν δέησιν, ἄπρακτον ἐξαπέστειλαν, ἑτέρου δὲ πεμφθέντος συντομωτέρου (ἤπειγε γὰρ ἀνάγκη τοὺς Χίους) ἔδοσαν· κενὸν γὰρ θύλακον αὐτοῖς οὗτος ἀνατείνας ἀλφίτων αὐτὸν ἔφη δεῖσθαι. ὅμως δ' οὖν καὶ τοῦτον ὡς ἀδολέσχην ἐμέμψαντο· ἀποχρώντως γὰρ κενὸς δειχθεὶς ὁ θύλακος ἐσήμηνε τὴν τῶν
24 Χίων αἴτησιν. ἔνθεν ὁ τραγικὸς Ἴων κινηθεὶς εἶπεν ἐπ' αὐτῶν

οὐ γὰρ λόγοις Λάκαινα πυργοῦται πόλις,
ἀλλ' εὖτ' Ἄρης νεοχμὸς ἐμπέσῃ στρατῷ,
βουλὴ μὲν ἄρχει, χεὶρ δ' ἐπεξεργάζεται,

ἅτε βουλευομένων μὲν τὰ κράτιστα, στυγούντων δὲ τὴν ῥητορικήν. ὅθεν εἰ μὴ τὰς τέχνας ἐκβάλλουσιν αἱ πόλεις, ἐκβεβλήκασι δὲ τὴν ῥητορικήν,
25 οὐκ ἂν εἴη τῶν τεχνῶν ἡ ῥητορική. τὸ μὲν γὰρ ἀναστρέφειν, καὶ λέγειν ὡς καὶ φιλοσόφους ἐξώρισάν τινες τῶν Ἑλληνίδων πόλεων, εὐηθές ἐστιν. πρῶτον μὲν γὰρ οὐκ ἂν ἔχοιεν τούτῳ παρασχεῖν

by vote as ambassador to Tissaphernes in order to oppose the Athenians, while the Athenians were making long and complicated harangues, drew two lines on the ground with his staff, the one straight and short, the other long and crooked, and said "Choose, O king, whichever of these two you please," with a riddling reference in the long and crooked line to the claptraps of rhetoric, and in the line that was at once both short and straight to simple
and concise straightforwardness of speech; and to 23
secure this they aim at speech void of superfluity not only amongst their own people but also amongst foreigners. And when the Chian ambassador made a request for the export of grain, they sent him away out of their assembly empty-handed because he made his request at great length; but when another man was sent who was more concise (for the Chians were hard pressed by necessity), they granted his request; for he held up before them an empty sack [a] and said, "This needs barley-meal." All the same, they censured this man, too, as a chatterbox; for the showing of the empty sack sufficiently indi-
cated the petition of the Chians. Hence, the tragic 24
poet Ion was moved to say about them,—

The battlements of Sparta are not words;
But when the War-god strikes its host afresh,
The head doth plan and rule, the hand perform,—

meaning that they are excellent in planning and detest rhetoric. Consequently, if the cities do not expel the arts but have expelled rhetoric, rhetoric will not be one
of the arts. For to reverse the argument and allege 25
that some of the Greek cities have also banished philosophers is stupid. For, firstly, they will not be able to

[a] *Cf.* Herod. iii. 46.

μαρτυρίαν ὥσπερ ἐπὶ ῥητορικῆς οἱ τοὐναντίον συναγαγόντες· ἔπειτα εἰ καὶ ἐξέβαλόν τινες τῶν πόλεων φιλοσοφίαν, οὐ κατὰ γένος πᾶσαν ἐξέβαλον ἀλλὰ τινὰς αἱρέσεις, οἷον τὴν Ἐπικούρειον ὡς ἡδονῆς διδάσκαλον, τὴν Σωκρατικὴν δὲ ὡς ἐκφαυλίζουσαν τὸ θεῖον. αἱ μέντοι γε προειρημέναι πόλεις οὐ τινὰ μὲν παρῃτήσαντο ῥητορικὴν τινὰ δὲ προσήκαντο, ἀλλὰ κοινῶς πᾶσαν περιέστησαν.

26 Πρός γε μὴν τοῖς εἰρημένοις, καὶ εἰ τέχνη πάντως ἐστὶν ἡ ῥητορική, ἤτοι τῷ ἔχοντι ἢ ταῖς πόλεσιν ἔσται χρειώδης ὡς καὶ αἱ λοιπαὶ τῶν τεχνῶν· οὔτε δὲ τῷ ἔχοντι οὔτε ταῖς πόλεσιν ἔστιν ὠφέλιμος, ὡς παραστήσομεν· οὐκ ἄρα τέχνη καθέ-
27 στηκεν. καὶ δὴ τῷ μὲν ἔχοντι οὐκ ἔστιν ὠφέλιμος, ἐπεὶ πρῶτον μὲν ἐν ἀγοραῖς καὶ γραμματοφυλακείοις ἀναγκαῖόν ἐστι καλινδεῖσθαι, κἄν τε θέλῃ κἄν τε μὴ θέλῃ, μετὰ μοχθηρῶν καὶ παλιμβόλων καὶ συκοφαντῶν διατρίβειν, εἰς τοὺς αὐτοὺς ἐκείνοις τόπους κατερχόμενον, εἶτα καὶ τῆς αἰδοῦς ὀλίγην ποιεῖσθαι φειδώ, ἵνα μὴ εὐκαταφρόνητος
28 εἶναι δοκῇ τοῖς πανουργοτέροις, θρασέως δὲ λέγειν καὶ τὴν τόλμαν ὥσπερ ὅπλον προβεβλῆσθαι, ἵνα φοβερὸς ᾖ τοῖς ἀντιδίκοις, ἀπατητικόν τε καὶ γόητα τυγχάνειν καὶ χειρίστοις ἐντεθραμμένον πράγμασι, μοιχείαις τε καὶ κλοπαῖς καὶ ταῖς πρὸς τοὺς γονεῖς ἀχαριστίαις, εἰς τὸ πραγματικῶς ταῦτα
29 διελέγχειν ὅτε δεῖ, καὶ πάλιν ἐπιθολοῦν, ἔχειν δὲ ἐχθροὺς πολλοὺς καὶ μῖσος πρὸς πάντας, τοὺς μὲν ὅτι ἀντηδικήθησαν, τοὺς δὲ εἰδότας ὅτι τοῦ μισθωσαμένου ἐστι, καὶ ὃ ἄλλους διέθηκε, τοῦτο καὶ αὐτούς ποτε πλείονι λήμματι δελεασθεὶς

furnish evidence for this, as do those who have come to the opposite conclusion in the case of rhetoric; and secondly, if indeed some of the cities did expel philosophy, they did not expel it all without exception but certain sects such as the Epicurean, as teaching pleasure, and the Socratic, as slighting the Divinity. But the cities mentioned above did not reject one sort of rhetoric and admit another but shunned it all without exception.

In addition to what has been said, if rhetoric is an 26
art at all, it will be of use either to its possessor or to the cities, like the rest of the arts; but it is not of use either to its possessor or to the cities, as we shall establish; therefore it is not an art.
Now to its possessor it is not useful, since, firstly, 27
he is compelled—whether he wishes or not—to spend his time at the assemblies and record offices, and to consort with knavish and double-dealing slanderers by going down to the same places as they; and, secondly, he has to be very sparing of modesty, lest he should seem to the more unscrupulous to be a
contemptible fellow; and he has to speak out boldly 28
and thrust his audacity before him like a spear, so as to be terrible in the eyes of his opponents, and to be a cheat and a juggler and reared up amidst the worst kinds of conduct,—adulteries and thefts and acts of ingratitude towards parents,—so as to expose them, when necessary, in a practical way, or again to be-
cloud them; he must also have many enemies and 29
hatred towards them all,—some because they have suffered retaliation, others because they are aware that it is the habit of one who has been hiring, when enticed by a larger fee, to treat themselves in the
same way that he has treated others. Besides all 30

30 διαθήσει, μετὰ τοῦ διὰ παντὸς ἀγωνιᾶν καὶ πειρατοῦ τρόπον ὁτὲ μὲν φεύγειν ὁτὲ δὲ διώκειν, ὥστε κοπούμενον νύκτωρ καὶ μεθ' ἡμέραν ὑπὸ τῶν πράγματα ἐχόντων ὀχλεῖσθαι, μεστὸν δὲ ἔχειν τὸν βίον θρήνων τε καὶ δακρύων, καὶ τινῶν μὲν εἰς δεσμωτήριον τινῶν δὲ ἐπὶ τύμπανον ἀπαγομένων. ὥστε τῷ μὲν ἔχοντι ἐπιβλαβὴς ἡ ῥητορική.

31 Καὶ μὴν οὐδὲ ταῖς πόλεσίν ἐστιν ὠφέλιμος· οἱ γὰρ νόμοι πόλεών εἰσι σύνδεσμοι, καὶ ὡς ψυχὴ σώματος ἐκφθαρέντος φθείρεται,[1] οὕτω νόμων ἀναιρεθέντων καὶ αἱ πόλεις διόλλυνται. παρὸ καὶ ὁ θεολόγος Ὀρφεὺς τὸ ἀναγκαῖον αὐτῶν ὑποφαίνων φησὶν

ἦν χρόνος ἡνίκα φῶτες ἀπ' ἀλλήλων βίον εἶχον
σαρκοδακῆ, κρείσσων δὲ τὸν ἥττονα φῶτα
δάιζεν—

32 μηδενὸς γὰρ ἐπιστατοῦντος νόμου ἕκαστος ἐν χερσὶ τὸ δίκαιον εἶχε, καὶ ὡς

ἰχθύσι ⟨μὲν⟩ καὶ θηρσὶ καὶ οἰωνοῖς πετεηνοῖς
ἐπιτέτραπται
ἔσθειν ἀλλήλους, ἐπεὶ οὐ δίκη ἔστι μετ' αὐτοῖς—

μέχρις ὅτου ὁ θεὸς οἰκτείρων μογοῦσιν αὐτοῖς θεσμοφόρους θεὰς ἐξαπέστειλεν, ἃς ἐπὶ τῷ τὴν ἀλληλοφάγον ἀνομίαν καταλῦσαι πλέον ἢ ἐπὶ τῷ καρποῖς ἡμερῶσαι τὸν βίον ἐθαύμασαν ἄνθρωποι.
33 ἐντεῦθεν καὶ οἱ Περσῶν χαρίεντες νόμον ἔχουσι βασιλέως παρ' αὐτοῖς τελευτήσαντος πέντε τὰς ἐφεξῆς ἡμέρας ἀνομίαν ἄγειν, οὐχ ὑπὲρ τοῦ δυστυχεῖν ἀλλ' ὑπὲρ τοῦ ἔργῳ μαθεῖν ἡλίκον κακόν

[1] ψυχῆς σώματος ἐκφθαρείσης ⟨ἄνθρωπος⟩ φθείρεται cj. Bekk.

this, he must be engaged continually in contests and, like a pirate, now be fleeing and now pursuing, so that he is wearied and worried night and day by those in trouble, and has his life filled with tears and lamentations, as some are led away to prison and others to the whipping-post. Thus rhetoric is injurious to its possessor.

Furthermore, it is not useful to cities either. For 31
the laws are what bind cities together, and as the soul perishes when the body has perished, so the cities are destroyed when the laws are abolished. Hence, the theologian Orpheus[a] hints at their necessity when he says,—

> There was a time when every man liv'd by devouring his fellow
> Cannibal-wise, and the stronger man did feast on the weaker,

(for when no law was in control each man maintained 32
his right by force of hand, even as it is permitted to

> Fishes and beasts of the wild and the winged ravens and vultures,
> Each to devour the other, for justice exists not among them),[b]

until God in his pity for their misery sent to them law-bearing goddesses, and men admired these for the way they stopped the lawless cannibalism more than for the way they civilized life by means of the fruits of the earth. Hence, too, the shrewd Persians have a 33
law that on the death of their king they must practise lawlessness for the next five days, not in order to be in a state of misery but in order to learn by experience

[a] *Cf. Adv. Phys.* i. 15.
[b] *Cf.* Hesiod, *Op.D.* 275 f.

ἐστιν ἡ ἀνομία, σφαγὰς καὶ ἁρπαγὰς καὶ εἴ τι χεῖρόν ἐστιν ἐπάγουσα, ἵνα πιστότεροι τῶν βασι-
34 λέων φύλακες γένωνται. ἀλλ' ἥ γε ῥητορικὴ κατὰ τῶν νόμων εἰσκεκύκληται. τεκμήριον δὲ παμμέγεθες τὸ παρὰ μὲν τοῖς βαρβάροις, παρ' οἷς ἢ οὐδ' ὅλως ἢ σπανίως ἔστι ῥητορική, τοὺς νόμους ἀσαλεύτους μένειν, παρὰ δὲ τοῖς προσιεμένοις αὐτὴν ὁσημέραι νεοχμοῦσθαι, ὥσπερ καὶ παρ'
35 Ἀθηναίοις, καθάπερ καὶ Πλάτων ὁ τῆς ἀρχαίας κωμῳδίας ποιητὴς λέγει· καὶ γὰρ τρεῖς ἐάν τις, φησίν, ἐκδημήσῃ μῆνας, οὐκέτι ἐπιγινώσκει τὴν πόλιν, ἀλλὰ παραπλησίως τοῖς νυκτὸς περιπατοῦσι παρὰ τὰ τείχη καθάπερ τινὰς ἀγγάρους κατάγεται,[1] τὸ ὅσον ἐπὶ τοῖς νόμοις μὴ τῆς αὐτῆς οὔσης
36 πόλεως. πρόδηλον δέ ἐστι τὸ κατὰ τῶν νόμων αὐτὴν ὑπάρχειν καὶ ἐξ ὧν ἐν ταῖς κακοτέχνοις τέχναις ὑποτίθενται. ὁτὲ μὲν γὰρ παραινοῦσι τῷ ῥητῷ καὶ ταῖς φωναῖς τοῦ νομοθέτου προσέχειν ὡς σαφέσι καὶ μηδεμιᾶς ἐξηγήσεως δεομέναις, ὁτὲ δὲ ἀναστρέψαντες μήτε τῷ ῥητῷ μήτε ταῖς
37 φωναῖς ἀλλὰ τῇ διανοίᾳ κατακολουθεῖν· οὐδὲ γὰρ ὁ κολάζειν ἀξιῶν τὸν ἐπανατεινάμενόν τινι σίδηρον τὸν ὁπωσοῦν ἐπανατεινάμενον, οἷον δακτύλιον, ἢ ὁποῖον, καθάπερ βελόνην, κολάζειν ἠξίωσεν, ἀλλ' ἐὰν τὴν διάνοιαν αὐτοῦ πολυπραγμονῶμεν, τὸν ἀνδροφονῆσαι τολμήσαντα τεθελήκέναι τιμωρεῖσθαι.
38 κελεύουσι δὲ ἐνίοτε καὶ κατὰ ἀποκοπὴν ἀναγινώσκειν τοὺς νόμους καὶ ἐκ τῶν λειπομένων ἕτερόν τι νόημα συντιθέναι. πολλάκις δὲ καὶ ἀμφιβόλους λέξεις διαστέλλουσι, πρόσφορον ἑαυτοῖς κατα-

[1] κατάγεται cj. Bekk. : κατάγεσθαι MSS., Bekk.

how great an evil lawlessness is, inflicting, as it does, murders and rapine and things which are, if possible, worse, so that they may become more trusty guardians of their kings. But rhetoric was brought to the front 34
in opposition to the laws. A very strong proof of this is the fact that amongst the barbarians, amongst whom there is either no rhetoric at all or very little, the laws remain unmoved, whereas amongst those who cultivate rhetoric they are altered daily, as is the case with the Athenians, as Plato, the poet of the 35
Old Comedy, affirms.[a] For he says that if a man has been absent abroad for three months he no longer recognizes the city, but on his return (like walkers by night) he goes past the walls, as foreign couriers might do, since so far as regards the laws it is not the same city. And that rhetoric is against the laws is 36
already plain from the statements they make in their mal-artful arts. For at one time they advise us to attend to the ordinance and words of the lawgiver as being clear and needing no explanation, at another time they turn round and advise us to follow neither the ordinance nor the words but the intention ; for 37
he who proposes to punish the man who holds an iron instrument over anyone does not propose to punish the man who holds it over him in any and every way, (as, for example, one might hold a ring), or of whatever form if may be (a needle, for instance), but, if we inquire closely as to his intention, it is his desire to take vengeance on the man who has dared to commit murder. And sometimes they bid us cut 38
out bits as we read the laws, and construct a different sense from what remains. Often, too, they make distinctions in ambiguous phrases and support the

[a] *Cf.* Meineke, *Com. Gr.* 3 p. 692.

σκευάζοντες τὸ σημαινόμενον, καὶ ἄλλα μυρία πρὸς ἀνατροπὴν τῶν νόμων ποιοῦσιν. ὅθεν καὶ ὁ Βυζάντιος ῥήτωρ ἐρωτηθεὶς πῶς ὁ Βυζαντίων
39 ἔχει νόμος εἶπεν "ὡς ἐγὼ θέλω." καθὰ γὰρ οἱ ψηφοπαῖκται τὰς τῶν θεωμένων ὄψεις δι' ὀξυχειρίαν κλέπτουσιν, οὕτως οἱ ῥήτορες διὰ πανουργίαν τὰς τῶν δικαστῶν διανοίας ἀμαυρώσαντες
40 τῷ νόμῳ συγκλέπτουσι τὰς ψήφους. τό γε μὴν τῶν παρανόμων ψηφισμάτων εἶδος οὐδεὶς ἐτόλμησε γράφειν ἀλλ' ἢ οἱ ῥήτορες. τὸν γοῦν γραφέντα κατὰ Κτησιφῶντος Δημοσθένης πολλὰ βοῶν καὶ τερατευόμενος ἥρπασεν. ὅθεν καὶ ὁ Αἰσχίνης "κακὸν ἔθος" φησὶν "εἰς τὰ δικαστήρια παρῆλθεν· ὁ μὲν γὰρ κατήγορος ἀπολογεῖται, ὁ δὲ φεύγων τὴν γραφὴν κατηγορεῖ, οἱ δὲ δικασταὶ ὧν μὴ εἰσὶ κριταί, περὶ τούτων ψηφοφορεῖν ἀναγκά-
41 ζονται." ἀλλ' εἰ κατὰ τῶν νόμων ἐστὶν ἡ ῥητορική, πρὸς τῷ μὴ χρησιμεύειν τι καὶ βλαβερὰ καθέστηκεν. οὐ μὴν ἀλλ' οὐδὲ οἱ δημαγωγοῦντες ῥήτορες ἐπ' ἀγαθῷ τῶν πόλεων προβαίνουσιν, ἀλλ' ὃν λόγον ἔχει φαρμακοπώλης πρὸς ἰατρόν,
42 τοῦτον ὁ δημαγωγὸς πρὸς τὸν πολιτικόν. κακοδιδασκαλεῖ γὰρ τοὺς πολλοὺς τὰ κεχαρισμένα λέγων, καὶ διαβολαῖς αὐτοὺς ἐξαλλοτριοῖ πρὸς τοὺς ἀρίστους. λόγῳ μὲν γὰρ καὶ τῷ δοκεῖν ὑπὲρ τοῦ κοινῇ συμφέροντος ὑπισχνεῖται πάντα ποιήσειν, ταῖς δὲ ἀληθείαις ἀπ' οὐδενὸς ὑγιοῦς τροφὴν πορίζεται, ἐοικότως ταῖς τίτθαις, αἳ μικρὸν τοῦ ψωμίσματος τοῖς παιδίοις διδοῦσαι τὸ ὅλον καταπίνουσιν.

43 Τοσαῦτα μὲν οὖν καὶ τοῖς Ἀκαδημαϊκοῖς ἐν καταδρομῆς μέρει λέγεται περὶ ῥητορικῆς, ὥστε

signification which suits themselves ; and they do thousands of other things which tend to the upsetting of the laws. Hence also, the Byzantine orator, when asked " How goes the Byzantians' law ? ", replied " As I choose." For just as jugglers deceive 39
the eyes of the beholders by their sleight of hand, so the orators by their low cunning blind the minds of the judges to the law and so steal away the votes. More- 40
over, no one except the orators has dared to publish decrees of a kind contrary to the laws. Yet Demosthenes by his loud shouting and talking marvels snatched away the law which indicted Ctesiphon. Hence Aeschines says[a]—" An evil custom has entered the law-courst ; for the accuser defends himself, while the defendant acts as accuser, and the jurors are obliged to cast their votes concerning matters of which they are not judges." But if 41
rhetoric is against the laws, it is not only of no use but actually harmful.—Furthermore, even the demagogic orators do not come forward for the good of the cities, but the demagogue bears to the statesman the relation which the druggist bears to the physician. For 42
he depraves the crowd by his doctrines, using flattering words, and sets them against the better class by his slanders. By word and in seeming he promises to do everything for the public benefit, but in reality he provides nourishment from no wholesome source, like nurses who offer the babes a morsel of the pap and then swallow the whole themselves.

Such, then, are the arguments used by the men of 43
the Academy[b] concerning rhetoric, by way of run-

[a] *Cf.* Aeschin. *Adv. Ctesiph.* 193. [b] *Cf.* § 20.

εἰ μήτε τῷ ἔχοντι μήτε τοῖς πέλας ἐστὶν ὠφέλιμος, οὐκ ἂν εἴη τέχνη. ἀλλὰ πρὸς ταῦτα ἀπολογούμενοι τινὲς μέν φασιν ὅτι διττῆς οὔσης ῥητορικῆς, τῆς μὲν ἀστείας καὶ ἐν σοφοῖς τῆς δὲ ἐν μέσοις ἀνθρώποις, τὴν κατηγορίαν γεγονέναι οὐ τῆς ἀστείας
44 ἀλλὰ τῆς τῶν μοχθηρῶν. τινὲς δὲ καὶ ὑποδείγμασι χρῶνται· ὡς γὰρ ὁ τὸν πατέρα τύπτων παγκρατιαστὴς οὐ διὰ τὴν παγκρατιαστικὴν τέχνην γίνεται πατροτύπτης ἀλλὰ διὰ τὴν τῶν τρόπων μοχθηρίαν, οὕτως ὁ ῥητορικὴν ἐξασκήσας, εἶτα πατρίδος αὐτῇ καὶ νόμων χρώμενος οὐ διὰ ῥητορικὴν τοιοῦτός ἐστιν ἀλλὰ διὰ τὴν ἰδίαν πονηρίαν.
45 λέληθε δὲ τοὺς μὲν πρώτους ὅτι ἄκοντες δεδώκασι τὴν ἀνυπαρξίαν τῆς ῥητορικῆς· μηδενὸς γὰρ εὑρισκομένου σοφοῦ, ἢ σπανίως γε εὑρισκομένου, δεήσει καὶ τὴν ἐν αὐτοῖς ῥητορικὴν ἢ ἀνύπαρκτον
46 ἢ σπάνιον εἶναι. πρὸς δὲ τοὺς δευτέρους ῥητέον ὅτι ἀνόμοιόν ἐστι τὸ παράδειγμα τοῖς ἐν χερσὶ ζητουμένοις· ἡ μὲν γὰρ ἄθλησις οὐχ ὑποδείκνυσι τὴν πρὸς τὸ κακὸν χρῆσιν αὐτῆς, οἷον τὴν πατροτυψίαν, ἡ δὲ ῥητορικὴ τοῦθ' ὡς προηγούμενον ἔργον διδάσκει, οἷον πῶς ἂν τὰ μικρὰ μεγάλα ποιήσαιμεν τὰ δὲ μεγάλα μικρά, ἢ πῶς ἂν τὰ μὲν
47 δίκαια ἄδικα φανείη τὰ δὲ ἄδικα δίκαια. καθόλου δὲ τῆς ῥητορικῆς ἐξ ἐναντίων συνισταμένης λόγων οὐκ ἐνδέχεται τὸν μὲν ἀστεῖον λέγειν ῥήτορα, τὸν δὲ μὴ τοιοῦτον οὐκέτι. ὁποῖος γὰρ ἂν ᾖ ὁ ῥήτωρ, πάντως τοὺς ἐναντίους ἐκμελετᾶν ὀφείλει λόγους, ἐν δὲ τοῖς ἐναντίοις ἔστι καὶ τὸ ἄδικον· πᾶς ἄρα ῥήτωρ καὶ τοῦ ἀδίκου συναγωνιστὴς ὢν ἄδικός ἐστιν.

48 **Ἀλλὰ ὅτι μὲν οὐ ῥητέον τὴν ῥητορικὴν τέχνην,**

ning it down, so that, if it is useful neither to its possessor nor to his neighbours, it will not be an art. But in reply to all this some assert that as there are two forms of rhetoric, the one refined and in use among the wise, the other in use among inferior people, the accusation is not made against the refined kind but against that of the baser class. Some of 44
them, too, use illustrations : just as the pancratiast who beats his father is a father-beater not because of his pancratiastic art but because of his bad morals, so the man who has practised rhetoric and then used it against his country and its laws, does so not because of rhetoric but because of his own wickedness. But 45
the first group of these apologists fail to notice that, against their will, they have granted the nullity of rhetoric ; for since the wise man is never, or at least rarely, found, it must follow that the rhetoric in use among the wise is likewise either non-existent or rare. And in reply to the second group one must 46
say that their example is not on a par with the matter now in question, for while athleticism does not suggest the employment of it for mischief, such as father-beating, rhetoric declares this to be its main task,—how, for instance, we are to make small things great and great things small, or how just things may be made to appear unjust, and the unjust just. And in 47
general, as rhetoric consists of opposite statements, one cannot say that the refined speaker is an orator, but the unrefined no longer an orator. For the orator, of whatever sort he may be, must certainly practise himself in contradictory speeches, and injustice is inherent in contradictions ; therefore every orator, being an advocate of injustice, is unjust.

From all this it is plain that we must not call 48

ἐκ τούτων συμφανές· τὸ δὲ μετὰ τοῦτο καὶ ἐκ τῆς ὕλης περὶ ἥν ἐστι σκοπῶμεν αὐτῆς τὸ ἀνυπόστατον. καίτοι προαποδέδοται ἡμῖν τὸ κεφάλαιον ἐν τῷ πρὸς τοὺς γραμματικούς· εἰ γὰρ περὶ λόγον ἡ ῥητορικὴ πονεῖται, οὔτε δὲ λέξις ἔστι τι οὔτε λόγος ἐκ λέξεων συγκείμενος, ὡς ἐπεδείξαμεν, διὰ τὸ οὗ τὰ μέρη μὴ ἔστιν ἀνύπαρκτον εἶναι, ἀκολουθήσει καὶ τὸ τὴν ῥητορικὴν ἀνυπόστατον
49 ὑπάρχειν. ὅμως δ' οὖν ῥητέον πρῶτον μὲν ὅτι οὐκ εἰ τὸν λόγον ἐξεπόνησεν ἡ ῥητορική, πάντως ἐστὶν ἔντεχνος, ἀλλ' εἰ τὸν συμφέροντα. καθὰ γὰρ φαρμάκων διαφόρων ὄντων, καὶ τῶν μὲν θανασίμων τῶν δὲ σωτηρίων, ἡ μὲν ἐν τοῖς θανασίμοις ἕξις καταγιγνομένη οὔτε τέχνη τίς ἐστιν οὔτε ἰατρική, ἡ δὲ περὶ τὰ σωτήρια καὶ τέχνη καὶ βιωφελής, οὕτω καὶ λόγων τῶν μὲν συμφερόντων τῶν δὲ βλαπτικῶν ὄντων, εἰ μὴ περὶ τοὺς συμφέροντάς ἐστιν ἡ ῥητορικὴ ἀλλὰ τοὺς βλαβερούς, πρὸς τῷ μὴ εἶναι τέχνη ἔτι καὶ κακοτεχνία γενήσεται. παρεστήσαμεν δέ γε πρότερον ὅτι βλαπτικωτάτοις ἐνυποδύεται λόγοις· τοίνυν οὐδὲ
50 τέχνη καθέστηκεν. καὶ μὴν εἴπερ ἡ συκοφαντικὴ καὶ ἡ ὀχλοκοπικὴ τὸ λέγειν ἐξήσκησαν καὶ οὐκ εἰσὶ τέχναι, δῆλον ὡς καὶ ἡ ῥητορικὴ κατὰ ψιλὸν τὸ ἐκπεπονηκέναι τὴν ἐν τῷ λέγειν δύναμιν ἐξεταζομένη οὐ γενήσεται τέχνη. ἀλλὰ μὴν ἡ συκοφαντικὴ καὶ ἡ ὀχλοκοπικὴ τὸ λέγειν ἐξήσκησαν καὶ οὐκ εἰσὶ τέχναι· τοίνυν οὐδὲ ἡ ῥητορική.
51 πρός γε μὴν τοῖς εἰρημένοις, οὐδὲ ἴδιον ῥητορικῆς ἐστὶ τοῦτο, ἀλλὰ τὸ κοινὸν παντὸς λογικοῦ μαθήματος· καὶ γὰρ ἰατρικὴ εὖ λέγει περὶ τῶν

rhetoric an art; and in the next place let us infer its unreality from the matter with which it deals. The sum of our argument, in fact, has been given already [a] in our treatise *Against the Grammarians*, for if rhetoric has to do with speech, but, as we have shown, neither the word is anything nor the speech composed of words, then, as that object whose parts do not exist is non-existent, it will follow that rhetoric also is non-existent. None the less, we must affirm, 49
firstly, that if rhetoric deals with speech it is not necessarily technical, but only if the speech be beneficial. For just as in the case of drugs, which differ in quality, some being deadly, others salutary, the skill which busies itself with the deadly ones is neither the art of medicine nor any art at all, whereas that which deals with the salutary drugs is both an art and useful for life, so also in the case of speeches, of which some are beneficial, others harmful, if rhetoric is not concerned with the beneficial but with the harmful, besides not being an art it will also be a mischievous artifice. And, in fact, we have already established [b] that it masks itself in most harmful speech; so it is not an art.—Moreover, if sycophancy and mob-courting 50
practise speaking but are not arts, it is evident that rhetoric, when examined, will not be an art by reason of the mere fact that it has gained by practice ability in speaking. But, in fact, sycophancy and mob-courting practise speaking and are not arts; neither, then, is rhetoric an art.—In addition to what has been 51
said, speech is not a property confined to rhetoric but is common to every branch of learning which uses words; for the art of medicine [c] speaks well con-

[a] *Cf. Adv. Gram.* 131 ff.

[b] *Cf.* §§ 41 ff.

[c] *Cf.* § 8.

ἑαυτῆς θεωρημάτων καὶ μουσικὴ περὶ μουσικῶν. διόπερ ὡς ἑκάστη τούτων οὐκ ἔστιν ἕνεκα τοῦ λέγειν ῥητορική, οὕτως οὐδὲ περὶ ἧς ἐστὶν ἡ ζήτησις.

52 Συνελόντι δὲ φάναι, οὐδὲ κατασκευάζει καλὴν λέξιν ἡ ῥητορική. οὐδὲ γὰρ ὑποδείκνυσιν ἡμῖν τὴν εἰς τοῦτο τεχνολογίαν, οἷον ὅτι καλῇ λέξει χρῆται πρῶτον μὲν ὁ μὴ ἐκκλίνων τὰ κατὰ τὴν συνήθειαν λεγόμενα, καθὼς καὶ ἐν τῷ πρὸς τοὺς γραμματικοὺς ὑπεδείξαμεν, εἶτα καὶ ὁ τοῦ νοουμένου πράγματος ἀσφαλῶς κρατῶν· ῥέμβεται γὰρ ἡ λέξις ἀγνοουμένων τούτων, παρὸ καὶ εἰς τοῦτο ἀποβλέποντες ἀγαθὸν λέγομεν εἶναι ῥήτορα ἕκαστον
53 τῶν ἰδίων ἐπιτηδευμάτων. σὺν δὲ τούτοις καὶ ὁ περιεσκεμμένος τίνες τῶν λέξεων κατὰ τὴν συνήθειαν[1] κεῖνται καὶ τίνες δοξαστῶς, τὸ ἑκάστῳ πρόσφορον ἀποδίδωσιν. βαλανεῖον μὲν γὰρ ἀνδρεῖον κατὰ τὴν συνήθειαν[1] εἴρηται ἀπὸ τοῦ ἄνδρας λούειν, ὁ δὲ πλούσιος μακάριος καὶ ὁ θάνατος κακὸν τῶν δοξαστῶν· τό τε γὰρ τὸν θάνατον τῶν κακῶν εἶναι καὶ τὸν πλοῦτον τῶν ἀγαθῶν ἄδηλον
54 καὶ δοξαστόν. χρῷτο δ' ἂν καλῶς λέξει καὶ ὁ κατειληφὼς τίνος ἕνεκα τὰς μεταλήψεις ποιούμεθα τῶν λέξεων, ἤτοι ὑπὲρ τοῦ μὴ εὐθυρρημονεῖν, προσκοπὴν φέροντος τοῦ εὐθέος ῥήματος, ἢ ὑπὲρ τοῦ τι σαφηνίζειν, ὡς ὅταν τὸ μὲν αἴτιον εἰς τὸ ποιοῦν μεταλαμβάνωμεν, τὸ δὲ σημεῖον εἰς τὸ
55 δηλοῦν. εἰ μὲν οὖν, ὡς ἔφην, ἐτεχνολογεῖτό τινα περὶ τούτων τοῖς ῥήτορσι, τάχα ἂν καὶ τὸ καλῶς

[1] ἀλήθειαν (bis) cj. Bekk.

cerning its own principles, and that of music concerning those of music. Hence, as none of these is rhetoric in virtue of its speech, neither is the subject of our inquiry.

52 And, to speak succinctly, rhetoric does not create good speech. For it does not suggest to us the technical rules for it, as for instance that the man who uses good speech is, firstly, he who does not pervert the language in common use (as we have pointed out in our book *Against the Grammarians*),[a] and secondly, he who is securely master of the subject in mind; for if the subjects are not understood, the language wanders off the point, and so, in view of this, we say that every man is a good orator about his own pursuits. Besides this, the person who has carefully 53 considered which of the words are employed according to common usage, and which have their source in (personal) opinion, attributes its due to each. Thus "a bath" is called ἀνδρεῖον according to common usage from the fact that it washes ἄνδρας (men)—, but when we say that the rich man is "blessed," or that death is "an evil," these descriptions have their source in opinion; for that death is one of the evils and wealth one of the goods is non-evident and a matter of opinion. He, too, will make a good use of language 54 who has grasped the reason why we substitute one word for another, whether in order to avoid straightforward speech, when the straightforward word causes offence, or in order to make a thing clear, as when we substitute "agent" for "cause," or "indication" for "sign." Now if, as I said, any tech- 55 nical rules on these matters were observed by the orators, they might possibly have derived from

[a] *Cf. Adv. Gram.* 189 ff.

λέγειν καὶ τὴν κεκαλλωπισμένην λέξιν ἐκ ῥητορικῆς εἶχον. νῦν δὲ ἐπεὶ ταύτης τῆς θεωρίας οὐ ψαύουσιν, ἢ εἰ ψαύοιεν, οὔ τοί γε κατὰ ῥητορικήν, λεκτέον μὴ ῥητορικῆς ἴδιον εἶναι τὸ καλλιλεκτεῖν.
56 ἥ τε λέξις καθ' ἑαυτὴν οὔτε καλή ἐστιν οὔτε μοχθηρά. τεκμήριον δὲ τὸ τῇ αὐτῇ ὑπὸ μὲν ἀστείου καὶ σεμνοῦ λεγομένῃ προσκόπτειν ἡμᾶς, ὑπὸ δὲ μίμου γελωτοποιοῦντος μηδαμῶς. διόπερ ὅταν λέγηται ὁ ῥήτωρ καλῆς λέξεως εἶναι κατασκευαστικός, ἤτοι κατὰ τοῦτο λέγεται καθὸ τὴν τὰ συμφέροντα πράγματα δηλοῦσαν λέξιν κατασκευάζει, ἢ τὴν οἷα ἐστὶν ἡ τοῦ ἑλληνίζειν, ἢ τὴν ἐναργῶς καὶ συντόμως καὶ ἐγκατασκεύως
57 δηλοῦσαν τὰ πράγματα. οὔτε δὲ καθὸ τὴν τὰ συμφέροντα πράγματα μηνύουσαν· οὐδὲν γὰρ ἴσασι περὶ τούτων τῶν πραγμάτων οἱ ῥήτορες. οὔτε καθὸ τὴν οἷα ἐστὶν ἡ τοῦ ἑλληνίζειν· κοινὸν γὰρ ἦν τοῦτο τῶν τῇ συνηθείᾳ καὶ ταῖς ἐλευθέραις τέχναις προσεχόντων. οὔτε καθὸ τὴν σαφῶς καὶ συντόμως καὶ ἐγκατασκεύως μηνύουσαν τὰ πράγματα· τοὐναντίον γὰρ περίοδον καὶ ἐπιφώνημα θέλοντες λέγειν οἱ ῥήτορες, καὶ μὴ φωνᾶεν φωνάεντι συγκρούειν, καὶ ὁμοιοτέλευτον διάνοιαν κατακλίνειν, ἐκκλείονται τῆς σαφοῦς ἅμα καὶ
58 συντόμου τῶν πραγμάτων ἑρμηνείας. οὐκ ἄρα ῥητορικῆς ἐστὶ τὸ κατασκευάζειν καλὴν λέξιν καὶ τὸ εὖ λέγειν. δοθέντος τε τούτου οὐκ ἄν τις ἕλοιτο τὴν τοιαύτην φράσιν, πρῶτον μὲν διὰ τὸ μὴ πίπτειν αὐτὴν εἰς τὴν κοινὴν τοῦ βίου χρῆσιν· οὐδεὶς γὰρ ἡμῶν οὕτω διαλέγεται ὡς οἱ ῥήτορες ἐπὶ τῶν δικαστηρίων, ἐπεὶ καταγελασθήσεται. καὶ αὐτοὶ δὲ ἐκεῖνοι πάντοτε ἐξελθόντες τῆς

rhetoric fine speaking and refined phraseology. But as it is, seeing that they do not touch this subject, or if they touch it it is not because of rhetoric, one must declare that fine speech is not peculiar to rhetoric.—
Again, language of itself is neither good nor bad. A 56
proof of this is the fact that the same language which offends us when spoken by a cultivated and solemn person does not offend us in the least when spoken by the jester cracking jokes. Hence, when the orator is said to be capable of framing fine speech, it is either for the reason that he frames speech which expresses profitable things, or speech which is a form of good Greek, or speech which indicates objects clearly,
concisely and competently. But it is not for the 57
reason that his speech indicates things of profit; for orators know nothing about those things. Nor is it because his speech is a form of good Greek; for this is also shared by those who keep to customary usage and the liberal arts. Nor is it because his speech indicates objects clearly, concisely and competently; on the contrary, in their desire to give out their well-rounded periods and concluding clauses, and not to let vowel collide with vowel, and to avoid similar sentence-endings, they preclude themselves from expounding things at once clearly and concisely.
Therefore it does not belong to rhetoric to produce 58
fine phrasing and good speaking. And, this being granted, no one will choose a style such as theirs, and that because, firstly, it does not fit in with the common practice; for none of us talks as do the orators in the law-courts, since he would be jeered at. And they themselves, when they come away

διατριβῆς καὶ τοῦ ἀγῶνος ἄλλῃ χρῶνται πρὸς
59 τοὺς πέλας ἑρμηνείᾳ. εἶτα καὶ ὡς ἔφην, προσκοπὴν ἐμποιεῖ τὸ μετὰ προσοχῆς καὶ ῥητορείας λαλεῖν. μετακτέον δὲ τὰ πρότερον πρὸς τοὺς ἀναλογιστικοὺς τῶν γραμματικῶν εἰρημένα, καὶ διδακτέον ὅτι τῇ συνηθείᾳ προσεκτέον μᾶλλόν ἐστι θέλοντας εὖ λέγειν ἤπερ τέχνῃ τινὶ περιεργοτέρᾳ.

60 Τὰ νῦν δὲ μετελθόντες καὶ ἀπὸ τοῦ τέλους τῆς ῥητορικῆς ποιώμεθα τὰς ἐνστάσεις. πάλιν τοίνυν λεκτέον ὡς εἰ μηδέν ἐστι ῥητορικῆς τέλος, οὐδέν ἐστι ῥητορικὴ διὰ τὸ πᾶσαν τεχνικὴν ἕξιν πρός τι τέλος λαμβάνειν τὴν ἀναφοράν. οὐχὶ δέ γε ἔστι τι ῥητορικῆς τέλος, ὡς δείξομεν· οὐκ ἄρα ἐστὶ
61 τέχνη ἡ ῥητορική. οἱ μὲν οὖν πλεῖστοι καὶ χαρίεντες ἔσχατον οἴονται τῆς ῥητορικῆς ἔργον εἶναι τὸ πείθειν. καὶ γὰρ οἱ περὶ τὸν Πλάτωνα εἰς τοῦτο ἀπιδόντες δύναμιν εἰρήκασιν αὐτὴν τοῦ διὰ λόγων πείθειν, καὶ οἱ περὶ τὸν Ξενοκράτην πειθοῦς δημιουργόν, καὶ Ἀριστοτέλης δύναμιν τοῦ θεωρεῖν τὸ ἐνδεχόμενον πιθανόν. καὶ Ἀρίστων ὁ Κριτολάου γνώριμος σκοπὸν μὲν ἐκκεῖσθαί φησιν αὐτῇ
62 τὴν πειθώ, τέλος δὲ τὸ τυχεῖν τῆς πειθοῦς. καὶ Ἑρμαγόρας τελείου ῥήτορος ἔργον εἶναι ἔλεγε τὸ τεθὲν πολιτικὸν ζήτημα διατίθεσθαι κατὰ τὸ ἐνδεχόμενον πειστικῶς. Ἀθήναιος δὲ λόγων δύναμιν προσαγορεύει τὴν ῥητορικὴν στοχαζομένην τῆς τῶν ἀκουόντων πειθοῦς, καὶ Ἰσοκράτης φησὶ μηδὲν ἄλλο ἐπιτηδεύειν τοὺς ῥήτορας ἢ
63 ἐπιστήμην πειθοῦς. ὅθεν καὶ ἡμεῖς στοιχοῦντες

from their business and the trial, always use another mode of expression towards their companions. And 59
secondly, talking in a studied and oratorical manner gives offence, as I have said. And one may transfer what has been said already against the Grammarians who rely on analogy,[a] and affirm that if we wish to speak well we should pay attention to common usage rather than to any superfluous art.

For the present let us pass on and base our objec- 60
tions on the "end"[b] of rhetoric. So we must state, once again, that if there is no "end" of rhetoric, rhetoric is nothing, because every technical activity has reference to some end. But, as we shall show, rhetoric has no end; therefore rhetoric is not an art.
Now most clever people suppose that the ultimate 61
task of rhetoric is persuasion. For Plato, with an eye to this, has said that it is "the ability to persuade by means of speech," and Xenocrates[c] that it is "the creator of persuasion," and Aristotle[d] that it is "the ability to discern the credible which is possible." Ariston, also, the friend of Critolaüs, declares that the professed aim of rhetoric is persuasion, and its end the securing persuasion. And Hermagoras used to say 62
that the task of the perfect orator is to settle the political question proposed as persuasively as possible. And Athenaeus calls rhetoric a power of speech which aims at the persuasion of the audience; and Isocrates asserts that orators pursue nothing else than the science of persuasion. Hence, we too, 63

[a] *Cf. Adv. Gram.* 179 ff.

[b] "End" in the sense of "aim" or "final cause."

[c] But this definition is attributed by Quintilian to Isocrates; and the definition ascribed to Xenocrates in § 6 is different.

[d] *Cf.* Arist. *Rhet.* I. ii. 1.

τῇ τούτων φορᾷ λέγομεν εὐθὺς ὅτι τὸ πιθανὸν προσαγορεύεται τριχῶς, καθ' ἕνα μὲν τρόπον ὅπερ ἐναργῶς τε ἀληθές ἐστι καὶ ἀληθοῦς ἐμποιοῦν φαντασίαν ἐπισπᾶται ἡμᾶς εἰς συγκατάθεσιν, καθ' ἕτερον δὲ ὅπερ ψεῦδός ἐστι καὶ ἀληθοῦς ἐμποιοῦν φαντασίαν ἐπισπᾶται ἡμᾶς εἰς συγκατάθεσιν (ὅπερ καὶ εἰκὸς ὀνομάζειν εἰώθασιν οἱ ῥήτορες ἀπὸ τοῦ ἐοικὸς εἶναι τῷ ἀληθεῖ), κατὰ δὲ τὸν τρίτον τρόπον τὸ κοινὸν τοῦ τε ἀληθοῦς
64 καὶ ψεύδους. τοσαυταχῶς δὴ λεγομένου τοῦ πιθανοῦ, ἄξιόν ἐστι πυθέσθαι τῶν ῥητόρων κατὰ τί τούτων τῶν πιθανῶν οἴονται τὴν ῥητορικὴν τοῦ πείθειν ἐφίεσθαι, καὶ περὶ ποῖον αὐτῶν τεχνιτεύειν αὐτὴν ἀξιοῦσιν, περὶ τὸ ἐναργῶς ἀληθὲς ἢ περὶ τὸ ἐοικὸς τούτῳ ψεῦδος ἢ ὃ περὶ τὴν ἀμφο-
65 τέρων κοινότητα στρέφεται. ἀλλὰ περὶ μὲν τὸ ἐναργῶς ἀληθὲς οὐχ οἷόν τε· τοῦτο γὰρ ἐξ αὑτοῦ πείθει καὶ ἐπισπᾶται ἡμᾶς πρὸς συγκατάθεσιν, ὥστε παρέλκειν τὴν ἐκ ῥητορικῆς ἐπ' αὐτοῦ συνισταμένην πειθώ. καὶ καθάπερ οὐδεμιᾶς δεόμεθα τέχνης πρὸς τὸ πείθεσθαι ὅτι νῦν ἡμέρα ἔστιν ἢ ὅτι νῦν ἐγὼ διαλέγομαι, πραγμάτων ὄντων ἐναργῶν καὶ αὐτοφωράτων, οὕτως οὐδὲ πρὸς τὸ συγκατατίθεσθαι τῷ ἀνδροφόνον εἶναι τὸν ἐπ' αὐτοφώρῳ ληφθέντα ἀνδροφόνον χρεία ῥητορικῆς.
66 καὶ ἄλλως, εἰ τοῦ προδήλως ἀληθοῦς, ᾗ πιθανόν ἐστι, θεωρητικὴ καθέστηκεν ἡ ῥητορική, πάντως καὶ τοῦ ἀπιθάνου γενήσεται θεωρητική· ταυτὶ γὰρ κατὰ τὴν ὡς πρὸς ἄλληλα σχέσιν λαμβάνεται, καὶ ᾧ λόγῳ ὁ τὸ ἀριστερὸν κατειληφὼς ἐξ ἀνάγκης ἐπιβάλλει καὶ τῷ οὗ ἀριστερόν ἐστιν, οὕτως ὁ τὸ πιθανὸν ἀληθὲς διακρίνων ἀπὸ τοῦ μὴ τοιούτου

following in their steps, affirm straightaway that the term " credible " (or " probable ") is used in three senses,[a]—in one sense of that which is plainly true and which, by implanting an impression of truth, draws us on to assent ; in another sense of that which is false, but which, by implanting an impression of truth, draws us on to assent (and this the orators are accustomed to call " likely," from the fact that it is like the truth) ; and in a third sense of that which shares in truth and falsehood alike. The term 64
" credible," then, being used in all these ways, it is proper to ask the orators in respect of which of these " credibles " do they suppose that rhetoric aims at persuasion, and to which of them do they claim that it addresses itself as an art,—is it to the obviously true, or to the false which is like the true, or to that which contains both together? But it cannot possibly be to 65
the obviously true ; for this persuades of itself and draws us to assent, so that persuasion regarding it produced by rhetoric is superfluous. And just as we need no art to be persuaded that " now it is day " or that " now I am talking," these being obvious and self-evident facts, so too there is no need of rhetoric to make us assent to the fact that the murderer taken in the act is a murderer.—And besides, if rhetoric 66
has for its object the obviously true, in so far as it is credible, it will certainly be also concerned with the incredible ; for these two are conceived as being relative the one to the other, and for the same reason that he who comprehends " left " necessarily conceives also that of which it is " left," he who discerns the " credible " which is true possesses knowledge also of the " incredible," from the fact of its not being

[a] *Cf. Adv. Log.* i. 174.

67 γνῶσιν ἔχει καὶ τοῦ ἀπιθάνου. ἐπεὶ οὖν πᾶν ἀληθές, ὁποῖόν ποτ' ἂν ᾖ, ἤτοι πιθανόν ἐστιν ἢ ἀπίθανον, ἀκολουθήσει τὴν ῥητορικὴν παντὸς ἀληθοῦς εἶναι θεωρητικήν. τῷ δὲ παντὸς ἀληθοῦς εἶναι θεωρητικὴν ἀκολουθήσει τὸ καὶ παντὸς ψεύδους· ᾧ γὰρ λόγῳ ὁ διακριτικὸς τοῦ πιθανοῦ ἐξ ἀνάγκης καὶ τοῦ ἀπιθάνου διακριτικὸς ἔσται, τῷ αὐτῷ καὶ ὁ παντὸς ἀληθοῦς ἐπιγνώμων συνεπιβαλεῖ[1] παντὶ τῷ ἀντικειμένῳ, τουτέστι τῷ ψεύδει. εἰ δὲ τοῦτο, ἔσται ἡ ῥητορικὴ γνῶσις ἀληθῶν τε καὶ ψευδῶν. οὐ πάνυ δέ γε τοῦτο· τοίνυν οὐδὲ τοῦ αὐτόθεν ἀληθοῦς ἐστὶ θεωρητική.
68 καὶ μὴν τοῖς ἀντικειμένοις συναγορεύειν ἐπαγγέλλεται, τὰ δὲ ἀντικείμενα οὐκ ἔστιν ἀληθῆ· οὐκ ἄρα τοῦ ἀληθοῦς ἐφίεται ἡ ῥητορική. καὶ μὴν οὐδὲ τοῦ ψεύδους· οὐδεμία γὰρ περὶ ψεῦδος ἵσταται τέχνη, ἀλλ' ἀναγκαῖόν ἐστι τὴν ῥητορικὴν τοῦτο μεταδιώκουσαν ἢ μὴ εἶναι τέχνην ἢ κακοτεχνίαν ὑπάρχειν, μετὰ τοῦ πάλιν τὰς αὐτὰς
69 ὑπαντιάζειν ἀπορίας. εἰ γὰρ περὶ τὸ πιθανὸν ψεῦδος καταγίνεται, πάντως εἴσεται καὶ τὸ ἀπίθανον. ἐπεὶ οὖν πᾶν ψεῦδος ἤτοι πιθανόν ἐστιν ἢ ἀπίθανον, παντὸς ψεύδους ἐπιστήμη γενήσεται, καὶ διὰ τοῦτο καὶ παντὸς ἀληθοῦς, ὥστ' αὐτὴν μὴ διαφέρειν τῆς διαλεκτικῆς· ὃ κατὰ πολλοὺς
70 τρόπους ἐστὶν ἄτοπον. οὐ μὴν ἀλλ' εἰ τῶν ἀντικειμένων συνήγορος καθέστηκε, τὰ δὲ ἀντικείμενα οὐκ ἔστι ψευδῆ, οὐκ ἂν εἴη ψεύδους θεωρητική. ἔτι εἴπερ εἰκός ἐστι τὸ τὰς πλείστας ἀφορμὰς εἰς τὸ ἀληθὲς εἶναι παρεχόμενον, καὶ παράλογον, ὃ δὴ τούτῳ ἀντίκειται, τὸ ὀλίγας ἀφορμὰς καὶ

[1] συνεπιβαλεῖ cj. Bekk. : συνεπιβάλλει mss., Bekk.

of the same character. Since, then, everything true, 67
of whatsoever sort it be, is either credible or incredible, it will follow that rhetoric is concerned with everything true. But from its concern with everything true will follow its concern with everything false as well; since, for the reason that he who can discern the credible will necessarily be able to discern also the incredible, for the same reason he who has knowledge of everything true will also perceive therewith everything of the contrary sort, that is to say, false. And if this is so, rhetoric will be the knowledge of things true and false. But this is certainly not so; hence, rhetoric has not the self-evidently true for its object.—Moreover, it professes 68
to advocate opposite causes, but opposites are not (both) true; therefore rhetoric does not aim at the true.—Nor yet at the false; for no art of the false exists, but it is necessary that rhetoric, if it pursues this, must either not be an art or be an evil technique, besides being confronted again with the same difficulties. For if it is concerned with 69
the credible falsehood, it will certainly know also the incredible. Since, then, every falsehood is either credible or incredible, it will be the knowledge of everything false and therefore also of everything true, so that it does not differ from dialectic; and this is in many ways absurd. If, however, it advo- 70
cates opposites, and opposites are not (both) false,[a] it will not have the false for its object.—Further, if the "likely" is that which furnishes the most numerous grounds for supposing it to be true, and its opposite, the "improbable," that which furnishes

[a] *i.e.* of contradictory propositions one must be true.

σπανίους ἔχον εἰς τὸ ἀληθὲς εἶναι, πάντως ἡ ῥητορικὴ εἰς τὸ ἐναντίον ἐπιχειροῦσα οὐ μᾶλλον
71 τοῦ εἰκότος ἢ τοῦ ἀντικειμένου στοχάζεται. καὶ μὴν οὐδὲ τὸ κοινὸν τοῦ τε ἀληθοῦς καὶ ψεύδους μεταδιώκει· ἐν τούτῳ γὰρ καὶ ψεῦδος κατεπέπλεκτο. ἄτοπόν τε καθεστήκει τὸ τέχνην ψευδέσι χρῆσθαι, σὺν τῷ κατὰ τὸν προϋποδεδειγμένον τρόπον ἀκολουθεῖν τὸ καὶ ἐπιστήμην αὐτὴν ἀληθῶν τε καὶ ψευδῶν γίνεσθαι, τοῦ πράγματος μὴ οὕτως ἔχοντος. ἀλλ' εἰ μήτε ἀληθὲς μήτε ψεῦδος μήτε τὸ κοινὸν ἀμφοτέρων θεωρεῖν δύναται ἡ ῥητορική, παρὰ δὲ ταῦτα οὐδέν ἐστι πιθανόν, οὐκ ἂν εἴη ῥητορικῆς τὸ πείθειν.

72 Ἡμεῖς μὲν οὖν ταύταις ἀξιοῦμεν ταῖς ἐνστάσεσι χρῆσθαι πρὸς τοὺς ῥήτορας, ἄλλοι δὲ καὶ τὰς λεχθησομένας εἰώθασι παραλαμβάνειν, αἷς ἔξεσται τῷ βουλομένῳ χρῆσθαι. φασὶ γάρ, ἤτοι τέχνη ἐστὶν ἡ ῥητορικὴ ἢ οὐκ ἔστιν. καὶ εἰ μὲν μὴ ἔστι, μηδὲ τέλος αὐτῆς ζητῶμεν· εἰ δὲ ἔστι, πῶς κοινὸν ἔχει τέλος καὶ τοῦ μὴ ῥήτορος; τὸ γὰρ πείθειν πολλοῖς πάρεστι διὰ πλοῦτον ἢ κάλλος ἢ δόξαν,
73 ὡς πρότερον ὑπεδείκνυμεν. ῥηθέντων τε πολλάκις τῶν λόγων καὶ ἐπ' αὐτοῖς πεπεισμένων τῶν δικαστῶν οὐδὲν ἧττον προσμένουσιν οἱ ῥήτορες, ἕτερόν τι ἀπεκδεχόμενοι τέλος, καὶ προσμένοντες δέονται. οὐκ ἄρα τὸ πείθειν ῥητορικῆς ἐστὶ τέλος,
74 ἀλλ' εἰ ἄρα, τὸ μετὰ τοῦτο ἐπακολουθοῦν. ἄλλως τε καὶ ἐναντίος ἐστὶν ὁ ῥητορικὸς λόγος πειθοῖ. πρῶτον μὲν γὰρ περίεργος καθέστηκεν, προσκό-
75 πτουσι δὲ οἱ πολλοὶ τῇ τοῦ λόγου περιεργίᾳ· εἶτα ὁ ἀσαφὴς λόγος οὐκ ἔστι πειστικός, ὁ δὲ τῶν

few and rare grounds for supposing it to be true, then rhetoric, since it argues both ways, certainly aims at the "likely" no more than at its opposite.—Nor 71
again does it pursue after that which shares in both truth and falsehood; for this implies falsehood. And that an art should employ falsehoods is absurd, besides the fact that—according to our previous demonstration—it becomes in consequence the science of things true and false, which is not the case. But if rhetoric cannot have for its object either the true or the false or that which includes both, and besides these there is no other "credible," then persuasion will not belong to rhetoric.

These, then, are the objections which we deem it 72
proper to use against the Orators, but others[a] are accustomed to adduce those which shall now be stated, and which anyone who likes will be able to use. They argue thus: Rhetoric either is or is not an art; and if it is not, let us not seek for its "end"; but if it is, how can it have an end which is shared by the non-rhetorician? For it is in the power of many to persuade by means of wealth or beauty or glory, as we previously indicated.[b]—Moreover, often, when 73
the speeches have been made and the judges in consequence persuaded of the facts, the Orators, none the less, wait on in expectation of gaining some further end, and as they wait they entreat. The end of rhetoric, therefore, is not persuasion but, if anything, that which follows after it.—
Again, rhetorical speech is opposed to persuasion. 74
For, firstly, it is superabundant, and most people are offended by its superabundance; and secondly, 75
speech which lacks clarity is not persuasive, but

[a] *e.g.* Plato (in the *Gorgias*) and Epicurus. [b] *Cf.* §§ 2, 5.

ῥητόρων λόγος ἐν περιόδοις κείμενος καὶ ἐνθυμή-
μασιν ἧττόν ἐστι σαφής. οὐκ ἄρα πειστικὸς ὁ
76 ἀπὸ τῆς ῥητορικῆς ἐστὶ λόγος. ὅ τε εὔνοιαν τοῖς
δικασταῖς ἐμποιῶν λόγος, οὗτός ἐστι πειστικός·
εὔνοιαν δὲ ἐμποιεῖ οὐχ ὁ ῥητορικὸς ἀλλ' ὁ ἀφελὴς
καὶ τὸν ἰδιωτικὸν ὑποφαίνων τύπον. τῷ μὲν γὰρ
τοῦ ῥήτορος ἀντίκεινται πάντες ταῖς ὑπεροχαῖς
φθονοῦντες· κἂν γὰρ δίκαια κατασκευάζῃ ὁ ῥήτωρ,
δοκοῦσι μὴ διὰ τὴν τῶν πραγμάτων φύσιν ἀλλὰ διὰ
τὴν τοῦ ῥήτορος πανουργίαν τὰ μὴ δίκαια τοιαῦτα
77 αὐτοῖς φαίνεσθαι· τῷ δὲ τοῦ ἰδιώτου ὡς ἀσθενεῖ
πᾶς τις συναγωνίζεται, καὶ τῷ ἧττον δικαίῳ
προσδοξάζει τὸ μᾶλλον δίκαιον διὰ τὸ ὑπὸ ἀφε-
λοῦς καὶ ἰδιώτου κατασκευάζεσθαι. παρ' ἣν
αἰτίαν Ἀθηναίοις τὸ παλαιὸν οὐκ ἐπετέτραπτο
συνήγορον παρίστασθαι τοῖς κρινομένοις ἐπὶ τῆς
ἐν Ἀρείῳ πάγῳ βουλῆς, ἀλλ' ἕκαστος ὡς εἶχε
δυνάμεως, ἀδιαστρόφως καὶ ἀπανούργως ὑπὲρ
78 ἑαυτοῦ τοὺς λόγους ἐποιεῖτο. καὶ μὴν εἴπερ
ἐπίστευον αὐτοῖς οἱ ῥήτορες ὅτι πειστικὴν ἔχουσι
δύναμιν, ἐχρῆν αὐτοὺς μήτε ἔλεον μήτε οἴκτους
μήτε ὀργὰς ἢ ἄλλα τινὰ τοιαῦτα κινεῖν, ἅπερ
πείθει μὲν οὐδαμῶς, παραλογίζεται δὲ τὴν τῶν
δικαστῶν γνώμην καὶ ἀντισκοτεῖ τῷ δικαίῳ.
Ἀλλ' ὅτι μὲν οὐκ ἐνδέχεται τὸ πείθειν τέλος
79 εἶναι ῥητορικῆς, δέδεικται· τινὲς δὲ τοῦτο μὲν οὐ
λέγουσιν αὐτῆς τέλος, τὸ δὲ τοὺς ἐνδεχομένους
εὑρεῖν λόγους, οἱ δὲ τὸ δόξαν ἐμποιεῖν τοῖς δικα-
σταῖς περὶ τῶν πραγμάτων οἵαν οἱ λέγοντες
θέλουσιν, ἄλλοι δὲ τὸ συμφέρον, τινὲς δὲ τὸ νικᾶν.
80 ὧν πρὸς μὲν τοὺς πρώτους ῥητέον ὡς εἴπερ τοὺς

the speech of the Orators, which consists of periods and enthymemes, is inferior in clarity. Therefore the speech derived from rhetoric is not persuasive.—Further, the speech which arouses goodwill 76
in the judges is that which is persuasive; but what arouses goodwill is not the rhetorical speech but that which is simple and reflects the ordinary style. For to the style of the Orator all those who detest airs of superiority are opposed. For even if the Orator maintains what is just, they imagine that unjust things seem to them just, not because of the real nature of the things but because of the trickery of the orator. But with the common man's speech 77
everyone sympathizes, feeling its weakness, and attributes greater justice to that which is less just because it is maintained by a plain and ordinary person. And for this reason the Athenians, in olden days, were not allowed to have an advocate to support those on trial at the court of the Areopagus, but each man, to the best of his ability, made a speech in his own defence without trickery or verbal jugglery.—Moreover, if the Orators believed their own statement 78
that they possess a power of persuasion, they ought not to excite pity or lamentation or indignation, or other feelings of that sort—things which do not persuade at all but pervert the minds of the judges and obscure justice.

Thus it has been shown that the "end" of rhetoric cannot be persuasion; and some say that 79
its end is not this but the discovery of appropriate words; others, that it is to implant in the judges an opinion about the facts such as the speakers wish; others, that it is the advantageous; and some that it is victory. In reply to the first of 80

ἐνδεχομένους εἰς τὰς ὑποθέσεις λόγους εὑρεῖν
ἐπαγγέλλεται ἡ ῥητορική, ἤτοι τοὺς ἀληθεῖς ἢ
δυνατοὺς ῥηθῆναι ἐπαγγέλλεται. οὔτε δὲ τοὺς
ἀληθεῖς· κανόνα γὰρ καὶ κριτήριον τῆς τῶν ἀληθῶν
καὶ ψευδῶν διαγνώσεως αὐτοὺς ἔχειν δεῖ, ὅπερ
οὐκ ἔχουσιν· οὔτε τοὺς δυνατοὺς ῥηθῆναι· ἀγνοοῦν-
τες γὰρ τοὺς ἀληθεῖς οὐδὲ τοὺς δυνατοὺς ῥηθῆναι
81 ἐπιγνώσονται. οὐκ ἄρα ῥητορικῆς ἐστὶ τὸ τοὺς
ἐνόντας καὶ δυνατοὺς εὑρίσκειν λόγους. ἥ τε
ῥητορικὴ οὐδὲν ἄλλο ἐστὶν ἢ τὸ τοὺς ἐνδεχομένους
εὑρεῖν λόγους, ὅθεν ὁ τοῦτο λέγων τέλος δυνάμει
τὴν ῥητορικὴν τέλος εἶναι λέγει τῆς ῥητορικῆς.
82 οὗ τε χάριν ἅπαντά φησι πράσσειν ὁ ῥήτωρ,
ἐκεῖνο ἂν τέλος εἴη· οὐχὶ δέ γε χάριν τῶν ἐν-
δεχομένων ἐπιχειρήσεων πάντα πράσσει ὁ ῥήτωρ,
ἀλλὰ τοῦ μετὰ τὰς ἐπιχειρήσεις ἐπακολουθοῦντος·
83 τοίνυν οὐκ ἂν εἴη τέλος ἐκεῖνο. καὶ μὴν οὗπερ
δεῖται τέλους τυχεῖν ὁ ῥήτωρ, τούτου καὶ ὁ μισθω-
σάμενος αὐτὸν ἰδιώτης· τοῦ δέ γε τοὺς ἐνδεχο-
μένους εὑρεῖν λόγους οὐ σπεύδει τυχεῖν ὁ ἰδιώτης,
ἀλλ' ἑτέρου τινός· ἐκεῖνο ἄρα τέλος γενήσεται,
84 καὶ οὐ τὸ τοὺς ἐνδεχομένους λόγους εὑρεῖν. καὶ
μὴν οὐδὲ τὸ δόξαν ἐμποιεῖν τοῖς δικασταῖς περὶ
τῶν πραγμάτων οἵαν οἱ λέγοντες θέλουσιν· τοῦτο
γὰρ οὐ διήνεγκε τοῦ πείθειν, ἐπείπερ ὁ πεπεικὼς
δόξαν ἐμπεποίηκε τοῖς δικασταῖς περὶ τῶν πραγ-
μάτων οἵαν οὗτος θέλει. ἐδείξαμεν δὲ ἡμεῖς ὅτι
οὐκ ἔστι τῆς ῥητορικῆς τέλος τὸ πείθειν, ὥστε
85 οὐδὲ τὸ δόξαν ἐμποιεῖν. ἀλλὰ μὴν οὐδὲ τὸ συμ-
φέρον, ὡς ἠξιώκασί τινες· ὅ τι γὰρ τοῦ μέρους

---

a *Cf. P.H.* ii. 14 ff., *Adv. Log.* i. 29 ff.

these we must say that if rhetoric professes to discover words pertinent to the assumptions, it professes to speak either true or possible words. But not such as are true; for they must possess a standard and criterion [a] for distinguishing the true and the false, and that they do not possess; nor yet such as it is possible to speak; for not knowing the true words, neither will they know those which it is
possible to speak. Therefore it does not belong to 81
rhetoric to discover words which are pertinent and possible.—Again, rhetoric is nothing else than the discovery of appropriate words; accordingly, he who states this "end" is virtually stating that rhetoric
is the end of rhetoric.—Also, that for the sake of 82
which the orator declares that he does everything will be his end; but the orator does not do everything for the sake of the appropriate arguments, but for the sake of that which follows after the arguments;
the former, then, will not be his end.—Moreover, the 83
end which the orator needs to attain is that which the private person who hired him also needs; but the private person is not eager to attain the discovery of appropriate words, but something else; this latter, then, and not the discovery of appropriate words, will
be the end.—Furthermore, neither will the end be 84
the implanting in the judges such an opinion regarding the case as the speakers desire; for this does not differ from persuasion, since he that has persuaded has implanted in the judges such an opinion regarding the case as he himself desires. But we have shown [b] that persuasion is not the end of rhetoric; neither,
then, is implanting an opinion.—Nor, again, is "the 85
advantageous," as some have assumed [c]; for that

[b] *Cf.* §§ 73 ff.

[c] *Cf.* § 79.

ἐστὶ τέλος, τοῦτο οὐκ ἂν εἴη τοῦ ὅλου τέλος· μέρους δέ γε τῆς ῥητορικῆς τοῦ συμβουλευτικοῦ τέλος λέγουσιν οἱ ῥήτορες εἶναι τὸ συμφέρον· οὐκ ἄρα τῆς ὅλης ῥητορικῆς ἐστὶ τέλος. ὅπερ τε πάσης τέχνης ἐστὶ τέλος κοινῶς, τοῦτο οὐκ ἂν εἴη τῆς ῥητορικῆς μόνης· τὸ συμφέρον δέ γε πάσης τέχνης ἐν τῷ βίῳ τέλος ἐστίν· οὐκ ἄρα τῆς ῥητορικῆς
86 ἰδιαίτερον καθέστηκεν. λείπεται οὖν τὸ νικᾶν αὐτῆς εἶναι τέλος. ὃ πάλιν ἀδύνατόν ἐστιν. ὁ γὰρ πολλάκις μὴ τυγχάνων τοῦ κατὰ γραμματικὴν τέλους οὐκ ἂν εἴη γραμματικός, καὶ ὁ πολλάκις μὴ τυγχάνων τοῦ κατὰ μουσικὴν τέλους οὐκ ἂν εἴη μουσικός. τοίνυν καὶ ὁ μὴ τυγχάνων πολλάκις τοῦ κατὰ ῥητορικὴν τέλους οὐκ ἂν εἴη ῥήτωρ.
87 ὁ δέ γε ῥήτωρ πλειονάκις[1] [μᾶλλον] ἢ νικᾷ νικᾶται, καὶ τοσούτῳ πλειονάκις[2] ὅσῳ δυναμικώτερός ἐστι, τῶν τὰ ἄδικα ἐχόντων πράγματα ἐπ᾽ αὐτὸν συντρεχόντων. οὐκ ἄρα ῥήτωρ ἐστὶν ὁ ῥήτωρ. ὅ τε μὴ τυχὼν τοῦ κατὰ ῥητορικὴν τέλους οὐκ ἂν ἐπαινοῖτο, ῥήτορα δὲ ἐνίοτε νικηθέντα ἐπαινοῦμεν· οὐκ ἄρα ῥητορικῆς τέλος ἐστὶ τὸ νικᾶν.

88 Ὥστε εἰ μήτε ὕλην ἔχει ἡ ῥητορικὴ περὶ ἣν τεχνιτεύει, μήτε τέλος ἐφ᾽ ὃ ἀνάγεται, οὐκ ἂν ὑπάρχοι ἡ ῥητορική· οὔτε δὲ ὕλην ἔχει οὔτε τέλος, καθὼς παρεστήσαμεν· οὐκ ἄρα ὑπάρχει ἡ ῥητορική.

89 Ἐπαπορήσειε[3] δ᾽ ἄν τις αὐτοῖς καὶ ἀπὸ τῶν μερῶν αὐτῆς. μέρη δὲ λέγουσι ῥητορικῆς τὸ δικανικόν τε καὶ συμβουλευτικὸν καὶ ἐγκωμιαστικόν, τούτων δὲ τοῦ μὲν δικανικοῦ τέλος εἶναι τὸ δίκαιον, τοῦ δὲ συμβουλευτικοῦ τὸ συμφέρον,

[1] πλειονάκις ἢ cj. Bekk.: πλεῖον μᾶλλον ᾧ mss., Bekk.
[2] πλειονάκις scripsi: πλεῖον mss., Bekk. (μᾶλλον cj. Bekk.).

which is the end of a part will not be the end of the whole; but the Orators say that "the advantageous" is the end of the deliberative part of rhetoric; therefore it is not the end of rhetoric as a whole.—Also, that which is the common end of all Art will not be the end of rhetoric alone; but "the advantageous" is the end of all art in our experience; therefore there is no end peculiar to
rhetoric.—It remains, then, to say that "victory" 86
is its end; but this again is impossible. For he who often fails to attain the end of the Art of Grammar will not be a grammarian, and he who often fails to attain the end of the Art of Music will not be a musician. So also he who fails often to attain the
end of the Art of Rhetoric will not be an orator. But 87
the orator is more often loser than victor, and the more so the more capable he is, as those who have an unjust case join in flocking to him. Therefore the orator is not an orator.—Also, he who fails to attain the end of rhetoric will not be praised; but we sometimes praise an orator when defeated; "victory," then, is not the end of rhetoric.

Consequently, if rhetoric has neither a subject- 88
matter[a] which it treats technically nor an end to which it is directed,[b] rhetoric will not exist; but, as we have established, it has neither subject-matter nor end; therefore rhetoric does not exist.

One may also raise difficulties for them based on 89
the parts of rhetoric. The parts of rhetoric, they say, are "the juridical, the deliberative and the laudatory," and of these "the juridical" has justice for its end, "the deliberative" has the advantageous,

[a] *Cf.* §§ 48 ff. [b] *Cf.* §§ 60 ff.

[1] ἐπαπορήσειε cj. Bekk.; ἐναπορήσειε mss., Bekk.

τοῦ δὲ ἐγκωμιαστικοῦ τὸ καλόν. ὅπερ εὐθέως
90 ἄπορόν ἐστιν. εἴπερ γὰρ ἄλλο τι καθέστηκεν ἡ δικανικὴ ὑπόθεσις καὶ ἄλλο τι ἡ συμβουλευτικὴ καὶ οὐ ταὐτὸν ἡ ἐγκωμιαστική, πάντως τὸ τέλος τῆς δικανικῆς οὐκ ἂν εἴη καὶ τῆς συμβουλευτικῆς τέλος, καὶ τὸ ταύτης οὐκ ἔσται τῆς ἐγκωμιαστικῆς, καὶ ἐναλλάξ. ἐπεὶ οὖν τῆς συμβουλευτικῆς τέλος ἐστὶ τὸ συμφέρον, οὐκ ἂν εἴη τοῦτο τῆς δικανικῆς τελος. ἦν δέ γε τῆς δικανικῆς τέλος τὸ δίκαιον·
91 οὐκ ἄρα τὸ δίκαιόν ἐστι συμφέρον. καὶ πάλιν, ἐπεὶ ὡς τὰ μέρη ταῦτα διαφέρει ἀλλήλων, οὕτω καὶ τὰ τέλη διοίσει, παρόσον τῆς ἐγκωμιαστικῆς τέλος ἐστὶ τὸ καλόν, τῆς δὲ δικανικῆς τὸ δίκαιον, ἐνδέχεται τὸ καλὸν εἶναι μὴ δίκαιον καὶ τὸ
92 δίκαιον μὴ καλόν[1]· ὅπερ ἄτοπον. καὶ μὴν εἰ τῆς ὅλης ῥητορικῆς τέλος ἐστὶ τὸ πείθειν, τοῦ δὲ δικανικοῦ τὸ δίκαιον καὶ τοῦ συμβουλευτικοῦ τὸ συμφέρον καὶ τοῦ ἐγκωμιαστικοῦ τὸ καλόν, οὐ πάντως τὸ δίκαιον ἔσται πιθανόν, οὐδὲ τὸ συμφέρον, οὐδὲ τὸ καλόν· ὅπερ μάχεται τῷ διὰ παντὸς τὴν ῥητορικὴν ἐφίεσθαι τοῦ πείθειν.

93 Ἄλλως τε ἐπὶ τοῦ δικανικοῦ ἤτοι διὰ δικαίων μόνον λόγων ἕλξει τοὺς δικαστὰς ἐπὶ τὸ τέλος ἡ ῥητορικὴ ἢ διὰ τῶν δικαίων ἅμα καὶ ἀδίκων. ἀλλ' εἰ μὲν διὰ τῶν δικαίων μόνον, ἀρετὴ γενήσεται· οὐχὶ δέ γε ἀρετή ἐστιν ἡ στοχαζομένη τῆς ὀχλικῆς πειθοῦς, ἐν ᾗ πολὺ τὸ εἰκαῖον καὶ ἐξαπατητικόν· οὐκ ἄρα διὰ τῶν δικαίων μόνον ἐπὶ
94 τὸ τέλος ἄγειν πέφυκε τοὺς ἀκούοντας. εἶτα οὐδὲ συστήσεται ἐξ ἐναντίας λόγος ἀεί ποτε τὸ δίκαιον

[1] τῆς δὲ . . . μὴ καλόν Herv. (simil. Fabr.): **μὲν εἶναι καὶ τὸ δίκαιον καλόν** MSS., Bekk.

and "the laudatory" has the noble. But this is at once open to doubt. For if the juridical purpose is 90
one thing, and the deliberative another, and the laudatory something different, certainly the end of the juridical will not also be the end of the deliberative, and the end of the latter will not be that of the laudatory, and *vice versa.* Since, then, the end of the deliberative is the advantageous, this will not be the end of the juridical. But the end of the juridical is the just; therefore the just is not advantageous.—And again, just as these parts differ, 91
so also the ends will differ; and therefore, inasmuch as the end of the laudatory is the noble and that of the juridical the just, it is possible for the noble not to be just and the just not to be noble; which is absurd.—Moreover, if persuasion is the end of rhetoric 92
as a whole, but the just of the juridical part, and the advantageous of the deliberative, and the noble of the laudatory, then certainly the just is not persuasive, nor is the advantageous, nor the noble; and this conflicts with the notion that rhetoric aims wholly at persuasion.

Furthermore, in the case of the juridical part, 93
rhetoric will draw the jurors towards its end either by just speeches only or by unjust as well as by just speeches. But if it is by just speeches only, it will be virtue; but what aims at the persuasion of the crowd is not virtue, for it contains much that is rash and misleading; it is not, therefore, of a nature to lead the hearers to its end by just speeches only.—Moreover, if it always pursues justice, no 94
contrary argument will be forthcoming, and when

ταύτης μεταδιωκούσης, τοῦ δ' ἐναντίου μὴ ὄντος λόγου οὐδὲ ῥητορική τις γενήσεται, ὥστε οὐδὲ ταύτῃ τοῖς δικαίοις μόνον χρήσεται λόγοις. καὶ μὴν οὐδὲ τοῖς ἀδίκοις, ἐπεὶ ἄδικος γενήσεται, καὶ πάλιν τοῦ ἐναντίου μὴ ὄντος λόγου ἀσύστατος ἔσται. λείπεται ἄρα δι' ἀμφοτέρων αὐτὴν βαδίζειν. ὃ πολλῷ τῶν προτέρων ἐστὶν ἀτοπώτερον· ἔσται γὰρ ἅμα ἀρετὴ καὶ κακία, ὃ τῶν ἀδυνάτων καθέστηκεν. οὐ τοίνυν ῥητέον μέρος εἶναί τι ῥητορικῆς δικανικὸν ὃ τέλος ἔχει τὸ δίκαιον.

95 Πρός γε μὴν τοῖς εἰρημένοις, εἴπερ ὑποδεῖξαι τὸ δίκαιον τοῖς δικασταῖς ἐν τῷ δικανικῷ μέρει τῆς ῥητορικῆς ὁ ῥήτωρ προτίθεται, ἤτοι αὐτόθεν ἐστὶ φαινόμενον καὶ ὁμόλογον τὸ δίκαιον ὅπερ ὑποδείκνυσιν, ἢ ἀμφισβητήσιμον. ἀλλὰ φαινόμενον μὲν οὐκ ἂν εἴποιεν· ἐπὶ τούτου γὰρ οὐ συνίσταται ὁ ῥητορικὸς λόγος, ἀναμφισβητήτου
96 καθεστῶτος. λείπεται οὖν τὸ ἀμφισβητήσιμον. ὃ πάλιν ἐστὶν ἄπορον· τοσοῦτον γὰρ ἀπέχουσιν οἱ εἰς τοὐναντίον ἐπιχειροῦντες τὴν ἀμφισβήτησιν λύειν ὡς καὶ ἐκ τῶν ἐναντίων αὐτὴν ἐπισφίγγειν, ἐπιθολοῦντες τὴν τῶν δικαστῶν γνώμην. καὶ τούτου πίστις ἡ περὶ Κόρακος φερομένη παρὰ
97 τοῖς πολλοῖς ἱστορία. νεανίας γὰρ πόθῳ ῥητορικῆς κατασχεθεὶς προσῆλθεν αὐτῷ τὸν ὁρισθησόμενον ὑπ' αὐτοῦ μισθὸν δώσειν ἐπαγγελλόμενος, ἂν τὴν πρώτην νικήσῃ δίκην. συμφωνίας δὲ γενομένης, καὶ τοῦ μειρακίου ἱκανὴν ἕξιν ἐμφαίνοντος ἤδη, ὁ μὲν Κόραξ ἀπῄτει τὸν μισθόν, ὁ δ' ἀντέλεγεν. ἀμφότεροι δὲ παρελθόντες εἰς τὸ δικαστήριον ἐκρίνοντο, ὅτε καὶ πρῶτόν φασι τὸν Κόρακα τοιαύτῃ τινὶ χρῆσθαι ἐπιχειρήσει, λέγοντα ὡς

there exists no contrary speech neither will rhetoric exist, so that on this showing, too, it will not use just speeches only.—Nor, again, will it use unjust speeches, for if so it will be unjust, and as the contrary speech is again non-existent, it will be non-existent. It remains, then, to say that it proceeds by means of both. But this is far more absurd than the former suppositions; for it will be at once virtue and vice, which is a thing impossible. One must, then, deny that there is a juridical part of rhetoric which has justice for its end.

In addition to what has been said, if the orator 95
proposes to explain to the judges what is just by the juridical part of rhetoric, the justice which he explains is either evident of itself and agreed or it is disputable. But they will not say that it is evident; for about this no rhetorical speech is composed, since
it is undisputed. It remains, then, to say that it is dis- 96
putable. But this again is open to doubt. For those who argue on opposite sides are so far from settling the dispute that they even intensify it by their contradictions and befog the minds of the judges. And there is evidence for this in the story commonly told about
Corax.[a] A young man seized with a desire for rhetoric 97
went to him and promised that he would pay him the fee he would charge, if he should win his first case. And when the compact was made, and the youth was now displaying sufficient skill, Corax demanded his fee, but the other said "No." Both then repaired to the court and had the case tried; and then, it is said, Corax first used an argument of this kind,—

[a] Noted Sicilian rhetorician, *circ.* 460 B.C. The same story is told of Protagoras by Diog. Laert. ix. 56.

ἐάν τε νικήσῃ ἐάν τε μή, λαβεῖν ὀφείλει τὸν μισθόν, νικήσας μὲν ὅτι ἐνίκησεν, λειφθεὶς δὲ κατὰ τὸν τῆς συμφωνίας λόγον· ὡμολόγησε γὰρ αὐτῷ ὁ ἀντίδικος ἀποδώσειν τὸν μισθὸν ἐὰν τὴν πρώτην νικήσῃ δίκην, ἣν αὐτόθεν νικήσας ὀφείλει τὴν
98 ὑπόσχεσιν χρεωλυτεῖν. θορυβησάντων δὲ τῶν δικαστῶν ὡς δίκαια λέγοντος αὐτοῦ, παραλαβὼν τοὺς λόγους ὁ νεανίας τῷ αὐτῷ ἐπιχειρήματι, μηδὲν μεταθείς, ἐχρῆτο· "ἐάν τε" γὰρ "νικήσω" φησὶν "ἐάν τε νικηθῶ, οὐκ ὀφείλω τὸν μισθὸν ἀποδοῦναι Κόρακι, νικήσας μὲν ὅτι ἐνίκησα, λειφθεὶς δὲ κατὰ τὸν τῆς συμφωνίας λόγον· ὑπεσχόμην γὰρ ἀποδώσειν τὸν μισθὸν ἐὰν τὴν πρώτην νικήσω δίκην, λειφθεὶς δὲ οὐκ ἀποδώσω."
99 εἰς ἐποχὴν δὲ καὶ ἀπορίαν ἐλθόντες οἱ δικασταὶ διὰ τὴν ἰσοσθένειαν τῶν ῥητορικῶν λόγων ἀμφοτέρους ἐξέβαλον τοῦ δικαστηρίου, ἐπιφωνήσαντες τὸ "ἐκ κακοῦ κόρακος κακὸν ᾠόν."

100 Οἷος δέ ἐστιν ὁ περὶ τοῦ δικανικοῦ μέρους λόγος, τοιοῦτος γένοιτ' ἂν καὶ ὁ περὶ τοῦ συμβουλευτικοῦ, ἵνα μὴ μακρηγορῶμεν. τὸ μὲν γὰρ ἐγκωμιαστικόν, σὺν τῷ ταῖς αὐταῖς ἀπορίαις ὑπάγεσθαι, ἔτι καὶ
101 ἀμέθοδόν ἐστιν. ἐπεὶ γὰρ οὔτε πάντες ἄνθρωποι ἐγκωμιάζεσθαι θέλουσιν οὔτε ἐπὶ τοῖς αὐτοῖς, δεῖ τὸν μέλλοντα καλῶς ἐγκωμιάζειν εἰδέναι τὴν τοῦ ἐγκωμιαζομένου διάθεσιν· οὐ πᾶν δέ γε τὸ καθ' ἕτερον κίνημα ληπτόν ἐστιν ἑτέρῳ. καὶ ἄλλως οἱ ῥήτορες οὐ παραδεδώκασί τινα μέθοδον δι' ἧς εἰσόμεθα τὸ πότε καὶ τίνα ἐγκωμιαστέον ἐστίν· οὐκ ἄρα δυνατὸν ἀπὸ ῥητορικῆς ὑγιῶς ἐγκωμιάζειν.
102 ἤτοι τε ἐπὶ τοῖς μὴ οὖσιν ἀγαθοῖς δοκοῦσι δὲ εἶναι,

that whether he won the case or lost it he ought to receive the fee ; if he won, because he had won, and if he lost, in accordance with the terms of the compact ; for his opponent had agreed to pay him the fee if he won his first case, so that if he did win it he was thereby bound to discharge the debt. And after 98
the judges had applauded him for speaking justly the young man in turn began his speech and used the same argument, altering nothing : "Whether I win," he says, "or whether I am beaten, I am not bound to pay Corax the fee ; if I win, because I have won ; and if I lose, in accordance with the terms of the compact ; for I promised to pay the fee if I should win my first case, but if I should lose I shall not pay." The judges 99
then, thrown into a state of suspense and perplexity owing to the equipollence of the rhetorical arguments, drove them both out of the court, crying "A bad egg from a bad crow !" [a]

To save a long account we may say that the argu- 100
ment concerning "the deliberative" part will be similar to that concerning the juridical. As to the "laudatory," or eulogistic, part, besides being subject to the same doubts, it is also lacking in method. For 101
since not all men wish to be eulogized, nor on the same grounds, he who is going to eulogize well must know the disposition of the person who is being eulogized ; but not every emotion in one man is perceptible by another. Moreover, the Orators have not furnished us with any method whereby we may know when and whom we ought to eulogize ; it is impossible, therefore, to eulogize properly by means of rhetoric.—Also, the orator will eulogize on account 102

[a] A play on the name Corax, which is the Greek for "carrion-crow" (or "daw"). *Cf.* the Latin saying "mali corvi malum ovum."

ἢ ἐπὶ τοῖς κατ' ἀλήθειαν οὖσιν ἐγκωμιάσει ὁ ῥήτωρ. οὔτε δὲ ἐπὶ τοῖς μὴ οὖσι, προσδιαφθείρει γὰρ τοὺς ἐγκωμιαζομένους, οὔτε ἐπὶ τοῖς οὖσιν· ἀγνοεῖ γὰρ ταῦτα, ὅτε καὶ τοῖς φιλοσόφοις διὰ τὴν ἀνεπίκριτον περὶ αὐτῶν μάχην ἐστὶν ἀκατάληπτα. οὐκ ἄρα δύναταί τινα ἐγκωμιάζειν ὁ
103 ῥήτωρ. οἵ τε μὴ εἰδότες ἐφ' οἷς ἐγκωμιαστέον ἐστὶν οὐδὲ ἐγκωμιάζειν δύνανται· οἱ δέ γε ῥήτορες οὐκ ἴσασιν ἐφ' οἷς ἐγκωμιάζειν δεῖ, καθὼς παραστήσομεν· τοίνυν οὐδὲ ἐγκωμιάζειν δυνήσονται. ἐγκωμιαστέον γάρ φασιν ἀπὸ γένους καὶ κάλλους καὶ πλούτου καὶ πολυτεκνίας καὶ τῶν ἐοικότων, ἀνάπαλίν τε ψεκτέον ἀπὸ δυσγενείας καὶ δυσμορ-
104 φίας καὶ πενιχρότητος. ὅπερ εὔηθες· δεῖ γὰρ ἡμᾶς ἀπὸ τῶν παρ' ἡμᾶς γινομένων τοὺς ἐπαίνους ἕλκειν καὶ ψόγους, εὐγένεια δὲ καὶ εὐτυχία κάλλος τε καὶ πολυτεκνία καὶ τὰ τοιαῦτα οὐκ ἔστι παρ' ἡμᾶς γινόμενα, ὥστε οὐκ ἐπαινετέον ἀπ' αὐτῶν, ἐπεί τοί γε εἰ ψιλῶς ἐπαινετέον τὴν εὐγένειάν ἐστι καὶ πολυτεκνίαν καὶ πᾶν τοιουτῶδες, ἐπαινετέος καὶ Βούσιρις καὶ Ἄμυκος καὶ Ἀνταῖος οἱ ξενοκτόνοι, ὅτι Ποσειδῶνος ἦσαν υἱεῖς, ἐπαινετὴ δὲ
105 καὶ ἡ Νιόβη, ὅτι πολύτεκνος. ἐναντίως τε εἰ ἡ ἀμορφία καὶ ἡ πενία ψεκτόν, ψεκτέος μὲν Ὀδυσσεύς, ὅτι χερνήτου λαβὼν σχῆμα

ἀνδρῶν δυσμενέων κατέδυ πόλιν,

ψεκτέος δὲ ὁ Διὸς Περσεύς, ὅτι πήραν περιηρτη-

[a] *Cf.* the encomiastic *Busiris* of Isocrates.

of things that are not good but seem so, or else on account of things which are really good. But he does not do so either on account of things which are not good,—for then he is further corrupting the persons eulogized,—or on account of those which are ; for of these he is ignorant, since even for philosophers they are non-apprehensible owing to the unsettled controversy about them. Therefore the Orator is not able
to eulogize anyone.—Also, those who do not know 103
on what account one should eulogize are unable to eulogize ; but the Orators do not know on what account one ought to eulogize, as we shall demonstrate ; hence they will not be able to eulogize. For they assert that one should eulogize on the ground of birth and beauty and wealth and abundance of children, and the like ; and conversely one should blame on the ground of low birth and ugliness and
poverty. But this is silly ; for we ought to attract 104
praise and blame from things inherent in ourselves, but noble birth and good fortune and beauty and abundance of children and things of that sort are not inherent in ourselves, so that we should not be praised on account of them ; for, to be sure, if we are to praise unconditionally noble birth and abundance of children and everything of that kind, we must praise those slayers of guests, Busiris [a] and Amycus and Antaeus, because they were sons of Poseidon, and also praise Niobê because she abounded in
children. Conversely, if ugliness and penury are to 105
be blamed, Odysseus is to be blamed because he took the form of a beggar [b] when

He entered the foemen's city,

and Perseus, the son of Zeus, is to be blamed because

[b] *Cf.* Hom. *Od.* iv. 244 ff.

μένος τὴν ἄνυδρον ὥδευε Λιβύην, καὶ Ἡρακλῆς, ὅτι λεοντῆν καὶ ξύλον ἐπὶ τοὺς ἄθλους ἐπήγετο.

106 Συνελόντι δὲ φάναι, δεδόσθω ταῦτα μέρη εἶναι τῆς ῥητορικῆς. ἀλλ' ἐπεὶ τὸ δίκαιον, ὅτι ἔστι δίκαιον, καὶ τὸ συμφέρον, ὅτι ἔστι συμφέρον, καὶ τὸ καλόν, ὅτι ἔστι καλόν, ἀποδείξει παρίσταται, καὶ οὐδέν ἐστιν ἡ ἀπόδειξις, οὐδὲ ῥητορική τι γενήσεται ἡ ἐπὶ τοιούτοις μέρεσι συνεστῶσα. ὅτι δὲ οὐδέν ἐστιν ἀπόδειξις, ἀκριβέστερον μὲν δείκνυται ἐν τοῖς σκεπτικοῖς ὑπομνήμασιν, ὑπο-
107 μνηστικώτερον δὲ καὶ νῦν παρασταθήσεται. εἰ γὰρ μηδὲν ὁ λόγος, οὐδὲ ἡ ἀπόδειξις ἔστι, ποιὸς λόγος οὖσα· οὐδὲν δέ γέ ἐστι λόγος, ὡς παρεστήσαμεν, διὰ τὸ μήτε ἐν φωναῖς μήτε ἐν ἀσωμάτοις λεκτοῖς ἔχειν τὴν ὑπόστασιν· οὐδὲ ἀπόδειξις ἄρα
108 ἔστιν. ἄλλως τε, εἰ ἔστιν, ἤτοι ἐναργὴς καθέστηκεν ἢ ἄδηλος. οὔτε δὲ ἐναργής ἐστιν· ἄδηλον γάρ τι περιέσχηκε, καὶ διὰ τοῦτό ἐστι διάφωνος, παντὸς τοῦ διαφωνουμένου πράγματος ἀδήλου
109 τυγχάνοντος. λείπεται ἄρα ἄδηλον αὐτὴν εἶναι. ἀλλ' εἰ τοῦτο, ἤτοι αὐτόθεν ληφθήσεται ἢ ἐξ ἀποδείξεως. οὔτε δὲ αὐτόθεν ληπτή ἐστιν (ἄδηλος γὰρ ἦν, τὸ δὲ ἄδηλον αὐτόθεν λαμβανόμενον ἄπιστον) οὔτε ἐξ ἀποδείξεως διὰ τὴν εἰς ἄπειρον
110 ἔκπτωσιν· οὐκ ἄρα τις ἔστιν ἀπόδειξις. μὴ οὔσης τε γενικῆς ἀποδείξεως οὐδὲ εἰδική τις ἔσται ἀπόδειξις, ὥσπερ καὶ μὴ ὄντος ζώου οὐδὲ ἄνθρωπος ἔστιν· γενικὴ δέ γε ἀπόδειξις οὐκ ἔστιν, ὡς παραστήσομεν· τοίνυν οὐδὲ ἄλλη τις γενήσεται τῶν ἐπ' εἴδους. ἐπεὶ γὰρ ἄδηλός ἐστιν, ὡς προεπελογισάμεθα, ὀφείλει διά τινος καταστῆναι. διὰ

he walked through waterless Libya with a wallet hung round him, and Heracles because he took with him on his labours a lion's skin and a club.

To speak briefly, then, let it be granted that these 106
are the parts of rhetoric. But since the fact that the just is just and that the advantageous is advantageous and that the noble is noble is established by proof, and proof is nothing, neither will rhetoric, which is composed of these parts, be anything. That proof is nothing is shown more precisely in our *Notes on Scepticism*,[a] and we shall now establish it more by way of a note or reminder. If speech is nothing, 107
neither does proof exist, being a kind of speech; but speech is nothing, as we have established,[b] because it subsists neither in utterance nor in incorporeal *lekta*; therefore proof does not exist.—Further, if it exists it 108
is either evident or non-evident.[c] But it is not evident; for it contains something non-evident, and because of this is disputable, as everything in dispute is non-evident. It remains then to say that it is non- 109
evident. But if so, it will be perceived either of itself or after proof. But it is not perceptible of itself (for it is non-evident, and the non-evident, if perceived of itself, is not to be trusted), nor after proof, because of the regress *ad infinitum*; therefore no proof exists.—Further, as generic proof[d] does not exist, no specific 110
proof will exist, just as if "animal" does not exist neither does "man" exist; but generic proof does not exist, as we shall establish; so no other, specific, proof will exist. For since it is non-evident, as we previously argued, it must be confirmed by some-

[a] *Cf. P.H.* ii. 134 ff., *Adv. Log.* ii. 299 ff.

[b] *Cf. Adv. Gram.* 135 ff.

[c] *Cf. P.H.* ii. 174 182.

[d] *Cf. P.H.* ii. 172

τίνος οὖν; ἤτοι γὰρ διὰ γενικῆς ἢ εἰδικῆς ἀπο-
111 δείξεως. οὔτε δὲ διὰ εἰδικῆς διὰ τὸ μήπω βέβαιον εἶναι τὴν τῆς γενικῆς ὕπαρξιν, οὔτε διὰ γενικῆς· αὕτη γάρ ἐστιν ἡ ἀμφισβητουμένη. οὐ τοίνυν ἔστι τις γενικὴ ἀπόδειξις. ᾧ ἕπεται τὸ μηδὲ τὴν εἰδικὴν ὑπάρχειν. καὶ ἄλλως, ἡ γενικὴ ἀπόδειξις εἰ μὲν λήμματά τινα ἔχει καὶ ἐπιφοράν, οὐδὲ γενική ἐστιν, εἰ δὲ οὐκ ἔχει, οὐδὲ κατασκευάσει
112 τι, πολὺ δὲ μᾶλλον οὐδὲ τὴν ἑαυτῆς ὕπαρξιν. ἥ τε τὴν ἀπόδειξιν πιστουμένη ἀπόδειξις ἤτοι ζητεῖται ἢ ἀζήτητός ἐστιν. ἀλλ' ἀζήτητος μὲν οὐκ ἂν εἴη διὰ τὰς ἔμπροσθεν εἰρημένας αἰτίας, ζητουμένη δὲ ὀφείλει ὑπ' ἄλλης κατασκευάζεσθαι, κἀκείνη πάλιν ὑπ' ἄλλης, καὶ τοῦτο εἰς ἄπειρον. οὐκ ἄρα ἔστι τις ἀπόδειξις.

113 Ἀλλὰ γὰρ καὶ πρὸς τὰ συνέχοντα θεωρήματα τῆς ῥητορικῆς ἀντειπόντες ἀπ' ἄλλης ἀρχῆς καὶ τῶν πρὸς τοὺς γεωμέτρας καὶ ἀριθμητικοὺς ἀποριῶν ἁπτώμεθα.

thing. By what then? Either by a generic or by a
specific proof. Not by a specific, since the existence 111
of the generic is as yet unconfirmed; nor yet by a
generic; for that is the matter in dispute. So then, no generic proof exists. From this it follows that neither does the specific proof exist.—And besides, if the generic proof has premisses and a conclusion it is not generic, and if it does not it will not prove anything, and what is much worse, it will not even prove
its own existence.—Also, the proof which confirms 112
the proof is either questioned or unquestioned. But, for the reasons stated above, it will not be unquestioned, and if questioned it must be proved by another proof, and this again by another, and so on *ad infinitum*. Therefore no proof exists.

So, now that we have refuted the main doctrines 113
of Rhetoric, let us make a fresh start and deal with the objections to be raised against the Geometers and the Arithmeticians.

Γ

ΠΡΟΣ ΓΕΩΜΕΤΡΑΣ

1 Ἐπεὶ οἱ γεωμέτραι συνορῶντες τὸ πλῆθος τῶν ἐπακολουθούντων αὐτοῖς ἀπόρων[1] εἰς ἀκίνδυνον εἶναι δοκοῦν καὶ ἀσφαλὲς πρᾶγμα καταφεύγουσι, τὸ ἐξ ὑποθέσεως αἰτεῖσθαι τὰς τῆς γεωμετρίας ἀρχάς, καλῶς ἂν ἔχοι καὶ ἡμᾶς τῆς πρὸς αὐτοὺς ἀντιρρήσεως ἀρχὴν τίθεσθαι τὸν περὶ τῆς ὑπο-
2 θέσεως λόγον. καὶ γὰρ ὁ Τίμων ἐν τοῖς πρὸς τοὺς φυσικοὺς τοῦτο ὑπέλαβε δεῖν ἐν πρώτοις ζητεῖν, φημὶ δὲ τὸ εἰ ἐξ ὑποθέσεώς τι ληπτέον. διόπερ καὶ ἡμᾶς οἰκεῖόν ἐστιν ἐκείνῳ στοιχοῦντας τὸ παραπλήσιον ποιεῖν ἐν τῇ πρὸς τοὺς ἀπὸ τῶν
3 μαθημάτων διεξόδῳ. τάξεως δὲ ἕνεκα προληπτέον ὅτι πολλαχῶς μὲν καὶ ἄλλως ὑπόθεσις προσαγορεύεται, τὰ νῦν δὲ ἀπαρκέσει τριχῶς λέγεσθαι, καθ᾿ ἕνα μὲν τρόπον ἡ δραματικὴ περιπέτεια, καθὸ καὶ τραγικὴν καὶ κωμικὴν ὑπόθεσιν εἶναι λέγομεν καὶ Δικαιάρχου τινὰς ὑποθέσεις τῶν Εὐριπίδου καὶ Σοφοκλέους μύθων, οὐκ ἄλλο τι καλοῦντες
4 ὑπόθεσιν ἢ τὴν τοῦ δράματος περιπέτειαν. καθ᾿ ἕτερον δὲ σημαινόμενον ὑπόθεσις προσαγορεύεται ἐν ῥητορικῇ ἡ τῶν ἐπὶ μέρους ζήτησις, καθὸ καὶ

[1] ἀπόρων cj. Bekk. ἀποριῶν MSS., Bekk.

BOOK III

AGAINST THE GEOMETERS

Since the Geometers, perceiving the multitude of 1
difficulties which beset them, take refuge in a method which seems to be free from danger and safe, namely, to beg by " hypothesis " the principles of geometry, it will be well for us, too, to begin our attack against them with the argument about " hypothesis." For 2
Timon, in his book *Against the Physicists*, assumed that one ought to raise this question first of all, —I mean, whether anything should be accepted from " hypothesis." Hence it is proper for us, in conformity with him, to do likewise in our treatise against these Mathematicians. And, for the sake of 3
due order, one must premise that the word " hypotheses " is used in a number of different senses ; but it will be enough now to mention three : in one sense it means the *peripeteia* (or " argument " or " plot ") of a drama, as we say that there is a tragic or a comic " hypothesis," and certain " hypotheses " of Dicaearchus [a] of the stories of Euripides and Sophocles, meaning by " hypothesis " nothing else than the *peripeteia* of the drama. And " hypothesis " is used 4
with another signification in rhetoric, as investigation of particulars, in which sense the sophists are wont

[a] A voluminous writer, disciple of Aristotle.

οἱ σοφισταὶ πολλάκις εἰώθασιν ἐν ταῖς διατριβαῖς
λέγειν "θετέον ὑπόθεσιν." οὐ μὴν ἀλλὰ καὶ κατὰ
τρίτην ἐπιβολὴν ὑπόθεσιν καλοῦμεν ἀρχὴν ἀπο-
δείξεων, αἴτησιν οὖσαν πράγματος εἰς κατασκευήν
5 τινος. οὕτω γοῦν τρισὶν ὑποθέσεσι κεχρῆσθαί
φαμεν τὸν Ἀσκληπιάδην εἰς κατασκευὴν τῆς τὸν
πυρετὸν ἐμποιούσης ἐνστάσεως, μιᾷ μὲν ὅτι νοητοί
τινές εἰσιν ἐν ἡμῖν πόροι, μεγέθει διαφέροντες
ἀλλήλων, δευτέρᾳ δὲ ὅτι πάντοθεν ὑγροῦ μέρη
καὶ πνεύματος ἐκ λόγῳ θεωρητῶν ὄγκων συν-
ηράνισται δι' αἰῶνος ἀνηρεμήτων, τρίτῃ δὲ ὅτι
ἀδιάλειπτοί τινες εἰς τὸ ἐκτὸς ἐξ ἡμῶν ἀποφοραὶ
γίνονται, ποτὲ μὲν πλείους ποτὲ δὲ ἐλάττους πρὸς
τὴν ἐνεστηκυῖαν περίστασιν.
6 Ἀλλὰ γὰρ τοσαυταχῶς νοουμένης τὰ νῦν τῆς
ὑποθέσεως, πρόκειται τὰ νῦν ζητεῖν οὐ μὰ Δία
περὶ τῆς δραματικῆς διατάξεως, οὐδὲ περὶ τῆς
παρὰ τοῖς ῥήτορσι ζητήσεως, ἀλλὰ περὶ τῆς ἐν
τέλει λεχθείσης ὑποθέσεως, ἣν ἀρχὴν ἀποδείξεως
συμβέβηκεν εἶναι· ταύτην γὰρ καὶ οἱ γεωμέτραι
λαμβάνουσι τὴν ὑπόθεσιν, βουλόμενοί τι γεω-
7 μετρικῶς ἀποδεῖξαι. διόπερ εὐθὺς ῥητέον ὅτι
καὶ ἐπεὶ οἱ ἐξ ὑποθέσεως λαμβάνοντές τι καὶ
χωρὶς ἀποδείξεως ψιλῇ μόνον ἀρκοῦνται φάσει
πρὸς τὴν ταύτης πίστιν, πεύσεταί τις αὐτῶν
8 τοιούτῳ τινὶ χρώμενος ἐπιλογισμῷ. ἤτοι ἰσχυρόν
ἐστι καὶ βέβαιον πρὸς πίστιν τὸ ἐξ ὑποθέσεώς τι
λαβεῖν ἢ ἄπιστόν τε καὶ ἀσθενές. ἀλλ' εἰ μὲν
ἰσχυρόν, καὶ τὸ ἀντικείμενον ἐξ ὑποθέσεως ληφθὲν
πιστὸν γενήσεται καὶ βέβαιον, ὥστε θήσομεν τὰ

to say often in their discourses, " One must posit the hypothesis." Moreover, in a third application we term the starting-point of proofs " hypothesis," it being the postulating something for the purpose of
proving something. Thus we say that Asclepiades [a] 5
made use of three " hypotheses " to demonstrate the initial condition which produces fever,—the first, that there exist in us certain intelligible (or " non-perceptible ") passages, differing from one another in size ; the second, that particles of moisture and air are collected from all sides out of corpuscles perceived by reason and eternally in motion ; the third, that certain unceasing effluvia are emitted from within us to the outside air, these being now more, now less, in number according to the condition prevailing at the moment.

Well then, " hypothesis " being now conceived in 6
these three ways, we certainly do not now propose to inquire about the arrangement in dramas, nor about oratorical questionings, but about the " hypothesis " in the sense mentioned last, which was " the starting-point of proof " ; for this is the " hypothesis " which the Geometers adopt when they wish to prove any-
thing geometrically. Consequently, we must state 7
at once that since those who assume a thing by hypothesis are satisfied with mere assertion, without proof, for its confirmation, one will interrogate them,
employing some such reasoning as this :—Assuming 8
a thing by hypothesis is either a strong and firm confirmation or unreliable and weak.[b] But if it is strong, its contrary, when assumed by hypothesis, will also be reliable and firm, so that we shall be posi-

[a] *Cf. P.H.* iii. 32, ii. 140.
[b] *Cf. P.H.* i. 173.

μαχόμενα. εἰ δὲ ἐπὶ τοῦ τὸ ἐναντίον ἐξ ὑποθέσεως λαμβάνοντος χωρὶς ἀποδείξεως ἄπιστόν ἐστιν ἡ ὑπόθεσις, ἄπιστος γενήσεται καὶ ἐπ' ἐκείνου, ὥστε οὐδέτερον αὐτῶν θήσομεν. οὐ τοίνυν ληπτέον
9 ἐστὶν ἐξ ὑποθέσεώς τι. καὶ μὴν τὸ ὑποτιθέμενον πρᾶγμα ἤτοι ἀληθές ἐστι καὶ τοιοῦτον ὁποῖον αὐτὸ ὑποτιθέμεθα ἢ ψεῦδος. ἀλλ' εἰ μὲν ἀληθές ἐστι, μηδὲ αἰτώμεθα αὐτό, εἰς πρᾶγμα ὑποψίας πλῆρες καταφεύγοντες, τὴν ὑπόθεσιν, ἀλλ' αὐτόθεν λαμβάνωμεν, ἐπείπερ οὐθεὶς τἀληθῆ καὶ ὄντα ὑποτίθεται, καθάπερ οὐδὲ τὸ νῦν ἡμέραν εἶναι ἢ ἐμὲ διαλέγεσθαι καὶ ἀναπνεῖν· ἡ γὰρ περιφάνεια τούτων τῶν πραγμάτων αὐτόθεν βέβαιον ἔχει τὴν θέσιν καὶ οὐ δισταζομένην τὴν ὑπόθεσιν. ὥστε εἰ ἀληθές ἐστι τὸ πρᾶγμα, μηδὲ αἰτώμεθα
10 αὐτὸ ὡς μὴ ὂν ἀληθές. εἰ δ' οὐκ ἔστι τοιοῦτο ἀλλὰ ψεῦδος καθέστηκεν, οὐδὲν ὄφελος ἀνακύψει ἐκ τῆς ὑποθέσεως· κἂν γὰρ μυριάκις αὐτὸ ὑποτιθώμεθα, σαθροῖς, ὥς φασί, θεμελίοις [οὐκ][1] ἀκολουθήσει τὸ συμπέρασμα τῆς ζητήσεως ἐξ
11 ἀνυπάρκτων ὁρμωμένης ἀρχῶν. οὐ μὴν ἀλλ' εἴ τις οἷς ἂν ὑποθῆται, τούτοις τὰ ἀκολουθοῦντα πιστὰ τυγχάνειν ἀξιώσει, μήποτε πᾶσαν ἀναιρεῖ ζήτησιν. εὐθέως γὰρ ὑποθήσεται ἕκαστος ἡμῶν τὸ τὰ τρία τέσσαρα εἶναι, καὶ τούτου δοθέντος συνάξει ὅτι καὶ τὰ ἓξ ὀκτώ ἐστιν· εἰ γὰρ τὰ τρία τέσσαρά ἐστι, τὰ ἓξ ὀκτὼ γενήσεται· ἀλλὰ μὴν τὰ τρία τέσσαρά ἐστιν, ὡς ἡ ὑπόθεσις δίδωσιν· τὰ
12 ἄρα ἓξ ὀκτώ ἐστιν. πάλιν τε αἰτήσομεν ὅτι μένει τὸ κινούμενον, καὶ συγχωρηθέντος τοῦ πράγματος συνάξομεν ὅτι ἡ φλὸξ ἠρεμεῖ· εἰ γὰρ τὸ κινούμενον

[1] [οὐκ] del. Heintz.

ting conflicting things simultaneously. But if the hypothesis is unreliable in the case of the man who assumes the contrary by hypothesis without proof, it will also be unreliable in the case of the other man, so that we shall posit neither of the things. Nothing, then, must be assumed by hypothesis.—Moreover, the 9
thing which is assumed is either true and such as we assume it to be, or false. But if it is true, let us not postulate it, fleeing for refuge to a thing which is highly suspicious—namely, hypothesis,—but let us accept it straightaway, for no one assumes *ex hypothesi* things true and actual, such as "Now it is day," or "I am talking and breathing"[a]; for the very obviousness of these facts does of itself make the statement firm and the assumption undisputed. So that if the thing is true, let us not postulate it as though it were not true. But if it is not true but is false, no 10
help will emerge from the hypothesis; for though we assume it a myriad times, on rotten foundations, as the saying goes, will follow the conclusion of the inquiry which starts from non-existent principles.—Moreover, if anyone shall maintain that the con- 11
clusions which follow from whatever assumptions are made are trustworthy, it is to be feared that he is destroying all inquiry. For example, each of us will assume that three is four, and, this being granted, will infer also that six is eight; for if three is four, six will be eight; but in fact, as the hypothesis grants, three is four; therefore six is eight. Again, we shall 12
postulate that what moves is at rest, and this being agreed we shall infer that the flame is stationary; for if what moves is at rest, the flame is stationary;

[a] *Cf. Adv. Log.* i. 242, ii. 144; *Adv. Rhet.* 65.

μένει, ἡ φλὸξ ἠρεμεῖ· τὸ δέ γε κινούμενον μένει· ἡ ἄρα φλὸξ ἠρεμεῖ. ἀλλ' ὃν τρόπον οἱ γεωμέτραι ἀτόπους ἐροῦσιν εἶναι ταύτας τὰς ὑποθέσεις (βέβαιον γὰρ εἶναι δεῖν τὸν θεμέλιον, ἵνα συνομολογηθῇ καὶ τὸ ἀκόλουθον), οὕτω καὶ ἡμεῖς πάντα τὰ ὑποθετικῶς αὐτοῖς λαμβανόμενα οὐ προσησόμεθα
13 *χωρὶς ἀποδείξεως. ἄλλως τε, εἰ βέβαιόν ἐστι καὶ πιστὸν τὸ ὑποτιθέμενον ᾗ ὑποτίθεται, μὴ ταῦτα ὑποτιθέσθωσαν ἐξ ὧν ἀποδείξουσίν τι, ἀλλ' αὐτὸ τὸ ἀποδεικνύμενον, τουτέστι μὴ τὰ λήμματα τῆς ἀποδείξεως ἀλλὰ τὴν ἐπιφοράν· ὃ γὰρ δύναται πρὸς πίστιν αὐτοῖς ἐπὶ τῶν ἐκκαλυπτόντων ἡ ὑπόθεσις, τοῦτο δυνήσεται καὶ ἐπὶ τῶν ἐκκαλυπτομένων ἐκ τῆς ἀποδείξεως πραγμάτων. εἰ δ' ἄπιστόν ἐστι, κἂν πολλάκις ὑποτεθῇ, τὸ τῆς ἀποδείξεως συμπέρασμα χωρὶς ἀποδείξεως, ἄπιστον γενήσεται καὶ τὸ εἰς κατασκευὴν τούτου λαμβανό-*
14 *μενον, εἰ μὴ δι' ἀποδείξεως διδάσκοιτο. νὴ Δί', ἀλλ' εἴπερ, φασί, τὸ ἀκολουθοῦν ταῖς ὑποθέσεσιν ἀληθὲς εὑρίσκεται, πάντως καὶ τὰ ὑποτεθέντα, τουτέστιν οἷς ἐπηκολούθησεν, ἀληθῆ γενήσεται. ὃ πάλιν ἐστὶν εὔηθες· πόθεν γὰρ ὅτι τὸ ἀκολουθοῦν τισὶν ἐν ἀποδείξει πάντως ἀληθές ἐστιν; ἢ γὰρ ἐξ αὐτοῦ μαθόντες ἐκείνου τοῦτ' ἐροῦσιν, ἢ ἐκ τῶν*
15 *οἷς ἠκολούθησε λημμάτων. ἀλλ' ἐξ αὐτοῦ μὲν οὐκ ἂν εἴποιεν. ἄδηλον γάρ ἐστι, τὸ δὲ ἄδηλον ἐξ αὐτοῦ πιστὸν οὐκ ἔστιν· ἀποδεικνύναι γοῦν τοῦτο ἐπιβάλλονται ὡς μὴ ἐν αὐτῷ τὴν πίστιν ἔχον. καὶ μὴν οὐδ' ἐκ τῶν λημμάτων· περὶ γὰρ τούτων ἐστὶν ἡ πᾶσα διαμάχη, καὶ μηδέπω αὐτῶν πεπιστωμένων*[1] *οὐδὲ τὸ ἀποδεικνύμενον ἐξ αὐτῶν*

[1] *πεπιστωμένων* cj. Bekk. : *πεπιστευμένων* mss., Bekk.

but what moves is at rest; therefore the flame is stationary. But just as the Geometers will say that these hypotheses are absurd (for the foundation must be firm in order that the inference which follows may be agreed), so too we shall refuse to accept any of their hypothetical assumptions without proof. More- 13
over, if the assumption, as assumed, is firm and trustworthy, let them not assume the things from which they will prove something, but the thing proved itself,—that is, not the premisses of the proof but its conclusion; for the power for confirmation which their hypothesis possesses in the case of the things which reveal, the same power it will possess in the case of the things revealed by the proof. But if the conclusion of the proof without proof is untrustworthy, though it be assumed many times over, that which is assumed in order to demonstrate it will also be untrustworthy unless it be taught by means of proof.—But in Heaven's name, they say, if what 14
follows the hypotheses is found to be true, certainly the things assumed—that is, the things which it has followed—will be true. But this again is silly; for how do we know that that which follows certain things in a proof is in all cases true? For they will assert this as having learnt it either from the thing itself or from the premisses which it followed. But they will 15
not assert it from itself. For it is non-evident, and the non-evident is not of itself trustworthy; at any rate they essay to prove it, as though it were not of itself convincing. Nor yet from the premisses; for the whole controversy is about these, and while they are still unconfirmed the thing which is being proved

16 βέβαιον εἶναι δύναται. ἔτι οὐδ' ἂν τὸ λῆγον ᾖ ἀληθές, εὐθὺς καὶ τὸ ἡγούμενόν ἐστι τοιοῦτον. ὥσπερ γὰρ τῷ ἀληθεῖ πέφυκεν ἀληθὲς ἐπακολουθεῖν καὶ ψεύδει ψεῦδος, οὕτως ἠξίωται καὶ ψεύδει ἀληθὲς συνεισάγεσθαι, καθάπερ [ἐν] τῷ πέτασθαι τὴν γῆν, ψεύδει ὄντι, τὸ εἶναι τὴν γῆν ἀληθὲς
17 ὑπάρχον εἵπετο. ὅθεν οὐκ εἰ τὸ λῆγόν ἐστιν ἀληθές, πάντως καὶ τὸ ἡγούμενον ἀληθές, ἀλλ' ἐνδέχεται τοῦ λήγοντος ἀληθοῦς ὄντος τὸ ἡγούμενον ὑπάρχειν ψεῦδος.

Καὶ δὴ ὅτι μὲν οὐκ εὖ ποιοῦσιν οἱ ἀπὸ τῶν μαθημάτων ἐξ ὑποθέσεως λαμβάνοντες τὰς ἀρχὰς τῆς ἀποδείξεως καὶ ἑκάστου θεωρήματος, ἐπιφθεγγόμενοι τὸ δεδόσθω, διὰ τούτων αὐτάρκως
18 κατεσκεύασται· μετελθόντες δὲ ἑξῆς διδάσκωμεν ὅτι ψευδεῖς καὶ ἀπιθάνους αὐτῶν συμβέβηκεν εἶναι τὰς ἀρχὰς τῆς τέχνης. καὶ δὴ πολλῶν εἰς τοῦτο δυναμένων λέγεσθαι, ὡς ἐναρχόμενοι τῆς ὑφηγήσεως εἴπομεν, τούτοις προσαχθήσεται τὰ τῆς ἀπορίας ὧν ἀναιρουμένων καὶ τὰ λοιπὰ συναναιρεθήσεται. ἐπεὶ οὖν τῶν ἀρχῶν διαβληθεισῶν οὐδὲ αἱ κατὰ μέρος ἀποδείξεις αὐτοῖς δύνανται προκόπτειν, λέγωμεν τὰ ἁρμόζοντα πρὸς τὰς ἀρχάς.

19 Εὐθέως τοίνυν ὡς πρῶτόν τι καὶ στοχειωδέστατον διδάσκουσιν ἡμᾶς ὅτι σῶμα μέν ἐστι τὸ τὰς τρεῖς ἔχον διαστάσεις, μῆκος πλάτος βάθος, ὧν πρώτη μὲν διάστασίς ἐστιν ἡ κατὰ μῆκος ἄνωθεν κάτω, δευτέρα δὲ ἡ κατὰ πλάτος ἀπὸ δεξιῶν ἐπ' ἀριστερά, τρίτη δὲ ἡ κατὰ βάθος ἀπὸ τῶν πρόσω εἰς τοὐπίσω. ὥστε τῶν τριῶν τούτων ἓξ γίνεσθαι παρατάσεις, δύο καθ' ἑκάστην, τῆς

by means of them cannot be firm.—Further, even if 16
the consequent is true, the antecedent is not inevitably true. For as the true naturally follows the true, and the false the false, so it is maintained that the true is a consequence of the false,—for example, that "the earth exists," which is true, follows "the earth flies," which is false. Hence, if the consequent is 17
true, the antecedent is not in all cases true, but when the consequent is true it is possible for the antecedent to be false.

So now, by these arguments it has been sufficiently established that the Mathematicians do no good by assuming *ex hypothesi* the principles of proof and of each theorem, repeating the formula "Let it be granted." Passing on, let us show in the next place 18
that the principles of their art are in fact false and incredible. Now many arguments can be used to prove this, as we said when commencing our exposition, but our doubts shall be cast on those principles the destruction of which will involve that of the rest. So, since their particular proofs cannot go forward when the principles are under suspicion, let us state suitable arguments against the principles.

To start with they tell us, as a primary and most 19
fundamental fact, that "body" is that which has three dimensions,[a]—length, breadth, depth,—and of these the first dimension, that of length, is up and down, the second, that of breadth, is from right to left, the third, that of depth, from before to behind. Thus there are six extensions of these three, two in

[a] *Cf. P.H.* iii. 39 f.; *Adv. Phys.* i. 367 ff.

μὲν πρώτης τὴν ἄνω καὶ κάτω, τῆς δὲ δευτέρας τὴν ἐν ἀριστερᾷ καὶ ἐν δεξιᾷ, τῆς δὲ τρίτης τὴν πρόσω καὶ ὀπίσω. στιγμῆς μὲν γὰρ ῥυείσης γραμμὴν γίνεσθαί φασι, γραμμῆς δ' ἐπιφάνειαν,
20 ἐπιφανείας δὲ στερεὸν σῶμα. παρὸ καὶ ὑπογράφοντες λέγουσι στιγμὴν μὲν εἶναι σημεῖον ἀμερὲς καὶ ἀδιάστατον ἢ πέρας γραμμῆς, γραμμὴν δὲ μῆκος ἀπλατὲς ἢ πέρας ἐπιφανείας, ἐπιφάνειαν
21 δὲ πέρας σώματος ἢ πλάτος ἀβαθές. τάξει οὖν ἀναλαβόντες περὶ στιγμῆς λέγωμεν πρῶτον, εἶτα περὶ γραμμῆς, τὸ δὲ μετὰ τοῦτο περὶ ἐπιφανείας καὶ σώματος· τούτων γὰρ ἀναιρουμένων οὐδ' ἡ γεωμετρία γενήσεται τέχνη, μὴ ἔχουσα τὰ ἐφ' οἷς ἡ σύστασις αὐτῆς δοκεῖ προκόπτειν.

22 Ἡ τοίνυν στιγμή, ἥν φασι σημεῖον ἀδιάστατον ὑπάρχειν, ἤτοι σῶμα νοεῖται ἢ ἀσώματον. καὶ σῶμα μὲν οὐκ ἂν εἴη κατ' αὐτούς· τὰ γὰρ μὴ ἔχοντα διάστασιν οὐκ εἶναι σώματα. λείπεται οὖν ἀσώματον αὐτὴν ὑπάρχειν, ὃ πάλιν ἐστὶν ἀπίθανον. τὸ μὲν γὰρ ἀσώματον οὐδενὸς νοεῖται γεννητικὸν ὡσανεὶ ἀθιγὲς καθεστώς, ἡ δὲ στιγμὴ νοεῖται τῆς γραμμῆς γεννητική· οὐ τοίνυν ἐστὶ
23 σημεῖον ἀδιάστατον ἡ στιγμή. καὶ μὴν εἴπερ ὄψις τῶν ἀδήλων ἐστὶ τὰ φαινόμενα, ἐπεὶ οὐ δυνατὸν ἐν τοῖς φαινομένοις λαβεῖν τινὸς σημεῖον καὶ πέρας ἀδιάστατον, δῆλον ὡς οὐδ' ἐν τοῖς νοητοῖς ληφθήσεταί τι τοιοῦτον. ἐν δέ γε τοῖς αἰσθητοῖς οὐδὲν ἔστιν ἀδιάστατον λαβεῖν, ὡς
24 παραστήσω· ὥστ' οὐδ' ἐν τοῖς νοητοῖς. πᾶν τοίνυν τὸ ἐν τοῖς αἰσθητοῖς ὑποπῖπτόν τινος πέρας

each case,—up and down of the first, right and left of the second, before and behind of the third. For they assert that the line is produced by the flow of the point, the surface by that of the line, and the solid body by that of the surface. So in describing these 20
they say that " the point is a sign without parts or dimensions," or " the limit of a line," " the line is length without breadth," or " the limit of a surface," and " the surface is the limit of a body," or " breadth without depth." Taking these, then, in order, let us 21
speak first about the point, next about the line, and after that about the surface and body ; for if these are destroyed Geometry will not be an Art, as not possessing the conditions upon which success in its construction seems to depend.

Now the point, which they say is " a sign without 22
dimensions," is conceived as either a body or incorporeal.[a] And according to them it will not be a body; for things which have no dimension are not bodies. It remains, then, to say that it is incorporeal ; but this again is incredible. For the incorporeal, as being impalpable, is conceived as generative of nothing, but the point is conceived as generative of the line ; so the point is not a sign without dimensions.—More- 23
over, if apparent things are " the vision of things non-evident," then, since in apparent things it is impossible to perceive a limit of anything or sign which is without dimensions, it is plain that no such thing will be perceived in intelligible things either. But in fact, as I shall establish, it is impossible to perceive in things sensible anything without dimensions ; so that it is also impossible in intelligibles. Now everything 24
which is perceived in sensibles as the limit and sign

[a] *Cf. Adv. Phys.* i. 377.

καὶ σημεῖον σὺν τούτῳ καταλαμβάνεταί τινος
ἄκρον, σὺν τῷ καὶ μέρος ἐκείνου, οὗπερ ἐστὶν
ἄκρον, ὑπάρχειν· ἐὰν γοῦν ἀφέλωμεν αὐτό, μειω-
θήσεται τὸ ἀφ' οὗ ἡ ἀφαίρεσις. τὸ δὲ μέρος τινὸς
ὑπάρχον εὐθὺς καὶ συμπληρωτικὸν αὐτοῦ καθέστη-
κεν, ὃ δέ ἐστί τινος συμπληρωτικόν, πάντως αὔξει
τὸ μέγεθος ἐκείνου, καὶ ὅ ἐστι μεγέθους αὐξη-
25 τικόν, τοῦτο ἐξ ἀνάγκης ἔχει μέγεθος. πᾶν ἄρα
τὸ ἐν τοῖς αἰσθητοῖς σημεῖόν τινος καὶ ἄκρον
μέγεθος ἔχον οὐκ ἔστιν ἀδιάστατον. ὅθεν εἰ καὶ
τὸ νοητὸν μεταβατικῶς ἀπὸ τοῦ αἰσθητοῦ νοοῦμεν,
σὺν τούτῳ καθεστὼς σημεῖον καὶ πέρας γραμμῆς
αὐτὸ νοήσομεν, σὺν τῷ καὶ συμπληρωτικὸν αὐτῆς
ὑπάρχειν, ὥστε καὶ αὐτὸ διάστασιν ἕξει πάντως,
26 ὅ γε διαστάσεώς ἐστι περιποιητικόν. ἄλλως τε
τὴν ἀπὸ τοῦ κέντρου ἐκβληθεῖσαν εὐθεῖάν φασι
περιαγομένην τῷ πέρατι ἑαυτῆς κυκλογραφεῖν τὴν
ἐπίπεδον. ἐπεὶ οὖν τὸ ἄκρον ταύτης τῆς εὐθείας
ἐστὶ σημεῖον, καὶ τοῦτο περιαγόμενον καταμετρεῖ
τὴν περιφέρειαν, ἔσται τοῦτο συμπληρωτικὸν τῆς
περιφερείας· ἡ δέ γε περιφέρεια διάστασιν εἶχεν·
τοίνυν καὶ τὸ συμπληρωτικὸν αὐτῆς σημεῖον ἕξει
27 τινὰ διάστασιν. ἥ γε μὴν σφαῖρα καθ' ἓν σημεῖον
ἀξιοῦται τῆς ἐπιπέδου ἅπτεσθαι, ἐκκυλιομένη τε
γραμμὴν ποιεῖν, δῆλον ὡς τῶν ἐπικαταπιπτόντων
σημείων τὴν ὅλην συντιθέντων γραμμήν. τοίνυν
εἰ τοῦ μεγέθους τῆς γραμμῆς συμπληρωτικόν
ἐστι τὸ σημεῖον, ἕξει καὶ αὐτὸ μέγεθος. συγκε-
χώρηται δὲ τοῦ μεγέθους τῆς γραμμῆς συμπληρω-
τικὸν αὐτὸ τυγχάνειν· καὶ αὐτὸ ἄρα μέγεθος ἕξει
καὶ οὐκ ἀδιάστατον γενήσεται.

of something is apprehended as being likewise the
extremity of something, and also as being part of
that whereof it is the extremity; if, then, we take
it away, that from which it is taken will be dimin-
ished. And that which is part of a thing clearly
helps to complete that thing, and that which helps
to complete a thing will certainly increase its mag-
nitude, and what serves to complete a magnitude
necessarily possesses magnitude. Therefore every 25
sign or extremity of anything in sensibles, as posses-
sing magnitude, is not without dimensions. Hence,
if we conceive the intelligible by transference from
the sensible, we shall conceive it as being the sign and
limit of the line, and also as helping to complete it, so
that it too will certainly possess a dimension since it
is productive of a dimension. Furthermore, they say 26
that the straight line drawn from the centre when it
revolves describes a circle in the plane with its limit.
Since then, the extremity of this straight line is a
sign, and this by revolving measures out the circum-
ference, it will help to complete the circumference;
but the circumference possesses a dimension; so the
sign, too, which helps to complete it will possess a
dimension.—Moreover, it is held that the sphere 27
touches the plane at one sign, and by rolling forward
makes a line, the signs which make contact succes-
sively composing, as is evident, the whole line. Then,
if the sign helps to complete the magnitude of the
line, it too will possess magnitude. But it has been
agreed that it does help to complete the magnitude
of the line; therefore it too will possess magnitude
and will not be without dimensions.

28 Ἀλλ᾽ εἰώθασι πρὸς τὰς τοιαύτας ἐπιχειρήσεις
ὑπαντῶντες οἱ περὶ τὸν Ἐρατοσθένη λέγειν ὅτι
τὸ σημεῖον οὔτε ἐπιλαμβάνει τινὰ τόπον οὔτε
καταμετρεῖ τὸ διάστημα τῆς γραμμῆς, ῥυὲν δὲ
ποιεῖ τὴν γραμμήν. ὅπερ ἐστὶν ἀδιανόητον. ῥεῖν
γὰρ νοεῖται τὸ ἀπό τινος τόπου εἴς τινα τόπον
ἐπεκτείνεσθαι, ὥσπερ τὸ ὕδωρ. εἰ δὴ τοιοῦτόν
τι φαντασιούμεθα τὸ σημεῖον, ἀκολουθήσει οὐχ
οἷον ἀμερὲς αὐτὸ τυγχάνειν, ἀλλ᾽ ἐκ τῶν ἐναντίων
πολυμερές.

29 Τοσαῦτα μὲν περὶ στιγμῆς· ἴδωμεν δὲ παρακει-
μένως καὶ τὰ περὶ γραμμῆς ὀφείλοντα λέγεσθαι·
αὕτη γὰρ μετὰ τὴν στιγμὴν ἐτέτακτο. τοίνυν κἂν
δοθῇ στιγμή τις ὑπάρχειν, οὐκ ἔσται ἡ γραμμή.
εἰ γὰρ αὕτη ῥύσις ἐστὶ σημείου καὶ μῆκος ἀπλατές,
ἤτοι ἕν ἐστι σημεῖον εἰς μῆκος ἐκτεταμένον ἢ
30 πολλὰ ἀδιάστατα[1] στοίχῳ κείμενα· οὔτε δὲ ἕν
ἐστιν εἰς μῆκος ἐκτεταμένον, ὡς παραστήσομεν,
οὔτε πολλὰ σημεῖα στοίχῳ κείμενα, καθὼς καὶ
τοῦτο ὑπομνήσομεν· οὐκ ἄρα ἔστι γραμμή. εἰ
γὰρ ἕν ἐστι [τὸ][2] σημεῖον, ἤτοι τοῦτο[3] τὸ σημεῖον
ἕνα μόνον ἐπέχει τόπον, ἢ μετατίθεται τόπον ἐκ
τόπου, ἢ ἐπεκτείνεται ἀπό τινος τόπου εἴς τινα
31 τόπον. ἀλλ᾽ εἰ μὲν ἑνὶ ἐμπεριέχεται τόπῳ, οὐκ
ἔσται γραμμὴ ἀλλὰ στιγμή· ῥυὲν γὰρ ἐνοεῖτο
γραμμή. εἰ δὲ τόπον ἐκ τόπου μετέρχεται, ἤτοι,
ὡς προεῖπον, ὃν μὲν ἀπολεῖπον ὃν δὲ ἐπιλαμβάνον
μετέρχεται, ἢ οὗ μὲν ἐχόμενον τόπου εἰς ὃν δὲ
ἐκτεινόμενον. ἀλλ᾽ εἰ ὃν μὲν ἀπολεῖπον ὃν δὲ
ἐπιλαμβάνον, πάλιν οὐκ ἔσται γραμμὴ ἀλλὰ

[1] ἀδιάστατα Heintz : διαστατὰ MSS., Bekk.
[2] [τὸ] del. Heintz.

But in answer to these objections Eratosthenes[a] 28
is accustomed to say that the sign neither occupies any space nor measures out the interval of the line, but by flowing makes the line. But this is inconceivable. For flowing is conceived as extension from a place to a place, as water extends. And if we shall imagine the sign to be something of that sort, it will follow that it is not like a thing without parts, but of the opposite sort, abounding in parts.

So much, then, concerning the point: in the next 29
place let us see what ought to be said concerning the line; for this comes next in order after the point. Now even though it be granted that a point exists, the line will not exist. For if it is "a flux of the sign" and "length without breadth," it is either a single sign extended in length or a number of signs placed in a row without intervals; but it is neither a single 30
sign extended in length, as we shall establish, nor a number of signs placed in a row, as we shall also show; therefore line does not exist. For if it is a single sign, this sign either occupies one place only or moves on from place to place, or is extended from a place to a place. But if it is contained in one place, it will not 31
be a line but a point; for the line was conceived as the result of flux. And if it moves from place to place, either it moves—as I said before—by quitting one place and occupying another, or by keeping to one place and extending to another. But if it is by quitting one place and occupying another, again it will not be a line but a point; for as it was conceived 32

[a] Eratosthenes of Cyrenê, Head of the Library at Alexandria, *circ.* 220 B.C., and specially noted for his writings on geography.

[1] τοῦτο Heintz: αὐτὸ MSS., Bekk.

32 στιγμή· ᾧ γὰρ λόγῳ τόπον πρῶτον ἐπεσχηκὸς
στιγμή τις ἀλλ' οὐ γραμμὴ ἐνοεῖτο, τῷ αὐτῷ καὶ
τὸν δεύτερον ἐπιλαμβάνον τόπον νοηθήσεται στιγ-
μή. εἰ δὲ οὗ μὲν ἐχόμενον τόπου εἰς ὃν δὲ ἐκτει-
νόμενον, ἤτοι μεριστῷ τόπῳ ἀντιπαρεκτείνεται ἢ
33 ἀμερίστῳ. καὶ εἰ μὲν ἀμερίστῳ τόπῳ ἀντιπαρεκ-
τείνεται, πάλιν οὐκ ἔσται γραμμὴ ἀλλὰ στιγμὴ καὶ
σημεῖον· τὸ γὰρ ἀμερῆ τόπον ἐπεσχηκὸς ἀμερές
ἐστιν, ὃ δέ ἐστιν ἀμερές, στιγμὴ καὶ οὐ γραμμὴ
καθέστηκεν. εἰ δὲ μεριστῷ, πάντως ἐπεὶ τὸ μερι-
στῷ ⟨παρεκτεινόμενον⟩[1] μέρη ἔχει, εἴγε ἅπαντι
παρεκτείνεται τῷ τόπῳ, τὸ δὲ μέρη ἔχον, οἷς
ἀντιπαρεκτείνεται τοῖς τοῦ τόπου μέρεσι, σῶμά
ἐστιν, ἔσται τὸ σημεῖον καὶ μεριστὸν καὶ σῶμα·
ὅπερ ἄτοπον. ὥστε οὐχ ἕν ἐστι σημεῖον ἡ γραμμή.
34 καὶ μὴν οὐδὲ [τὰ][2] πολλὰ σημεῖα στοιχηδὸν κεί-
μενα. ταῦτα γὰρ τὰ σημεῖα ἤτοι ψαύοντα ἀλλήλων
νοεῖται ἢ οὐ ψαύοντα. καὶ εἰ μὲν οὐ ψαύοντα
ἀλλήλων, μεσολαβούμενα τόποις τισὶ διορισθή-
σεται, καὶ τόποις διοριζόμενα οὐκέτι ποιήσει μίαν
35 γραμμήν. εἰ δὲ ψαύοντα ἀλλήλων νοοῖτο, ἤτοι
ὅλα ὅλων ἅψεται ἢ μέρεσι μερῶν. καὶ εἰ μὲν
μέρεσι μερῶν ἅψεται, οὐκ ἔσται αὐτὰ ἔτι ἀδιά-
στατα καὶ ἀμερῆ· τὸ γὰρ μέσον δυοῖν σημείων
λόγου χάριν νοούμενον σημεῖον ἄλλῳ μὲν μέρει
ἅψεται τοῦ ἔμπροσθεν σημείου, ἄλλῳ δὲ τοῦ
ὄπισθεν, οὐ τῷ αὐτῷ δὲ τῆς ἐπιπέδου, διαφέροντι
δὲ τοῦ ἄλλου τόπου, ὥστε μηκέτ' αὐτὸ ταῖς
36 ἀληθείαις ἀμερὲς τυγχάνειν ἀλλὰ πολυμερές. εἰ
δὲ ὅλα ὅλων ἅψαιτο σημεῖα, δῆλον ὡς ἐν σημείοις

[1] μεριστῷ ⟨παρεκτεινόμενον⟩ scripsi : μεριστὸν MSS., Bekk. (cf. *Adv. Phys.* i. 385).

as a point but not a line when it occupied the first place, so, by the same reasoning, it will be conceived as a point when it occupies the second place. And if it is keeping to one place and extending to another, it extends over place which is either divisible or indivisible. But if it extends over indivisible 33
place, once again it will not be a line but a point or sign, for that which occupies indivisible place is indivisible, and that which is indivisible is a point and not a line. And if it extends over divisible place, then,—since ⟨that which extends over⟩ the divisible has parts,—since it is extended over all the place, and that which has parts wherewith it extends over the parts of the place is body, the sign will certainly be both divisible and corporeal; which is absurd. Consequently, the line is not one single sign.—Nor yet is it 34
a number of signs placed in a row. For these signs are conceived either as touching one another or as not touching. If as not touching one another, being intercepted they will be separated by certain spaces, and being separated by spaces they will no longer form one line. And if they are conceived as touching 35
one another, they will either touch wholes as wholes or parts with parts.[a] But if they shall touch parts with parts, they will no longer be without dimensions and without parts; for the sign which is conceived—shall we say?—as midway between two signs will touch the sign in front with one part, and that behind with another, and the plane with a different part, and the other place with yet another, so that in very truth it is no longer without parts but with many parts. And if the signs as wholes should touch wholes, it is 36

[a] *Cf. P.H.* iii. 45 f.

[1] [τὰ] del. Heintz.

σημεῖα περισχεθήσεται καὶ τὸν αὐτὸν ἐφέξει τόπον· ταύτῃ τε οὐκ ἔσται στοιχηδὸν κείμενα, ἵνα γένηται γραμμή, ἀλλ' εἰ τὸν αὐτὸν ἐπέσχηκε τόπον, μία καταστήσεται στιγμή. εἴπερ οὖν, ἵν' ἐπινοηθῇ γραμμή, δεῖ προεπινοεῖσθαι[1] τὸ σημεῖον ἐξ οὗ λαμβάνει τὴν νόησιν, ἐπιδέδεικται δὲ μήτε σημεῖον οὖσα μήτε ἐκ σημείων σύνθετος, οὐδὲν ἔσται γραμμή.

37 Καὶ μὴν πάρεστιν ἀποστάντας τῆς τοῦ σημείου νοήσεως προηγουμένως ἀναιρεῖν τὴν γραμμὴν καὶ τὸ ἀνεπινόητον αὐτῆς διδάσκειν. γραμμὴ γάρ ἐστιν, ὡς αὐτῶν πάρεστιν ἀκούειν τῶν γεωμετρῶν, μῆκος ἀπλατές, σκεψάμενοι δὲ ἡμεῖς ἀκριβῶς οὔτε ἐν τοῖς νοητοῖς οὔτε ἐν τοῖς αἰσθητοῖς εὑρήσομεν
38 δυνάμενόν τι ληφθῆναι μῆκος ἀπλατές. καὶ ἐν μὲν τοῖς αἰσθητοῖς, ἐπείπερ ὃ ἂν λάβωμεν αἰσθητὸν μῆκος, τοῦτο πάντῃ τε καὶ πάντως σὺν ποσῷ
39 πλάτει ληψόμεθα· ἐν δὲ τοῖς νοητοῖς, καθόσον ἕτερον μὲν ἑτέρου στενώτερον δυνάμεθα νοῆσαι μῆκος, ὅταν δὲ τὸ αὐτὸ μῆκος κατ' ἰσότητα φυλάττοντες σχίζωμεν ταῖς ἐπινοίαις τὸ πλάτος καὶ ἄχρι τινὸς τοῦτο[2] ποιῶμεν, ἔλαττον μὲν τὸ πλάτος καὶ ἔλαττον γινόμενον νοήσομεν, ἐπειδὰν δὲ ἅπαξ φθάσωμεν στερῆσαι τοῦ πλάτους τὸ μῆκος, οὐκέτι οὐδὲ μῆκος φαντασιούμεθα, ἀλλ'
40 ἀναιρεῖται καὶ ἡ τοῦ μήκους ἐπίνοια. καθόλου τε πᾶν τὸ νοούμενον κατὰ δύο τοὺς πρώτους ἐπινοεῖται τρόπους· ἢ γὰρ κατὰ περίπτωσιν ἐναργῆ ἢ κατὰ τὴν ἀπὸ τῶν ἐναργῶν μετάβασιν, καὶ ταύτην τρισσήν· ἢ γὰρ ὁμοιωτικῶς ἢ ἐπισυνθετικῶς ἢ ἀναλογιστικῶς. ἀλλὰ κατὰ μὲν περιπτωτικὴν

[1] προεπινοεῖσθαι Fabr. : προσεπινοεῖσθαι MSS., Bekk.
[2] τοῦτο cj. Bekk. : τὸ αὐτὸ MSS., Bekk.

plain that signs will be contained in signs and will occupy the same place ; and thus they will not be placed in a row, so as to form a line, but if they occupy the same place they will form one point. If, then, in order that the line may be conceived it is necessary that the sign, from which the notion of it is derived, should first be conceived, and it has been shown that the line is neither a sign nor composed of signs, then the line will be nothing.

Moreover, leaving aside the notion of the sign we 37
can destroy the line directly and show its inconceivability. For the line, as one may learn from the Geometers themselves, is " length without breadth," but when we have examined the matter closely, we shall not find either amongst intelligibles or amongst sensibles anything that is capable of being perceived
as length without breadth. Not amongst sensibles, 38
since whatever sensible length we perceive we shall in every case perceive it as combined with a certain
amount of breadth ; nor amongst intelligibles, inas- 39
much as we can conceive one length as narrower than another, but when we keep the same length invariably and in thought cut slices from its breadth and keep doing this up to a point, we shall conceive the breadth as growing less and less, but when we reach the point of finally depriving the length of breadth we shall no longer be imagining even length, but even
the notion of length will be destroyed.—In general, 40
also, everything conceived is conceived in two main ways, either by way of clear impression or by way of transference from things clear, and this way is threefold,—by similarity, or by composition, or by analogy.

ἐνάργειαν νοεῖται τὸ λευκὸν καὶ τὸ μέλαν καὶ γλυκὺ καὶ πικρόν, κατὰ δὲ τὴν ἀπὸ τῶν ἐναργῶν μετάβασιν ὁμοιωτικῶς μὲν νοεῖται καθάπερ ἀπὸ
41 τῆς Σωκράτους εἰκόνος Σωκράτης αὐτός, ἐπισυνθετικῶς δὲ καθάπερ ἀπὸ ἵππου καὶ ἀνθρώπου ἱπποκένταυρος· ἵππεια γὰρ καὶ βρότεια μίξαντες μέλη ἐφαντασιώθημεν τὸν μήτε ἄνθρωπον μήτε ἵππον ἀλλ' ἐξ ἀμφοτέρων σύνθετον ἱπποκένταυρον.
42 ἀναλογιστικῶς δέ τι νοεῖται πάλιν κατὰ δύο τρόπους, ὁτὲ μὲν αὐξητικῶς ὁτὲ δὲ μειωτικῶς, οἷον ἀπὸ τῶν κοινῶν ἀνθρώπων,

> οἷοι νῦν βροτοί εἰσιν,

παραυξητικῶς μὲν ἐνοήσαμεν Κύκλωπα, ὃς οὐκ ἐῴκει

> ἀνδρί γε σιτοφάγῳ ἀλλὰ ῥίῳ ὑλήεντι,

μειωτικῶς δὲ τὸν πυγμαῖον ἄνθρωπον, ὃς οὐχ
43 ὑπέπεσεν ἡμῖν περιπτωτικῶς. τοσούτων δὴ νοήσεως ὄντων τρόπων, εἴπερ νοεῖταί τι[1] ἀπλατὲς μῆκος, ἐξ ἀνάγκης ὀφείλει ἤτοι κατὰ περιπτωτικὴν ἐνάργειαν νοεῖσθαι ἢ κατὰ τὴν ἀπὸ τῶν ἐναργῶν μετάβασιν. ἀλλὰ κατὰ περιπτωτικὴν μὲν ἐνάργειαν οὐκ ἂν νοοῖτο· οὐδενὶ γὰρ περιεπέσομεν
44 μήκει χωρὶς πλάτους. λείπεται οὖν κατὰ τὴν ἀπὸ τῶν ἐναργῶν μετάβασιν λέγειν αὐτὸ νενοῆσθαι· ὃ πάλιν τῶν ἀδυνατωτάτων. εἰ γὰρ οὕτως ἐνοήθη, ἤτοι πάντως κατὰ ὁμοίωσιν ἢ κατὰ ἐπισύνθεσιν ἐνοεῖτο ἢ κατὰ ἀναλογίαν· κατ' οὐδένα δὲ τούτων τῶν τρόπων εἰς ἔννοιαν ἐλθεῖν πέφυκεν, ὡς παραστήσομεν· οὐκ ἄρα νοεῖταί τι μῆκος ἀπλατές.

[1] τι Heintz : τὸ mss., Bekk.

Thus, by clear impression are conceived the white, the black, the sweet and the bitter, and by transference from things clear are concepts due to similarity, —such as Socrates himself from a likeness of Socrates, and those due to composition,—such as the hippo- 41
centaur from horse and man, for by mixing the limbs of horse and man we have imagined the hippocentaur which is neither man nor horse but a compound of both. And a thing is conceived by way of analogy 42
also in two ways, sometimes by way of increase, sometimes by decrease ; for instance, from ordinary men—

Such mortals as now we see—

we conceive by way of increase the Cyclops who was

Less like a corn-eating man than a forest-clad peak of the mountains [a] ;

and by way of decrease we conceive the pygmy whom we have not perceived through sense-impressions. Now the modes of conception being so many, if length 43
without breadth is conceived it must necessarily be conceived either by way of clear sense-impression or by way of transference from clear things ; but it will not be conceived by way of clear sense-impression ; for we have had no impression of any length without breadth. It remains, then, to say that it is conceived 44
by way of transference from clear things ; but this again is most impossible. For if it was conceived in this way, it was certainly conceived either through similarity or through composition or through analogy ; but in none of these ways can it naturally be conceived, as we shall establish ; therefore no length without breadth is conceived. For it is obviously im- 45

[a] *Cf.* Hom. *Od.* ix. 191 *Adv. Phys.* i. 45, 395.

45 αὐτίκα γὰρ κατὰ μὲν τὴν ὁμοιότητα τῶν ἀμηχάνων
ἐστὶ νοεῖν τι μῆκος ἀπλατές. οὐδὲν γὰρ ἔχομεν
ἐν τοῖς φαινομένοις μῆκος χωρὶς πλάτους [νοού-
μενον],[1] ἵνα νοήσωμεν ὅμοιόν τι τούτῳ μῆκος
46 ἀπλατές. τὸ γάρ τινι ὅμοιον γινωσκομένῳ πάντως
ἐστὶν ὅμοιον, τῷ δὲ μὴ γινωσκομένῳ οὐδὲ ὅμοιον
εὑρεῖν δυνατόν. ἐπεὶ οὖν οὐκ ἔχομεν ἐναργῶς
ὑποπῖπτον ἡμῖν μῆκος χωρὶς πλάτους, οὐδὲ
47 ὅμοιόν τι αὐτῷ δυνησόμεθα νοεῖν. καὶ μὴν οὐδὲ
κατ' ἐπισύνθεσιν οἷόν τέ ἐστι προβαίνειν τοῖς
γεωμέτραις αὐτοῦ τὴν ἐπίνοιαν· εἰπάτωσαν γὰρ
ἡμῖν τίνα τῶν ἐκ περιπτώσεως ἐναργῶς γινωσκο-
μένων μετὰ τίνων συνθέντες νοήσομεν τὸ ἀπλατὲς
μῆκος, ὡς πρότερον ἐπ' ἀνθρώπου καὶ ἵππου
48 ποιοῦντες ἐφαντασιούμεθα ἱπποκένταυρον. λεί-
πεται οὖν ἐπὶ τὸν κατὰ ἀναλογιστικὴν αὔξησιν ἢ
μείωσιν τὸν ⟨τρίτον⟩[2] τῆς νοήσεως αὐτοῖς τρόπον
49 συμφεύγειν· ὃ πάλιν τῶν ἀπόρων θεωρεῖται. τὰ
γὰρ κατὰ ἀναλογίαν νοούμενα ἔχει τι κοινὸν πρὸς
τὰ ἀφ' ὧν νοεῖται, οἷον ἀπὸ τοῦ κοινοῦ μεγέθους
τῶν ἀνθρώπων παραυξητικῶς ἐνοήσαμεν τὸν
Κύκλωπα καὶ μειωτικῶς τὸν πυγμαῖον, ὥστε
εἶναί τι κοινὸν τοῖς κατὰ ἀναλογίαν νοουμένοις
πρὸς ἐκεῖνα τὰ ἀφ' ὧν νοεῖται. οὐδὲν δ' ἔχομεν
κοινὸν τοῦ τε ἀπλατοῦς καὶ τοῦ σὺν πλάτει νοου-
μένου μήκους, ἵν' ἀπ' ἐκείνου ὁρμηθέντες νοήσωμεν
50 τὸ ἀπλατὲς μῆκος. μὴ ἔχοντες δέ τι κοινὸν αὐτῶν
οὐδὲ κατὰ ἀναλογίαν ποιήσασθαι τὴν τοῦ ἀπλατοῦς
μήκους νόησιν ἰσχύσομεν. ὅθεν εἰ ἕκαστον τῶν
νοουμένων κατὰ τοὺς ἐκκειμένους νοεῖται τρόπους,
δεδίδακται δὲ κατὰ μηδέν' αὐτῶν νοούμενον τὸ

[1] [νοούμενον] del. Heintz.

[2] ⟨τρίτον⟩ add. cj. Bekk.

possible to conceive a length without breadth by way of similarity. For we have no length without breadth amongst things apparent by means of which we might conceive a similar length without breadth. For what 46
is similar to anything is certainly similar to a thing known, and it is impossible to find a thing similar to what is not known. Since, then, we possess no clear impression of a length without breadth, we shall not be able to conceive anything similar to it.—Nor yet is 47
it possible for the Geometers to get the notion of it by way of composition ; for let them tell us which of the things clearly known from sense-impression are we to compound with which so as to conceive length without breadth, as we did before, in the case of man and horse, when we imagined the hippocentaur. It remains, then, for them to take refuge in 48
the third mode of conception, that of analogy, by way of increase or decrease ; but this again is seen to be hopeless. For things conceived by analogy have 49
something in common with the things wherefrom they are conceived, as for instance from the common size of men we conceived by way of increase the Cyclops and by way of decrease the pygmy, so that things conceived by analogy have something in common with the things wherefrom they are conceived. But we find nothing in common between the length that is without breadth and that conceived along with breadth, so that by setting out from the latter we might conceive length without breadth. But if we find nothing common to them both we shall 50
not be able to form the conception of length without breadth by analogy. Hence, if each of the concepts is conceived according to the modes described, and it has been shown that length without breadth is

ἀπλατὲς μῆκος, ἀνεπινόητόν ἐστι τὸ ἀπλατὲς
μῆκος.
51 Ὅμως δ' οὖν καὶ πρὸς τοὺς οὕτως ἐναργεῖς
ἐλέγχους πειρῶνται κατὰ τὸ δυνατὸν ἀνδριζόμενοι
λέγειν οἱ γεωμέτραι ὅτι κατ' ἐπίτασιν νοεῖται τὸ
52 ἀπλατὲς μῆκος. λαβόντες γοῦν ὁποιονδήποτε
μῆκος σὺν ποσῷ πλάτει, φασὶν ὅτι μειοῦμεν κατ'
ἐπίτασιν τουτὶ τὸ πλάτος ἀεὶ καὶ μᾶλλον τὴν
στενότητα ἐπιτείνοντες, εἶθ' οὕτως τὸ κατ' ἐπί-
τασιν νοούμενον ἀπλατὲς εἶναι μῆκος λέγομεν·
εἰ γὰρ κατ' ὀλίγον ἐλαττοῦται στενούμενον τὸ
πλάτος κατ' ἐπίτασιν, ἐλεύσεταί ποτε καὶ εἰς
ἀπλατὲς μῆκος, καταληξάσης εἰς τοῦτο τῆς νοή-
53 σεως. ἀλλὰ μὴν ἐδείξαμεν, ἐρεῖ τις, ὅτι ἡ παν-
τελὴς στέρησις τοῦ πλάτους ἀναίρεσίς ἐστι καὶ
τοῦ μήκους. εἶτα τὸ κατ' ἐπίτασίν τινος νοού-
μενον οὐχ ἕτερόν ἐστι τοῦ προεπινοηθέντος, ἀλλ'
54 αὐτὸ ἐκεῖνο ἐπιτεταμένον. ἐπεὶ οὖν ἀπὸ τοῦ
ποσὸν ἔχοντος πλάτος κατ' ἐπίτασιν στενότητος
νοῆσαί τι θέλομεν, τὸ μὲν πάντῃ πάντως ἀπλατὲς
μῆκος οὐκ ἐπινοήσομεν (ἑτερογενὲς γάρ ἐστι),
στενὸν δέ τι ληψόμεθα πλάτος, ὥστε τὴν κατάληξιν
τῆς νοήσεως ἐν ἐλαχιστοτάτῳ γενέσθαι πλάτει,
πλὴν ὅμως πλάτει, τὸ δὲ μετὰ τοῦτο γενέσθαι τὴν
ἐπιβολὴν τῆς διανοίας εἰς ἑτερογενές, καὶ ὃ μήτε
55 μῆκός ἐστι μήτε πλάτος. εἴπερ τε δυνατόν ἐστι
μῆκός τι νοήσαντας σὺν ποσῷ πλάτει στερήσει
τοῦ πλάτους λαβεῖν μῆκος ἀπλατές, ἐνδέξεταί
ποτε κατὰ τὸν ὅμοιον τρόπον καὶ σάρκα σὺν
τρωτῷ ἰδιώματι νοήσαντας στερήσει τοῦ τρωτοῦ

---

[a] *Cf. Adv. Phys.* i. 403 ff.

[b] *Cf.* § 39.

conceived according to none of them, then length without breadth is inconceivable.

Notwithstanding, even to arguments so clear as 51
these the Geometers manfully endeavour to reply, as best they can, saying that length without breadth is conceived by way of "intension."[a] Thus, when we 52
have taken any given length along with a certain amount of breadth, they say that we diminish this breadth by "intension," intensifying ever more and more its narrowness, and so in the end we say that what is thus conceived by way of intension is length without breadth; for if the breadth is lessened little by little by being narrowed through intension, at some time it will come to be a length without breadth, the conception ending up in this. But surely, some- 53
one will say, we have proved that complete privation of breadth is also the abolition of length.[b] Also, that which is conceived through the intension of something is not different from the thing preconceived but just that thing intensified. Since, then, we desire to con- 54
ceive a thing by way of intension of its narrowness from that which has a certain amount of breadth, we shall not conceive length which is entirely without breadth (for that is different in kind), but we shall apprehend a narrow breadth, so that the conception ends up in the very least amount of breadth, but still breadth all the same, and after this the notion in the mind passes into something different in nature, which is neither length nor breadth. And if it is possible to 55
apprehend length without breadth by privation of the breadth when we have conceived a certain length along with a certain amount of breadth, then it will be feasible in like manner when we have conceived flesh with the quality of vulnerability to conceive

ἰδιώματος νοῆσαι ἄτρωτόν τε καὶ ἀπαθῆ σάρκα,
56 δυνατὸν δὲ ἔσται καὶ σῶμα νοήσαντας μετὰ ἀντιτύπου ἰδιώματος στερήσει τῆς ἀντιτυπίας λαβεῖν τι μὴ ἀντιτυποῦν σῶμα. ὅπερ τελέως ἐστὶν ἀδύνατον καὶ παρὰ τὴν κοινὴν τῶν ἀνθρώπων ἔννοιαν· τὸ γὰρ ἄτρωτον νοούμενον ἡμῖν οὐκέτι ἔστι σάρξ, ἐπείπερ σὺν τῷ τρωτῷ ἰδιώματι ἡ σὰρξ ἐνοεῖτο ὡς σάρξ, καὶ τὸ μὴ ἀντιτυποῦν σῶμα οὐκέτι νοεῖται σῶμα· σὺν γὰρ τῷ ἀντιτύπῳ ἰδιώματι ἐνοεῖτο τὸ σῶμα, καθό ἐστι σῶμα. ὅθεν καὶ τὸ νοούμενον μῆκος χωρὶς πλάτους οὐκ ἂν εἴη μῆκος· σὺν γὰρ τῷ ποσὸν ἔχειν πλάτος τὸ μῆκος ὡς μῆκος νοεῖται.

57 Ἀλλ' ὅ γε Ἀριστοτέλης, καίπερ ποικίλως κατασκευασθείσης τῆς τοῦ πράγματος ἀνεπινοησίας καὶ οὐκ ἐν ὀλίγῳ κειμένων ταράχῳ τῶν γεωμετρῶν, φησὶ μὴ ἀδιανόητον εἶναι τὸ ὑπὸ τούτων λεγόμενον μῆκος ἀπλατές, ἀλλὰ δύνασθαι χωρὶς πάσης περισκελείας εἰς ἔννοιαν ἡμῖν ἐλθεῖν. ἵστησι δὲ τὸν λόγον ἐπί τινος ἐναργεστέρου ὑποδείγματος καὶ
58 σαφοῦς. τὸ γοῦν τοῦ τοίχου μῆκος, φησί, λαμβάνομεν μὴ συνεπιβάλλοντες αὐτοῦ τῷ πλάτει, διόπερ ἐνέσται καὶ τὸ παρὰ τοῖς γεωμέτραις λεγόμενον μῆκος χωρὶς πλάτους τινὸς ἐπινοεῖν, ἐπείπερ ὄψις τῶν ἀδήλων ἐστὶ τὰ φαινόμενα, πλανώμενος ἢ τάχα κατασοφιζόμενος ἡμᾶς. ὅταν γὰρ τὸ τοῦ τοίχου μῆκος χωρὶς πλάτους νοῶμεν, οὐ χωρὶς παντὸς πλάτους αὐτὸ νοοῦμεν, ἀλλὰ χωρὶς τοῦ περὶ τῷ τοίχῳ καθεστῶτος πλάτους.
59 ὅθεν καὶ ἐνδέχεται συμπλέξαντας τὸ τοῦ τοίχου

[a] *Cf. P.H.* iii. 40. [b] *Cf. Adv. Phys.* i. 412.
[c] *Cf.* § 23.

invulnerable and impassive flesh by privation of the quality of vulnerability; and it will be possible by 56
conceiving body with the quality of solidity,[a] and by privation of the solidity, to perceive a non-solid body. But this is perfectly impossible and contrary to the common notion of mankind. For that which we conceive as invulnerable is no longer flesh, for flesh includes the quality of vulnerability when conceived as flesh, and the non-solid body is no longer conceived as body, for body, *qua* body, is conceived as including the quality of solidity. Hence, the length conceived without breadth will not be length, for length, as length, is conceived as including the quality of having a certain amount of breadth.

But although the inconceivability of the thing has 57
been established in a variety of ways, and the Geometers are in a state of no little confusion, yet Aristotle affirms that the length without breadth they talk of is not inconceivable but can come into our minds without any difficulty.[b] He bases his argument on an obvious and clear example. Thus we perceive 58
the length of a wall, he says, without thinking simultaneously of its breadth, and therefore it will be possible also to conceive of the "length without any breadth" talked of by the Geometers, seeing that "things evident are the vision of things non-evident"[c]; but he is in error, or perhaps humbugging us. For whenever we conceive the length of the wall without breadth, we do not conceive it as wholly without breadth but without the breadth which belongs to the wall. And thus it is possible for 59
us by combining the length of the wall with a certain

μῆκός τινι πλάτει καὶ ὁτῳδηποτοῦν νόησιν αὐτοῦ ποιεῖσθαι· ὥστε μῆκος λαμβάνεσθαι τὰ νῦν οὐ χωρὶς παντὸς πλάτους, καθάπερ ἀξιοῦσιν οἱ ἀπὸ τῶν μαθημάτων, ἀλλὰ χωρὶς τοῦδέ τινος πλάτους. προύκειτο δὲ τῷ Ἀριστοτέλει παραστῆσαι οὐχ ὅτι τινὸς πλάτους ἀμοιρεῖ τὸ κατὰ τοὺς γεωμέτρας λεγόμενον μῆκος, ἀλλ' ὅτι παντὸς ἐστέρηται πλάτους· ὅπερ οὐκ ἀπέδειξεν.

60 Ταῦτα μὲν οὖν περὶ τούτων· ἐπειδὴ δὲ οἱ γεωμέτραι καὶ πέρας ἐπιφανείας εἶναι λέγουσι τὴν γραμμήν, ὅ ἐστι μῆκος ἀπλατές, φέρε κοινότερον περὶ γραμμῶν ἅμα καὶ ἐπιφανειῶν διαπορῶμεν· οὕτω γὰρ εὐδιάβλητος καὶ ὁ ἐπὶ τὸ σῶμα γενήσεται
61 λόγος. εἰ γὰρ ἡ γραμμὴ πέρας ἐστὶν ἐπιφανείας, μῆκος ἀπλατὲς καθεστῶσα, δῆλον ὡς ὅταν ἐπιφάνεια ἐπιφανείᾳ παρατεθῇ, ἤτοι παράλληλοι γενήσονται δύο γραμμαὶ ἢ μία ἀμφότεραι. καὶ εἰ μὲν μία αἱ δύο γραμμαὶ γίνονται, ἐπεὶ ἡ γραμμὴ πέρας ἐστὶν ἐπιφανείας, ἡ δὲ ἐπιφάνεια πέρας σώματος, τῶν μὲν δυοῖν γραμμῶν μιᾶς ἅμα γινομένων γενήσονται καὶ αἱ δύο ἐπιφάνειαι μία ἐπιφάνεια, τῶν δὲ δυοῖν ἐπιφανειῶν μιᾶς ἐπιφανείας γενηθεισῶν ἐξ ἀνάγκης ἔσται καὶ τὰ δύο σώματα ἓν σῶμα, τῶν δὲ δυοῖν σωμάτων ἑνὸς γενομένων ἡ παράθεσις οὐκ ἔσται παράθεσις ἀλλ'
62 ἕνωσις. ὅπερ ἐστὶν ἀδύνατον. ἐπὶ τινῶν μὲν γὰρ σωμάτων δύναται ἡ παράθεσις ἕνωσις γίνεσθαι, καθάπερ ὕδατος καὶ τῶν ἐοικότων τούτῳ, ἐπὶ τινῶν δὲ οὐδαμῶς· καὶ γὰρ λίθος λίθῳ παρατιθέμενος καὶ σίδηρος σιδήρῳ καὶ ἀδάμας ἀδάμαντι κατὰ γραμμὴν οὐχ ἑνοῦνται. ὥστε οὐκ ἂν γένοιντο αἱ δύο γραμμαὶ μία γραμμή. καὶ γὰρ

amount, however small, of breadth to form a conception of it; so that in this case the length is perceived not without any breadth at all, as the Mathematicians claim, but without this particular breadth. But Aristotle's problem was to prove not that the length talked of by the Geometers is devoid of a certain breadth, but that it is wholly deprived of breadth; and this he has not proved.

So much then, concerning these matters; and see- 60
ing that the Geometers declare that the line, which is "length without breadth," is also "the limit of the plane," come and let us raise doubts in a more general way concerning both lines and planes; for thus the statement about body will become easy to refute. If, 61
then, the line, being length without breadth, is the limit of the plane, it is evident that when a plane is set beside a plane [a] either the two lines will be parallel or both will become one. And if the two lines become one, since the line is the limit of the plane, and the plane the limit of the body, as the two lines become one the two planes also will simultaneously become one plane, and when the two planes have become one plane the two bodies also will necessarily be one body, and when the two bodies have become one, the juxtaposition will not be juxtaposition but unification.[b] But this is impossible. For while 62
juxtaposition can become unification in some cases, as in that of water and things like it, in some cases it cannot; for when stone is set beside stone and iron beside iron and adamant beside adamant they are not unified in respect of their lines. Consequently, the two lines will not become one line.—Moreover,

[a] *Cf.* §§ 29 ff.

[b] *Cf. P.H.* iii. 42; *Adv. Phys.* i. 260 ff.

ἄλλως, εἰ ἕνωσίς ἐστι τῶν δύο γραμμῶν μιᾶς γενομένων καὶ σύμφυσις τῶν σωμάτων, ἐχρῆν τὸν χωρισμὸν γίνεσθαι μὴ κατὰ τὰ αὐτὰ αὐτῶν πέρατα ἀλλὰ κατὰ ἄλλα καὶ ἄλλα μέρη ἀποσπωμένων, ὥστε καὶ φθορὰν συμβαίνειν. οὐχὶ δέ γε τοῦτο γινόμενον θεωρεῖται, ἀλλὰ τὰ πέρατα τῶν σωμάτων καὶ πρὸ τῆς παραθέσεως καὶ μετὰ τὸν χωρισμὸν τοιαῦτά ἐστιν ὁποῖα καὶ ἐν τῇ παραθέσει ὄντα πρότερον ἐφαίνετο.[1] οὐ τοίνυν αἱ δύο γραμμαὶ
63 μία γίνονται. οὐ μὴν ἀλλ' εἴπερ αἱ δύο γραμμαὶ μία γίνονται, δεήσει τὰ παρατιθέμενα ἀλλήλοις σώματα ἑνὶ ἄκρῳ ἐλάσσω εἶναι· γεγόνασι γὰρ αἱ δύο μία, ἥτις ἓν ἔχειν ὀφείλει πέρας τε καὶ ἄκρον. οὐχὶ δέ γε τὰ παρατιθέμενα ἀλλήλοις σώματα ἑνὶ ἄκρῳ γίνεται ἐλάσσονα, ὥστε αἱ δύο γραμμαὶ
64 οὐκ ἂν γένοιντο μία γραμμή. εἰ δὲ παράλληλοι γίνονται δύο γραμμαὶ κατὰ παράθεσιν δυοῖν σωμάτων, τὸ ἐκ τῶν δυοῖν γραμμῶν μεῖζον ἔσται τῆς μιᾶς γραμμῆς. εἰ δὲ τὸ ἐκ τῶν δυοῖν γινόμενον γραμμῶν μεῖζόν ἐστι τῆς μιᾶς γραμμῆς, ἔχοι ἂν ἑκατέρα αὐτῶν πλάτος, ὃ μεθ' ἑτέρας μείζονα ποιεῖ τὴν διάστασιν, καὶ οὕτως οὐκ ἔστιν ἀπλατὲς μῆκος ἡ γραμμή. δυοῖν οὖν θάτερον, ἢ ἀναιρεῖν δεῖ τὴν ἐνάργειαν, ἢ μενούσης ταύτης ἀθετεῖν τὴν τῶν γεωμετρῶν ἐπίνοιαν, καθ' ἣν οἴονται τὴν γραμμὴν μῆκος ἀπλατὲς ὑπάρχειν.

65 Καὶ δὴ ταῦτα μὲν προηγουμένως ῥητέον ἐστὶν ἡμῖν πρὸς τὰς τῆς γεωμετρίας ἀρχάς· μεταβάντες δὲ διδάσκωμεν ὅτι καὶ κατὰ τὰς ἐκείνων αὐτῶν ὑποθέσεις οὐχ οἷόν τε προβαίνειν τὴν ζήτησιν. ἀρέσκει τοίνυν αὐτοῖς τὴν εὐθεῖαν γραμμήν, ὡς

[1] ἐφαίνετο cj. Bekk. : ἐφαίνοντο MSS., Bekk.

if there is unification of the two lines which have become one and natural junction of the bodies, the separation ought to take place when they are pulled asunder, not at the same limits but now at one part and now at another, so that as a result they perish. This, however, is not found to occur, but the limits of the bodies both before the juxtaposition and after the separation are just the same as they originally appeared to be during the juxtaposition. So the two lines do not become one.—If, however, the two 63
lines do become one, the bodies set beside each other will have to be less by one extremity; for the two have become one, and this must have one limit and extremity. But the bodies set beside each other do not become less by one extremity, so that the two lines will not become one line.—And if the two lines 64
are parallel in the juxtaposition of two bodies, that which results from the two lines will be greater than the one line. But if that which results from the two lines is greater than the one line, one of the two will have breadth, which along with the other will make the dimension greater, and thus the line is not length without breadth.—Of two things one, then,—we must either do away with the evidence of the senses, or, if this remains unshaken, we must disallow the notion of the Geometers which leads them to suppose that the line is "length without breadth."

This, then, is what we have primarily had to say 65
against the Geometers' principles; so now let us pass on and show that on their own assumptions it is not possible for their investigation to go forward. Thus, they are fain to believe, as we said above,[a] that the

[a] *Cf.* § 26.

καὶ ἀνώτερον ἐλέγομεν, στρεφομένην πᾶσιν αὑτῆς τοῖς μέρεσι κύκλους γράφειν· ᾧπερ θεωρήματι ὄντι συνεκτικωτάτῳ μαχόμενόν ἐστι τὸ τὴν γραμ-
66 μὴν μῆκος ἀπλατὲς ὑπάρχειν. ζητῶμεν δὲ τὸν τρόπον τοῦτον. εἰ γὰρ κατ' αὐτοὺς πᾶν μέρος τῆς γραμμῆς ἔχει σημεῖον, τὸ δὲ σημεῖον στρεφόμενον κύκλον γράφει, δεήσει κατ' αὐτούς, ὅταν ἡ εὐθεῖα γραμμὴ στρεφομένη καὶ πᾶσι τοῖς ἑαυτῆς μέρεσι κυκλογραφοῦσα τὸ διάστημα καταμετρῇ τὸ τῆς ἀπὸ τοῦ κέντρου μέχρι τῆς ἐξωτάτω περιφερείας ἐπιπέδου, τότε ἤτοι συνεχεῖς ἀλλήλοις ὑπάρχειν τοὺς καταγραφομένους κύκλους ἢ δι-
67 εστῶτας ἀπ' ἀλλήλων. ἀλλ' εἰ μὲν διεστᾶσιν ἀπ' ἀλλήλων, ἀκολουθήσει μέρος τι εἶναι τῆς ἐπιπέδου τὸ μὴ κυκλογραφούμενον, καὶ τῆς εὐθείας μέρος τὸ κατὰ τούτου μὲν φερόμενον τοῦ διαστήματος, μὴ κυκλογραφοῦν δέ. ὅπερ ἐστὶν ἄτοπον. ἢ γὰρ οὐκ ἔχει κατὰ τοῦτο τὸ μέρος σημεῖον ἡ εὐθεῖα γραμμή, ἢ ἔχουσα οὐ καταγράφει κύκλον, ὧν ἑκάτερον παρὰ τὸν γεωμετρικόν ἐστι λόγον· καὶ πᾶν γὰρ μέρος τῆς γραμμῆς σημεῖον ἔχειν φασί,
68 καὶ πᾶν σημεῖον στρεφόμενον κυκλογραφεῖν. εἰ δὲ συνεχεῖς ἀλλήλοις ὑπάρχειν οἴονται τοὺς κύκλους, ἤτοι οὕτως εἰσὶ συνεχεῖς ὡς τὸν αὐτὸν ἐπέχειν τόπον, ἢ ὥστε ἄλλον παρ' ἄλλον τετάχθαι μηδενὸς σημείου μεταξὺ πίπτοντος· πᾶν γὰρ σημεῖον τὸ μεταξὺ κατ' ἐπίνοιαν πῖπτον ὀφείλει καὶ αὐτὸ κύκλον γράφειν. καὶ εἰ μὲν τὸν αὐτὸν ἐπέχουσι τόπον πάντες, εἷς γενήσεται κύκλος, καὶ διὰ τοῦτο τῷ ἐλαχίστῳ κύκλῳ καὶ πρὸς τῷ κέντρῳ καθεστῶτι ὁ μείζων καὶ ἐξωτάτω καὶ πάντων περι-
69 ληπτικὸς καθεστὼς κύκλος ἴσος γενήσεται· εἰ γὰρ

straight line by revolving describes circles with all its parts ; but the view that the line is length without breadth is in conflict with this most convincing
theorem. Let us probe the matter in this way. If, 66
as they say, every part of the line has a sign, and the sign as it revolves describes a circle, then, whenever the straight line by revolving and describing circles with all its parts measures off the distance from the centre to the outermost circumference of the plane, it will be necessary, according to them, that the circles described should be either continuous with one an-
other or separate from one another. But if they are 67
separate from one another it will follow that there is a certain part of the plane which is not encircled, and a part of the straight line which moves over this interval but does not describe a circle. But this is absurd. For either the straight line has no sign in this part, or having one does not describe a circle ; but each of these alternatives is contrary to geometrical doctrine ; for they assert that every part of the line has a sign, and also that every sign when
revolving describes a circle. And if they suppose 68
that the circles are continuous with one another, they are continuous either in such a way as to occupy the same place or so as to be ranged in order one beside another with no sign falling between ; for every sign which is conceived as falling between must of itself describe a circle. But if they all occupy the same place, there will be one circle, and therefore the circle which is greater and outermost and inclusive of them all will be equal to the smallest circle which is at

ὁ μὲν ἐξωτάτω κύκλος καὶ πρὸς αὐτῇ τῇ περι-
φερείᾳ μεῖζον ἐπέχει διάστημα καὶ ὁ ἐσωτάτω
πρὸς τῷ κέντρῳ κύκλος μικρὸν ἐπέχει διάστημα,
πάντες δὲ οἱ κύκλοι τὸν αὐτὸν ἐπέχουσι τόπον, ὁ
τὸ μεῖζον ἐπέχων διάστημα ἴσος γενήσεται τῷ
⟨τὸ⟩[1] ἐλάχιστον ἐπέχοντι διάστημα· ὅπερ ἐστὶν
ἀπεμφαῖνον. οὐ τοίνυν οὕτως εἰσὶ συνεχεῖς οἱ
70 κύκλοι ὡς τὸν αὐτὸν τόπον ἐπέχειν. εἰ δὲ παράλ-
ληλοι τυγχάνουσιν ὥστε μεταξύ τι μὴ πίπτειν
ἀμερὲς σημεῖον, συμπληρώσουσι τὸ ἀπὸ τοῦ κέν-
τρου μέχρι τῆς περιφερείας πλάτος. εἰ δέ γε
συμπληρώσουσιν, ἐπέχουσί τι πλάτος. ἦσαν δέ
γε οὗτοι γραμμαί. αἱ ἄρα γραμμαὶ ἔχουσί τι
πλάτος καὶ οὐκ ἀπλατεῖς καθεστήκασιν.
71 Ἀπὸ δὲ τῆς αὐτῆς δυνάμεως ὁρμώμενοι ὁμοιό-
τροπον τῇ προαποδοθείσῃ συνθήσομεν ἐπιχείρησιν.
ἐπεὶ γάρ φασι τὴν κυκλογραφοῦσαν εὐθεῖαν γραμ-
μὴν δι' ἑαυτῆς τὸν κύκλον καταγράφειν, συνερω-
τῶντες αὐτοὺς φήσομεν, εἰ ἡ κυκλογραφοῦσα
εὐθεῖα γραμμὴ δι' ἑαυτῆς τὸν κύκλον καταγράφειν
πέφυκεν, οὐκ ἔστι μῆκος ἀπλατὲς ἡ γραμμή· ἀλλὰ
μὴν ἡ κυκλογραφοῦσα εὐθεῖα γραμμὴ δι' ἑαυτῆς
τὸν κύκλον καταγράφει, ὡς ἐκεῖνοι φασίν· οὐκ
ἄρα μῆκός ἐστιν ἀπλατὲς ἡ γραμμή, ὡς ἡμεῖς
72 τοῦτο ἀκόλουθον ὂν ἐκείνοις διδάξομεν. ὅταν
γὰρ ἡ ἀπὸ τοῦ κέντρου ἀγομένη εὐθεῖα στρέφηται
καὶ δι' ἑαυτῆς καταγράφῃ κύκλον, τότε ἤτοι κατὰ
πάντων τῶν μερῶν τοῦ ἐντὸς τῆς περιφερείας
πλάτους φέρεται ἡ εὐθεῖα γραμμή, ἢ οὐ κατὰ
πάντων ἀλλὰ κατὰ τινῶν. καὶ εἰ μὲν κατὰ τινῶν
φέρεται, οὐδὲ καταγράφει κύκλον, καθ' ὧν μὲν
μερῶν φερομένη καθ' ὧν δὲ οὔ. εἰ δὲ κατὰ πάντων

the centre; for if the outermost circle, that which is on 69
the very circumference, occupies a greater distance, and the innermost circle at the centre occupies a little distance, and all the circles occupy the same place, then that which occupies the greater distance will be equal to that which occupies the least distance,—which is absurd. So, then, the circles are not continuous in such a way as to occupy the same place.
And if they are parallel so that no indivisible sign falls 70
between, they will fill up the breadth from the centre to the circumference. But if they fill it up, they occupy some breadth. Yet these circles are lines. Lines, therefore, possess a certain breadth and are not "without breadth."

Starting with the same theory we shall construct 71
a confutation similar to that already stated. Since they assert that the straight line which describes a circle describes the circle of itself, we shall reply with the objection,—if the straight line which describes a circle is by nature such as to describe the circle of itself, the line is not length without breadth; but in fact, as they assert, the straight line which describes a circle does describe the circle of itself; therefore the line is not length without breadth, this being the
consequence of their theory, as we shall show. For 72
when the straight line drawn from the centre revolves and of itself describes a circle, the straight line then either moves over all the parts of the breadth within the circumference, or not over all but over some. And if it moves over some, it does not describe a circle, as it moves over some parts but not over others.

---

[1] ⟨τὸ⟩ add. cj. Bekk.

φέρεται, ὅλον τὸ τῆς περιφερείας καταμετρήσει
73 πλάτος, πλάτος δὲ καταμετροῦσα ἕξει πλάτος· τὸ
γὰρ τοῦ πλάτους καταμετρητικὸν ὀφείλει πλάτος
ἔχειν, ᾧ καταμετρεῖ. εἰ ἄρα ἡ[1] εὐθεῖα γραμμὴ
κυκλογραφοῦσα[2] ὅλον καταμετρεῖ τὸ πλάτος ⟨καὶ
ἔχει πλάτος⟩,[3] [καὶ] οὐκ ἔστι μῆκος ἀπλατὲς ἡ
γραμμή.

74 Τὸ δ' αὐτὸ σαφέστερον δειχθήσεται καὶ ὅταν λέγωσιν οἱ γεωμέτραι τὴν πλάγιον τοῦ τετραγώνου πλευρὰν καταγομένην τὸ παραλληλόγραμμον ἐπίπεδον καταμετρεῖν. εἴπερ γὰρ μῆκος ἀπλατές ἐστιν ἡ πλάγιος πλευρὰ τοῦ τετραγώνου καταγομένη, οὐ καταμετρήσει τὸ παραλληλόγραμμον ἐπίπεδον τοῦ τετραγώνου δι' ἑαυτῆς· τὸ γὰρ καταμετρητικὸν πλάτους ὀφείλει πλάτος ἔχειν. εἰ δὲ καταμετρεῖ, πάντως πλάτος ἔχει. ὥστε πάλιν ἢ τοῦτο τὸ θεώρημα ψεῦδος εἶναι τοῖς γεωμέτραις, ἢ μηδὲν ὑπάρχειν τὸ νοούμενον μῆκος ἀπλατές.

75 Τόν τε κύλινδρόν φασι κατ' εὐθεῖαν γραμμὴν
ἅπτεσθαι τῆς ἐπιπέδου καὶ ἐκκυλιόμενον τῇ ἀνὰ
μέρος ἄλλων καὶ ἄλλων εὐθειῶν θέσει καταμετρεῖν
τὴν ἐπίπεδον. ἀλλ' εἰ καὶ κατ' εὐθεῖαν ἅπτεται
τῆς ἐπιπέδου ὁ κύλινδρος καὶ κυλιόμενος τῇ ἀνὰ
μέρος ἄλλων καὶ ἄλλων εὐθειῶν θέσει καταμετρεῖ
τὴν ἐπίπεδον, πάντως καὶ ἡ ἐπίπεδος συνέστηκεν
ἐξ εὐθειῶν καὶ ἡ ἐπιφάνεια τοῦ κυλίνδρου πάλιν
76 ἐξ εὐθειῶν πεπλήρωται. ὅθεν ἐπεὶ ἔχει καὶ ἡ
ἐπίπεδος πλάτος καὶ ἡ ἐπιφάνεια τοῦ κυλίνδρου
ὁμοίως, καὶ οὐκ ἔστιν ἀπλατής, τὸ δὲ πλάτους

[1] εἰ ἄρα ἡ Heintz : ἡ ἄρα MSS., Bekk.
[2] κυκλογραφοῦσα Heintz : κυκλοφοροῦσα MSS., Bekk.
[3] ⟨καὶ ἔχει πλάτος⟩ addidi.

And if it moves over all, it will measure out all the breadth of the circumference, and measuring out 73
breadth it will possess breadth; for that which is capable of measuring out breadth must possess breadth wherewith it measures. Therefore, if the straight line in describing a circle measures out all the breadth and possesses breadth, the line is not "length without breadth."

The same thing will be shown more clearly when 74
the Geometers state that when the downward side of the square is drawn it measures out the plane bounded by the parallel lines. For if it is length without breadth, the downward side of the square when drawn will not of itself measure out the plane surface of the square bounded by the parallel lines; for that which is capable of measuring out a breadth must possess breadth. And if it measures out, it certainly possesses breadth. So that, once again, either this theorem of the Geometers is false, or the concept "length without breadth" is nothing.

Also, they say that the cylinder touches the plane 75
along a straight line and when rolling forward, by the placing of straight lines in turn, one after another, measures out the plane. But if the cylinder touches the plane along a straight line and when rolling measures out the plane by placing its straight lines in turn, one after another, the plane certainly is composed of straight lines and the surface of the cylinder, too, is made up of straight lines. Hence, 76
since the plane possesses breadth, and the surface of the cylinder likewise is not without breadth, and

ποιητικὸν ὀφείλει καὶ αὐτὸ πλάτος ἔχειν, δῆλον οὖν ὡς ὅτι καὶ αἱ εὐθεῖαι γραμμαὶ συμπληρωτικαὶ οὖσαι τοῦ πλάτους ἐξ ἀνάγκης πλάτος ἔχουσιν, ὥστε μηδὲν εἶναι μῆκος ἀπλατές, διὰ δὲ τοῦτο μηδὲ γραμμήν.

77 Εἰ δὲ καὶ δοίημεν τὴν γραμμὴν μῆκος εἶναι ἀπλατές, τὰ ἀκόλουθα τούτοις ἔτι τούτων ἀπορώτερα. ὥσπερ γὰρ τὸ σημεῖον ῥυὲν ποιεῖ γραμμήν, οὕτω καὶ ἡ γραμμὴ ῥυεῖσα ποιεῖ ἐπιφάνειαν κατ' αὐτούς, ἥτις ἐστί, φασί, πέρας σώματος δύο
78 ἔχον διαστάσεις, μῆκος καὶ πλάτος. εἴπερ οὖν ἡ ἐπιφάνεια πέρας ἐστὶ σώματος, τό γε σῶμα πάντως πεπερασμένον ἐστίν· καὶ εἰ τοῦτο, ὅτε παρατίθεται δύο σώματα ἀλλήλοις, τότε ἤτοι τὰ πέρατα τῶν περάτων ἢ τὰ πεπερατωμένα τῶν πεπερατωμένων ἅψεται, ἢ καὶ τὰ πεπερατωμένα τῶν πεπερατωμένων καὶ τὰ πέρατα τῶν περάτων, οἷον ἐπὶ τοῦ ἀμφορέως, εἰ νοήσαιμεν πέρας μὲν τὸ ἔξωθεν ὄστρακον πεπερατωμένον δὲ τὸν ἐν
79 αὐτῷ οἶνον. δυοῖν οὖν ἀμφορέων παραβληθέντων ἀλλήλοις ἤτοι τὸ ὄστρακον τοῦ ὀστράκου ἅψεται ἢ ὁ οἶνος τοῦ οἴνου ἢ καὶ τὸ ὄστρακον τοῦ ὀστράκου καὶ ὁ οἶνος τοῦ οἴνου. καὶ εἰ μὲν τὰ πέρατα τῶν περάτων ἅπτεται, τὰ πεπερατωμένα ἀλλήλων οὐχ ἅψεται, τουτέστι τὰ σώματα, ὅπερ ἦν ἀπεμφαῖνον. εἰ δὲ τὰ πεπερατωμένα μὲν ἀλλήλων ἅψεται, τουτέστι τὰ σώματα, τὰ πέρατα δὲ ἀλλήλων οὐχ ἅψεται, ἐκτὸς ἔσται τὰ σώματα τῶν οἰκείων
80 περάτων. εἰ δὲ καὶ τὰ πέρατα τῶν περάτων ἅπτεται καὶ τὰ πεπερατωμένα τῶν πεπερατωμένων, ἐπισυνθήσομεν τὰς ἀπορίας· ᾗ μὲν γὰρ τὰ πέρατα

what is productive of breadth must itself possess breadth, it is plain that the straight lines too, as they serve to fill up the breadth, necessarily possess breadth, so that no " length without breadth " exists, and consequently no line.

And even if we should grant that the line is 77
" length without breadth," the consequences of this will be even more hopeless than those stated. For as the sign when it has flowed [a] makes the line, so also the line when it has flowed makes, according to them, the plane, which is, they say, " the limit of the body," possessing two dimensions, length and
breadth. If, then, the plane is the limit of the body, 78
the body certainly is limited ; and if so, when two bodies are set beside each other, then either the limits will touch the limits or the things limited the things limited, or the things limited will touch the things limited and also the limits the limits,—as though, in the case of a jar, we were to conceive the external earthenware as the limit, and the wine
within it as the thing limited.[b] When, then, two 79
jars are set beside each other, either the ware will touch the ware or the wine the wine, or the ware will touch the ware and also the wine the wine. But if the limits touch the limits the things limited (that is, the bodies) will not touch each other, which is absurd. And if the things limited (that is, the bodies) shall touch each other, and the limits shall not touch each other, the bodies will be outside their own limits.
And if both the limits touch the limits and the things 80
limited the things limited, we shall be multiplying the difficulties ; for where the limits touch each other,

[a] *Cf.* § 29 ; *Adv. Phys.* i. 376.
[b] *Cf. Adv. Phys.* i. 431.

ἀλλήλων ἅπτεται, τὰ πεπερατωμένα οὐχ ἅψεται ἀλλήλων, ᾗ δὲ τὰ πεπερατωμένα, ἐκτὸς ἔσται τὰ σώματα τῶν οἰκείων περάτων, ἐπεὶ πέρας μέν
81 ἐστιν ἡ ἐπιφάνεια, πεπερασμένον δὲ τὸ σῶμα. τά τε πέρατα σώματά ἐστιν ἢ ἀσώματα. καὶ εἰ μὲν σώματά ἐστι, ψεῦδος ἔσται τοῖς γεωμέτραις τὸ ἀβαθῆ εἶναι τὴν ἐπιφάνειαν. εἰ γὰρ σῶμά ἐστιν, ἐξ ἀνάγκης ἕξει καὶ βάθος· πᾶν γὰρ σῶμα ὀφείλει βάθος ἔχειν. εἶτα οὐδὲ ἅψεταί τινος, ἀλλὰ πᾶν ἔσται ἀπειρομέγεθες. εἰ γὰρ σῶμά ἐστιν, ἐπεὶ πᾶν σῶμα πέρας ἔχει, κἀκεῖνο τὸ πέρας σῶμα ὂν ἕξει πέρας, κἀκεῖνο ὁμοίως, καὶ τοῦτ' εἰς
82 ἄπειρον. εἰ δὲ ἀσώματόν ἐστι τὸ πέρας, ἐπεὶ τὸ ἀσώματον οὐδενὸς δύναται θιγεῖν οὐδὲ θιχθῆναι, τὰ πέρατα οὐχ ἅψεται ἀλλήλων, τούτων δὲ μὴ ἁπτομένων οὐδὲ τὰ πεπερατωμένα ἅψεται ἀλλήλων. κἂν δῶμεν οὖν εἶναι μῆκος ἀπλατὲς τὴν γραμμήν, ὁ περὶ τῆς ἐπιφανείας λόγος ἄπορος ἐστίν. οἷς, κἂν ἡμεῖς μὴ λέγωμεν, ἀπόροις οὖσι συναπορεῖται καὶ τὸ στερεὸν σῶμα, ἐκ τούτων σύνθετον καθεστώς.

83 Σκοπῶμεν δὲ καὶ οὕτως. εἰ γὰρ σῶμά ἐστιν, ὥς φασὶν οἱ γεωμέτραι, τὸ τὰς τρεῖς ἔχον διαστάσεις, μῆκος πλάτος βάθος, ἤτοι χωριστόν ἐστι τούτων τὸ σῶμα, ὥστε ἄλλο μὲν εἶναι τὸ σῶμα ἄλλο δὲ τὸ μῆκος τοῦ σώματος πλάτος τε καὶ βάθος, ἢ ὁ ἀθροισμὸς τούτων ἐστὶ τὸ σῶμα. ἀλλὰ χωρίζεσθαι μὲν τούτων τὸ σῶμα οὐ πιθανόν ἐστιν· ὅπου γὰρ μήτε μῆκος ἔστι μήτε πλάτος μήτε
84 βάθος, ἐκεῖ οὐχ οἷόν τε νοῆσαι σῶμα· εἰ δὲ ὁ

the things limited will not touch each other, and where the things limited touch, the bodies will be outside their own limits, since the surface is the limit and the body the thing limited.—Also, the limits are 81
either bodies or incorporeal. But if they are bodies, the Geometers will find that it is false that the surface is without depth. For if it is corporeal, it will of necessity have depth; for every body must have depth. Then, too, it will not touch anything but will all be infinite in magnitude. For if it is body, since every body has a limit, that limit too, being a body, will have a limit, and likewise this last one, and so on *ad infinitum*. And if the limit is incorporeal, since 82
the incorporeal cannot touch or be touched by anything,[a] the limits will not touch each other, and as they do not touch neither will the things limited touch each other. So, even if we grant that the line is "length without breadth," the account given of the plane surface is dubious. And these things being dubious, along with them doubt is cast—even if we do not affirm it—on the solid body, seeing it is composed of these.

Let us also consider the matter in this way:—If 83
body is, as the Geometers assert, that which has the three dimensions, length, breadth and depth, either the body is separable from these, so that the body is one thing and the length, breadth and depth of the body something different, or else the aggregation of these is the body. But that the body should be separated from these is not credible; for where neither length nor breadth nor depth exists, there it is impossible to conceive body; and if the aggrega- 84

[a] *Cf. Adv. Phys.* i. 255; "tangere enim et tangi nisi corpus nulla potest res," Lucret. i. 304.

ἀθροισμὸς τούτων νοεῖται σῶμα καὶ ἄλλο παρὰ ταῦτα οὐδὲν ὑπάρχει, ἐξ ἀνάγκης, ἐπεὶ ἕκαστον τούτων ἀσώματόν ἐστι, καὶ ἡ κοινὴ τῶν ἀσωμάτων σύνοδος γενήσεται ἀσώματος. ὥσπερ γὰρ ἡ σύνθεσις τῶν στιγμῶν καὶ ἡ σύνοδος τῶν γραμμῶν ἀσωμάτων φύσει καθεστηκυιῶν οὐ ποιεῖ στερεὸν καὶ ἀντίτυπον σῶμα, οὕτω καὶ ἡ τοῦ πλάτους καὶ ἡ τοῦ μήκους ἔτι δὲ καὶ ἡ τοῦ βάθους συνέλευσις ἀσώματος οὖσα οὐκ ἂν ποιήσαι στερεὸν καὶ ἀντίτυπον σῶμα. εἰ δὲ μήτε χωρὶς τούτων ἐστὶ τὸ σῶμα μήτε ταῦτ' ἐστίν, ἀνεπινόητον, ὅσον ἐπὶ τοῖς γεωμέτραις, γίνεται τὸ σῶμα.

85 πρὸς τούτοις, εἴπερ μήκους καὶ πλάτους καὶ βάθους σύνοδος ποιεῖ σῶμα, ἤτοι πρὶν τῆς συνόδου ἕκαστον τούτων νοεῖται περιέχον ἐν ἑαυτῷ τὴν σωματότητα καὶ τοὺς ὥσπερ σωματικοὺς λόγους, ἢ μετὰ τὴν τούτων συνέλευσιν ἐπισυνέστη τὸ σῶμα. καὶ εἰ μὲν ἕκαστον τούτων πρὶν τῆς συνόδου νοεῖται περιεκτικὸν τῆς σωματότητος, ἔσται τούτων ἕκαστον σῶμα καὶ οὐ μετὰ τὴν σύνοδον

86 αὐτῶν ἐκεῖνο γενήσεται. εἶτ' ἐπεὶ τὸ σῶμα οὐ μῆκος μόνον ἐστίν, οὐδὲ πλάτος κατ' ἰδίαν, οὐδὲ βάθος κατὰ περιγραφήν, ἀλλ' ὁμόσε τὰ τρία, καὶ μῆκος καὶ πλάτος καὶ βάθος, τούτων τε ἕκαστον περιεῖχε τὴν σωματότητα, ἕκαστον αὐτῶν ἕξει τὰ τρία, καὶ τὸ μῆκος οὐ μόνον ἔσται μῆκος ἀλλὰ καὶ πλάτος καὶ βάθος, καὶ τὸ πλάτος οὐ μόνον ἔσται πλάτος ἀλλὰ καὶ μῆκος καὶ βάθος, καὶ τὸ βάθος ὁμοίως ἔσται καὶ μῆκος καὶ πλάτος. ὅπερ

87 τελέως ἐστὶν ἀλογώτατον. εἰ δὲ συνελθόντων τούτων τότε νοεῖται ἡ σύστασις τοῦ σώματος, ἤτοι συνελθόντων αὐτῶν μένει ἡ ἀρχῆθεν φύσις

tion of these is conceived as body, and there is nothing else besides these, then, since each of these is incorporeal, the united assemblage of these incorporeals will necessarily be incorporeal. For just as the combination of the points and the conjunction of the lines, which are by nature incorporeal, do not make a solid and resistant body, so too the union of breadth and length, and depth as well, being incorporeal, will not make a solid and resistant body. But if the body is neither separate from these nor identical with these, the body is—so far as the Geometers' account goes—inconceivable.—Further- 85
more, if the conjunction of length and breadth and depth makes body, either each of these is conceived as containing in itself corporeality and what we may call " the corporeal reasons " before the conjunction, or else body is constructed after these have come together. But if each of these is conceived as containing corporeality before the conjunction, each of these will be body, and body will not come into being after the conjunction.—Moreover, since body is not 86
length alone, nor breadth by itself, nor exclusively depth, but the three together, length and breadth and depth, and each of these includes corporeality, each of them will possess the three, and the length will be not length only but also breadth and depth, and the breadth will be not breadth only but also length and depth, and similarly the depth will also be length and breadth. But this is most completely illogical.—And if the composition of body is conceived as taking 87
place after these have come together, then either the original nature of those things which have come

τοῦ μὲν μήκους ὡς μήκους, τοῦ δὲ πλάτους ὡς πλάτους, τοῦ δὲ βάθους ὡς βάθους, ἢ μεταβέ-
88 βληκεν εἰς τὴν σωματότητα. καὶ εἰ μὲν μένει ἡ ἀρχῆθεν αὐτῶν φύσις, ἐπεὶ ἀσώματά ἐστιν, οὐδὲ διάφορον ποιήσει σῶμα, ἀλλὰ καὶ μετὰ τὴν σύνοδον ἀσώματα μενεῖ, τὴν φύσιν ὄντα ἀσώματα.
89 εἰ δὲ συνελθόντα μεταβάλλει εἰς τὴν σωματότητα, ἐπεὶ τὸ ἐπιδεχόμενον μεταβολὴν εὐθέως ἐστὶ σῶμα, ἕκαστον τούτων καὶ πρὶν τῆς εἰς ταὐτὸ συνόδου ἔσται σῶμα, οὕτω τε καὶ τὸ ἀσώματον γενήσεται σῶμα. ὥσπερ τε τὸ σῶμα μεταβαλὸν ἄλλην μὲν ἀντ' ἄλλης ἔχει ποιότητα, μένει δὲ οὐδὲν ἧσσον σῶμα, οἷον τὸ λευκόν, ἵνα μέλαν γένηται, καὶ τὸ γλυκύ, ἵνα πικρόν, καὶ ὁ οἶνος, ἵνα ὄξος, καὶ ὁ μόλιβδος, ἵνα ψιμμύθιον, καὶ ὁ χαλκός, ἵνα ἰός, ἄλλην μὲν ἀντ' ἄλλης ἀναδέχεται ποιότητα, οὐκ
90 ἐκβαίνει δὲ τοῦ σώματα εἶναι, ἀλλὰ καὶ τὸ μέλαν, ὅτε ἐκ λευκοῦ γέγονε μέλαν, καὶ τὸ πικρόν, ὅτε ἐκ τοῦ γλυκέος γέγονε πικρόν, καὶ τὸ ὄξος, ὅτε ἐκ τοῦ οἴνου γέγονεν ὄξος, μένει σώματα, οὕτω καὶ ταῦτα, εἴπερ μεταβάλλει [εἰς σώματα],[1] ἄλλα μὲν ἀντ' ἄλλων ἔσται ἀσώματα, ἀσώματα[2] δὲ οὐδὲν ἧττον· οὐ γὰρ ἐκβήσεται τὴν ἰδίαν φύσιν. εἰ οὖν οὔτε πρὶν τῆς συνελεύσεως τούτων ἔστι νοῆσαι τὸ σῶμα οὔτε μετὰ τὴν συνέλευσιν αὐτῶν, παρὰ δὲ ταῦτα οὐκ ἔστιν ἄλλως ἐπινοῆσαι, οὐδέν
91 ἐστι σῶμα. πρὸς δὲ τούτοις, εἰ μηδέν ἐστι μῆκος μηδὲ πλάτος μηδὲ βάθος, οὐδὲ τὸ κατὰ μετουσίαν τούτων νοούμενον ἔσται σῶμα· οὐχὶ δέ γε μῆκος ἔστιν οὐδὲ πλάτος οὐδὲ βάθος, ὡς διὰ τῶν ἐμ-

[1] εἰς σώματα seclusi.
[2] ἀσώματα, ἀσώματα Heintz : σώματα, σώματα MSS., Bekk.

together remains,—that of length as length, of breadth as breadth, of depth as depth,—or it is changed to corporeality.[a] But if their original 88
nature remains, since they are incorporeal they will not form a different body, but even after their conjunction they will remain incorporeal, being incorporeal by nature. And if after coming together they 89
change to corporeality, then, since that which admits of change is *ipso facto* corporeal, each of these will be body even before their coming together, and thus too the incorporeal will be body.—Also, just as the body when it has changed exchanges one property for another, but none the less remains body,—for example, white to become black, and sweet to become bitter, and wine to become vinegar, and lead to become white lead, and bronze to become rust, exchange one property for another yet do not cease to be bodies, but the black, when from being white it 90
has become black, and the bitter, when from being sweet it has become bitter, and the vinegar, when from being wine it has become vinegar, all remain bodies,—so these dimensions also, if they change, will change from one sort of incorporeal to another, but none the less ⟨will remain⟩ incorporeal; for they will not go out of their own proper nature.—If, then, it is not possible to conceive the body either before the coming together of these dimensions or after their coming together, and besides these no other alternative can be conceived, body is nothing.—And 91
further, if neither length is anything, nor breadth, nor depth, that which is conceived as participating in these will not be body; but length is not anything, nor is breadth, nor depth, as we have already pointed

[a] *Cf. Adv. Phys.* i. 371 ff.

προσθεν παρεμυθησάμεθα· οὐδὲ τὸ κατὰ μετουσίαν ἄρα τούτων νοούμενον ἔσται σῶμα.

92 Τὰς μὲν οὖν γεωμετρικὰς ἀρχὰς οὕτω συμβέβηκεν ἀνυποστάτους εἶναι· τούτων δὲ ἀναιρουμένων οὐδὲ ἄλλο τι γεωμετρικὸν θεώρημα συστῆναι δύναται. ὁποῖον γὰρ ἂν ᾖ τοῦτο, γραμμικῶς ὀφείλει ἀποδείκνυσθαι, ἐδείξαμεν δὲ ἡμεῖς ὅτι οὐδέν ἐστιν ἡ γενικὴ γραμμή, ᾧ ἀκολουθεῖ μηδὲ τῶν ἐπ' εἴδους τινὰ ὑπάρχειν, ἐάν τε εὐθεῖάν τις ταύτην ὑποτίθηται ἐάν τε κεκλασμένην ἐάν τε
93 ἄλλως πως ἔχουσαν. ὅθεν ἤρκει μὲν ἴσως ἐν τούτοις περατοῦν τὴν πρὸς τοὺς γεωμέτρας ἀντίρρησιν· ὅμως δὲ ἐπαγωνιζόμενοι πειρασόμεθα διδάσκειν ὅτι κἂν τῶν ἀρχῶν ἀποστῶμεν τῶν τῆς γεωμετρίας, οὐ δύνανται θεώρημα συστῆσαι
94 οἱ γεωμέτραι οὐδ' ἀποδεῖξαι. καίτοι πρὶν τούτων καὶ πρὸς τὰς ὑποβεβηκυίας αὐτῶν ἀρχὰς οὐκ ὀλίγα δυνατόν ἐστι λέγειν, οἷον ὅταν φῶσιν εὐθεῖαν εἶναι γραμμὴν τὴν ἐξ ἴσου τοῖς ἑαυτῆς μέρεσι κειμένην. ἵνα γὰρ τὰ ἄλλα παρῶμεν, ἐκεῖνο μὲν συμφανές ἐστιν ὅτι τῆς γενικῆς γραμμῆς μὴ οὔσης οὐδὲ εὐθεῖα γραμμὴ γένοιτ' ἄν· ὡς γὰρ ζώου μὴ ὄντος οὐδὲ ἄνθρωπος ἔστι καὶ ἀνθρώπου μὴ ὄντος οὐδὲ Σωκράτης ἔστιν, οὕτω τῆς γενικῆς ἀναιρουμένης γραμμῆς συνανῄρηται καὶ ἡ ἐπίπεδος εὐθεῖα
95 γραμμή. εἶτα καὶ τὸ ἴσον λέγεται διχῶς, κατὰ ἕνα μὲν τρόπον τὸ ἰσομέγεθες καὶ μήτε ὑπερέχον ἐκείνου τοῦ ᾧ λέγεται ἴσον μήτε ὑπερεχόμενον, καθὸ καὶ τὸ πηχυαῖον ξύλον ἴσον εἶναι λέγομεν τῷ πηχυαίῳ, καθ' ἕτερον δὲ τὸ ἔχον ἐξ ἴσου τὰ μέρη κείμενα, τουτέστι τὸ ὁμαλόν· οὕτω γοῦν τὸ

[a] *Cf.* § 86.

[b] *Cf.* §§ 37 ff.

out [a]; therefore that which is conceived as participating in these will not be body.

Thus, as regards the principles of geometry, the 92
result is that they are unfounded; and as these are abolished no other geometrical theorem can subsist. For the theorem, of whatever sort it be, must be proved by a diagram, but we have shown [b] that the generic line is nothing, and from this it follows that none of the specific lines exist, whether one assumes a straight one, or a curved one, or one of some other
form. Hence, it might, no doubt, have sufficed to 93
finish at this point our confutation of the Geometers; however, we shall contend against them further and try to show that, even if we disregard the principles of geometry, the Geometers are unable to construct
or prove a theorem. Before this, however, no little 94
can be said against their underlying principles,—as, for instance, when they declare that "a straight line is that which is equally placed with its parts." For, to pass over all other objections, this one is obvious,—that the generic line being non-existent, the straight line will not exist; for just as "man" does not exist if "animal" is non-existent, and "Socrates" does not exist if "man" is non-existent, so if the generic line is destroyed the plane straight line is destroyed
along with it.—Moreover, the term "equal" is used 95
in two senses, in one sense as "equal in magnitude" and neither exceeding nor being exceeded by that to which it is said to be equal (as we say that the staff of a cubit's length is equal to a cubit's length), in another sense of "that which has its parts placed equally," that is to say, "the even"; thus, for

96 ἴσον ἔδαφος καλοῦμεν ἀντὶ τοῦ ὁμαλόν. διχῶς οὖν τοῦ ἴσου προσαγορευομένου, ὅταν οἱ γεωμέτραι τὴν εὐθεῖαν γραμμὴν ὑπογράφοντες φῶσιν "εὐθεῖά ἐστι γραμμὴ ἡ ἐξ ἴσου τοῖς ἑαυτῆς μέρεσι κειμένη," ἤτοι τὸ κατὰ τὸ πρῶτον σημαινόμενον λαμβάνουσιν ἴσον ἢ τὸ κατὰ τὸ δεύτερον. ἀλλ' εἰ μὲν τὸ κατὰ τὸ πρῶτον, τελέως εἰσὶν ἀνόητοι· οὐδένα γὰρ ἔχει νοῦν τὸ εὐθεῖαν εἶναι γραμμὴν τὴν ἰσομεγέθη τοῖς ἑαυτῆς μέρεσι καὶ μήτε ὑπερέχουσαν ταῦτα μήτε ὑπερεχομένην ὑπὸ
97 τούτων. εἰ δὲ τὸ κατὰ τὸ δεύτερον, δι' αὐτοῦ τὸ ζητούμενον[1] διδάξουσιν, εἴγε ὅτι μὲν ἔστιν εὐθεῖα παριστᾶσιν ἐκ τοῦ ὁμαλῶς τε καὶ ἐπ' εὐθείας ἔχειν κείμενα τὰ μέρη, τὸ δὲ ἐπ' εὐθείας τι κεῖσθαι οὐκ
98 ἔστι μαθεῖν μὴ ἐπιβαλόντας τῇ εὐθείᾳ. πολλῷ δὲ ἀτοπώτατοι τυγχάνουσι κἀκείνως ὁριζόμενοι "εὐθεῖά ἐστιν ἥτις ἐξ ἴσου τοῖς ἑαυτῆς πέρασι στρέφεται" ἢ οὕτως "ἥτις περὶ τὰ ἑαυτῆς πέρατα στρεφομένη πᾶσι τοῖς ἑαυτῆς μέρεσιν ἅπτεται τοῦ ἐπιπέδου." πρῶτον μὲν γὰρ καὶ αὗται αἱ ἀποδόσεις ὑποπίπτουσι ταῖς πρότερον εἰρημέναις ἡμῖν ἀπορίαις· εἶτα, καθὼς καὶ οἱ Ἐπικούρειοί φασιν, ἡ τοῦ κενοῦ εὐθεῖα εὐθεῖα μέν ἐστιν, οὐ στρέφεται δὲ διὰ τὸ καὶ αὐτὸ τὸ κενὸν μήτε ὅλον
99 μήτε κατὰ μέρος κίνησιν ἐπιδέχεσθαι. ἡ μὲν γὰρ ἐπὶ τέλει ἀπόδοσις καὶ εἰς τὸν δι' ἀλλήλων ἐμπίπτει τρόπον, ὅς ἐστι μοχθηρότατος. τό τε γὰρ ἐπίπεδον διὰ τῆς εὐθείας διδάσκουσι τήν τ' εὐθεῖαν διὰ τοῦ ἐπιπέδου· εὐθεῖαν μὲν γὰρ εἶναί φασιν ἥτις εἰς πάντα τὰ μέρη τοῦ ἐπιπέδου ἅπτεται,

[1] **δι' αὐτοῦ τὸ ζητούμενον** Heintz : **δι' αὐτοῦ τοῦ ζητουμένου** MSS., Bekk.

instance, we call a pavement "equal" instead of "even" (or "level"). The term "equal," then, 96
being applied in two ways, when the Geometers in describing the straight line say that "a straight line is that which lies equally with its parts," they are taking the term "equal" either in the first signification or in the second. But if it is in the first, they are perfectly senseless; for there is no sense in saying that the straight line is of equal magnitude with its parts, neither exceeding these nor being exceeded by these. And if it is in the second sense, they will be 97
proving the matter in question by means of itself, seeing that they establish the fact that it is straight from the fact that it has its parts lying evenly and in a straight line, whereas it is not possible to learn that a thing lies in a straight line without having sensed the straight line. But they are far more absurd when 98
they give the following definition,—"A straight line is that which revolves equally with its limits," or this—"which in revolving round its limits touches the plane with all its parts." For, firstly, these descriptions are subject to the doubts already expressed by us; and secondly, as the Epicureans affirm, the straight line of the void is, indeed, straight, but does not revolve because the void itself does not admit of motion either as a whole or in part. And the last de- 99
scription falls also into the vice of circular reasoning,[a] which is most unsound. For they both explain the plane by means of the straight line and the straight line by means of the plane; for they say that the straight line is that which touches the plane with all its

[a] *Cf. P.H.* i. 117.

ἐπίπεδον δὲ τυγχάνειν δι᾽ οὗ ἡ καταγομένη εὐθεῖα
πᾶσι τοῖς μέρεσιν ἅπτεται, ὥσθ᾽ ἵνα μὲν τὴν
εὐθεῖαν μάθωμεν, πρῶτον τὸ ἐπίπεδον μαθεῖν δεῖ,
ἵνα δὲ τοῦτο, ἀναγκαῖον προεγνωκέναι τὴν εὐθεῖαν.
ὅπερ ἄτοπον. καθόλου τε ὁ διὰ τοῦ ἐπιπέδου
τὴν εὐθεῖαν διδάσκων οὐδὲν ἄλλο ποιεῖ ἢ δι᾽
εὐθείας τὴν εὐθεῖαν παρίστησιν, ἐπείπερ τὸ ἐπίπεδον
πολλαί εἰσιν εὐθεῖαι κατ᾽ αὐτούς.

100 Οἷος δέ ἐστιν ὁ περὶ τῆς εὐθείας λόγος, τοιοῦτος
γένοιτ᾽ ἂν καὶ ὁ περὶ τῆς γωνίας. πάλιν γὰρ ὅταν
ὑπογράφοντες λέγωσιν ὅτι γωνία ἐστὶ δυοῖν εὐ-
θειῶν μὴ κατάλληλα κειμένων τὸ ὑπὸ τὴν κλίσιν
ἐλάχιστον, ἤτοι ἐλάχιστον λέγουσι τὸ ἀμερὲς σῶμα
101 ἢ τὸ κατ᾽ αὐτοὺς σημεῖον καὶ στιγμήν. ἀλλὰ τὸ
μὲν ἀμερὲς σῶμα οὐκ ἂν εἴποιεν, ἐπείπερ τοῦτο
μὲν οὐδ᾽ εἰς δύο μέρη δύναται διαιρεῖσθαι, ἡ δὲ
γωνία κατ᾽ αὐτοὺς ἐπ᾽ ἄπειρον τέμνεται. καὶ
ἄλλως, τῆς γωνίας ἣν μὲν μείζονά φασιν εἶναι
ἣν δὲ μικροτέραν· τοῦ δὲ ἐλαχίστου σώματος οὐδέν
ἐστι βραχύτερον, ἐπεὶ ἐκεῖνο ἀλλ᾽ οὐ τοῦτο γενή-
102 σεται ἐλάχιστον. λείπεται ἄρα τὸ κατ᾽ αὐτοὺς
σημεῖον εἶναι λέγειν· ὃ καὶ αὐτὸ τῶν ἀπόρων. εἰ
γὰρ πάντῃ πανταχῶς ἀδιάστατόν ἐστι τὸ σημεῖον,
οὐ διαιρεθήσεται ἡ γωνία. καὶ μὴν οὐδὲ μείζων
τις ἔσται ἢ ἐλάσσων γωνία· ἐν γὰρ τοῖς μηδεμίαν
ἔχουσι διάστασιν οὐκ ἂν εἴη τις κατὰ μέγεθος
103 διαφορά. ἄλλως τε, εἰ μεταξὺ τῶν εὐθειῶν πίπτει
τὸ σημεῖον, διορίζει τὰς εὐθείας, διορίζον δὲ οὐκ
104 ἔσται ἀδιάστατον. νὴ Δία, ἀλλ᾽ εἰώθασί τινες ἐξ

parts, and the plane is that which, when the straight line is drawn over it, it touches it with all its parts, so that in order to get to know the straight line we must first get to know the plane, and in order to do this, we must necessarily know beforehand the straight line; which is absurd. And, in sum, he who explains the straight line by means of the plane is doing nothing else than establishing the straight line by means of the straight line, since, according to them, the plane is many straight lines.

The argument about the angle will be of much the 100
same kind as that about the straight line. For again, when in describing the angle they say that the angle is "the minimum under the inclination of two lines which do not lie parallel," they mean by "minimum" either the indivisible body or what they call the sign or point. But they will not mean the 101
indivisible body, since this cannot be divided into two parts, whereas, according to them, the angle is divided to infinity. And besides, in the case of the angle, one, they say, is greater, another lesser; but nothing is smaller than the minimal body, for if so it, and not the body, would be the minimum. It remains 102
then to say that it is what they call the sign; but this itself is also dubious. For if the sign is in every way wholly without dimensions, the angle will not be divided.—Moreover, no angle will be greater or lesser; for in things which have no dimension there will be no difference in respect of magnitude. Be- 103
sides, if the sign falls between the straight lines, it divides the straight lines, and as dividing it will not be without dimensions.—But, in sooth, some of them 104

αὐτῶν γωνίαν λέγειν τὸ ὑπὸ τὴν κλίσιν πρῶτον διάστημα. πρὸς οὓς

ἁπλοῦς ὁ μῦθος τῆς ἀληθείας ἔφυ.

ἤτοι γὰρ ἀμερές ἐστι τὸ διάστημα τοῦτο ἢ μεριστόν. ἀλλ' εἰ μὲν ἀμερές, αἱ προειρημέναι τῶν ἀποριῶν ἀκολουθήσουσιν αὐτοῖς, εἰ δὲ μεριστόν, οὐδὲν ἔσται πρῶτον· τοῦ γὰρ ὑποσταθέντος πρώτου ἕτερον εὑρεθήσεται πρότερον διὰ τὴν ἀρεσκο-
105 μένην αὐτοῖς εἰς ἄπειρον τῶν ὄντων τομήν. ἐῶ λέγειν ὅτι καὶ ἄλλῃ τινὶ τεχνολογίᾳ μάχεται ἡ τοιαύτη τῶν γωνιῶν νόησις. διαιρούμενοι γάρ φασι τῆς γωνίας τὴν μέν τινα εἶναι ὀρθὴν τὴν δὲ ἀμβλεῖαν τὴν δὲ ὀξεῖαν, καὶ τῆς μὲν ἀμβλείας ἄλλην καὶ ἄλλην μᾶλλον ἀμβλυτέραν εἶναι, ὡσαύτως δὲ
106 καὶ τῆς ὀξείας. εἰ δὴ γωνίαν φαμὲν τὸ ἐλάχιστον ὑπὸ τὴν κλίσιν διάστημα, οὐ σωθήσονται αἱ τοιαῦται τῶν γωνιῶν διαφοραί, παρόσον ὑπερέχουσί τε ἀλλήλας καὶ ὑπερέχονται ὑπ' ἀλλήλων. ἢ εἴπερ σώζονται, ἀναιρεῖται ἡ γωνία, μὴ ἔχουσα ἑστηκὸς μέτρον ᾧ διαγνωσθήσεται.

Περὶ μὲν οὖν εὐθείας γραμμῆς καὶ γωνίας
107 τοιαῦτα ῥητέον πρὸς αὐτούς· ὁριζόμενοι δὲ καὶ τὸν κύκλον φασὶ "κύκλος ἐστὶ σχῆμα ἐπίπεδον ὑπὸ μιᾶς γραμμῆς περιεχόμενον, πρὸς ἣν αἱ ἀπὸ τοῦ κέντρου προσπίπτουσαι εὐθεῖαι ἴσαι εἰσὶν ἀλλήλαις," ματαιάζοντες· τοῦ γὰρ σημείου καὶ τῆς γραμμῆς καὶ τῆς εὐθείας καὶ ἔτι τοῦ ἐπιπέδου καὶ τῆς γωνίας ἀνῃρημένων οὐδὲ κύκλος ἐπινοηθῆναι δύναται.

108 Ἀλλ' ἵνα μὴ δοκῶμεν σοφιστικοί τινες εἶναι καὶ τὴν σύμπασαν τῆς ἀντιρρήσεως κατασκευὴν

are wont to say that the angle is "the first interval under the inclination." Against whom

By nature simple is the tale which truth doth tell.[a]

For this interval is either without parts or with parts. But if it is without parts, they will find themselves beset in consequence with the difficulties already stated; and if it has parts, none of them will be "first"; for another will be found to be prior to that assumed to be "first" because of the division of existents *ad infinitum* which is approved by them. I forbear to argue 105
that such a notion of the angles is in conflict with another piece of their technology. For in their classification they say that one class of angle is "right," another "obtuse," another "acute"; and that, of the obtuse angles, some are more obtuse than others, and so likewise with the acute angles. But if we affirm that 106
the angle is "the least interval under the inclination," such differences in angles will not be preserved, in so far as they both exceed one another and are exceeded by one another. Or, if they are preserved, the angle is destroyed, not possessing a fixed standard by which it can be distinguished.

Such, then, are the arguments we must use against them with respect to the straight line and the angle; and in defining the circle they say "The circle is a 107
plane figure enclosed by one line, and the straight lines from the centre which fall on this are equal to one another," talking idly; for when the sign and the line and the straight line, and the plane, too, and the angle are destroyed, the circle cannot be conceived.

But in order that we may not seem to be sophistical 108
people and to expend all the reasoning in our refuta-

[a] *Cf.* Eurip. *Phoeniss.* 469.

ἐν μόναις καταναλίσκειν ταῖς τῆς γεωμετρίας ἀρχαῖς, φέρε μετελθόντες, ὡς πρότερον ὑπεσχόμεθα, καὶ τὰ μετὰ τὰς ἀρχὰς αὐτῶν θεωρήματα
109 ἐπισκεψώμεθα. ὅταν οὖν λέγωσι τὴν δοθεῖσαν εὐθεῖαν δίχα τεμεῖν, ἤτοι τὴν ἐπὶ τοῦ ἄβακος διδομένην λέγουσι διχοτομεῖν ἢ τὴν ἀπὸ ταύτης κατὰ μετάβασιν νοουμένην. οὔτε δὲ τὴν ἐπὶ τοῦ ἄβακος δοθεῖσαν διχοτομεῖν ἐροῦσιν· αὕτη μὲν γὰρ μῆκος καὶ πλάτος αἰσθητὸν ἔχειν φαίνεται, ἡ δὲ κατ' αὐτοὺς εὐθεῖα γραμμὴ μῆκός ἐστιν ἀπλατές, ὥστε μὴ ⟨γραμμὴ⟩[1] οὖσα κατ' αὐτοὺς ἡ γραμμὴ ἡ ἐπὶ τοῦ ἄβακος οὐδὲ δίχα τμηθήσεται ὡς γραμμή.
110 καὶ μὴν οὐδὲ ἡ ἀπὸ ταύτης κατὰ μετάβασιν νοουμένη. ὑποκείσθω γὰρ λόγου χάριν ἐξ ἐννέα στιγμῶν συνεστῶσα, ἀφ' ἑκατέρου μὲν τῶν ἄκρων τεσσάρων καὶ τεσσάρων ἀριθμουμένων, μιᾶς δὲ τὰς δύο τετράδας μεσολαβούσης στιγμῆς. οὐκοῦν εἰ δίχα τέμνεται ἡ ὅλη γραμμή, ἤτοι μεταξὺ ταύτης τῆς πέμπτης στιγμῆς καὶ τῆς ἑτέρας τετράδος ἐνεχθήσεται τὸ τέμνον, ἢ κατ' αὐτῆς τῆς
111 πέμπτης, ὥστε καὶ αὐτὴν διχάζειν. τὸ μὲν οὖν μεταξὺ τῆς πέμπτης στιγμῆς καὶ τῆς ἑτέρας τετράδος φέρεσθαι τὸ τέμνον τῶν ἀλόγων· γενήσεται γὰρ ἄνισα τὰ τμήματα, καὶ τὸ μὲν ἐκ τεσσάρων στιγμῶν συγκείμενον τὸ δὲ ἐκ πέντε. τὸ δὲ αὐτὴν διχάζειν τὴν στιγμὴν πολλῷ τοῦ προτέρου ἀλογώτερον· οὐκέτι γὰρ ἀδιάστατον ἀπολείψουσι τὸ σημεῖον, ὅ γε διχάζεται πρὸς τοῦ
112 τέμνοντος. ὁ δὲ αὐτὸς λόγος καὶ ἐπειδὰν φῶσι τὸν κύκλον εἰς ἴσα τέμνειν. εἰ γὰρ ὁ κύκλος εἰς ἴσα τέμνεται, πάντως ἐπεὶ μεσαίτατον ἔχει τὸ

[1] ⟨γραμμὴ⟩ addidi.

tion on the principles of geometry alone, come and let us pass on, as we previously promised,[a] and investigate the theorems which come after their principles. When, then, they say that they will " bisect the given 109
straight line," [b] they mean that they are bisecting either that given on the board or that which is conceived by transference from it. But they will not mean that they are bisecting that given on the board ; for this appears to possess sensible length and breadth, whereas, according to them, the straight line is " length without breadth," so that the line on the board, not being a line according to them, will not be bisected like a line. Nor, indeed, will the line which 110
is conceived by transference from that on the broad. For let us assume, for the sake of argument, that it is composed of nine points, four being numbered from the one extremity and four from the other and one point occupying the middle place between the two sets of four. Then, if the whole line is bisected, the secant will strike either between this fifth point and one of the sets of four or on the fifth point itself so as to divide it in two. That the secant should 111
strike between the fifth point and one of the sets of four is, however, illogical ; for the sections will be unequal, one being made up of four points and the other of five. But the dividing the point itself into two is much more illogical than the former alternative ; for they will no longer be leaving the sign without dimensions,[c] as it is divided into two by the secant.—And the argument is the same when 112
they say they are cutting the circle into equal parts.[d] For if the circle is cut into equal parts, then, since it

[a] *Cf.* § 93.
[b] *Cf. Adv. Phys.* i. 282 ff.
[c] *Cf.* § 20.
[d] *Cf. Adv. Phys.* i. 284 ff.

κέντρον, ὃ καὶ αὐτό ἐστι σημεῖον, ἤτοι τῷδε τῷ τμήματι ἢ τῷδέ [τινι] προσμερισθήσεται, ἢ καὶ αὐτὸ δίχα τμηθήσεται. ἀλλὰ τὸ μὲν τῷδε ἢ τῷδε προσμερισθῆναι ἄνισον τὴν διχοτόμησιν ποιεῖ, τὸ δὲ καὶ αὐτὸ διχοτομεῖσθαι μαχόμενόν ἐστι τῷ
113 ἀδιάστατον καὶ ἀμερὲς ὑπάρχειν τὸ σημεῖον. τό τε τέμνον τὴν γραμμὴν ἤτοι σῶμά ἐστιν ἢ ἀσώματον. οὔτε δὲ σῶμα δύναται τυγχάνειν· ἀμερὲς γάρ τι καὶ ἀσώματον καὶ μὴ ὑποπῖπτον αὐτῷ οὐκ ἂν τέμοι· οὔτε ἀσώματον. τουτὶ γὰρ πάλιν εἰ μὲν στιγμή ἐστι, τῷ ἀμερὴς εἶναι καὶ κατὰ ἀμεροῦς πίπτειν οὐκ ἂν τέμοι, εἰ δὲ γραμμή, πάλιν ἐπεὶ τῷ πέρατι ἑαυτῆς ὀφείλει τέμνειν, τὸ δὲ πέρας αὐτῆς
114 ἐστὶν ἀμερές, οὐ τέμνει. ἄλλως τε τὸ τέμνον πέρας ἤτοι μέσον τῶν δυοῖν στιγμῶν πῖπτον διχοτομεῖ τὴν γραμμήν, ἢ κατὰ μέσου φερόμενον τοῦ σημείου. ἀλλὰ τὸ μὲν κατὰ μέσου φέρεσθαι τοῦ σημείου τῶν ἀδυνάτων· δεήσει γάρ, ὡς πρότερον ἐλέγομεν, τὸ καθ' οὗ φέρεται μεριστόν εἶναι καὶ μηκέτι ἀδιάστατον τυγχάνειν.[1]
115 μεταξὺ δὲ τῶν δυοῖν στιγμῶν φέρεσθαι πολλῷ ἀλογώτερον. πρῶτον μὲν γὰρ οὐδὲν δύναται μεταξὺ συνεχείας μέσον πεσεῖν πέρας· εἶτα κἂν δοθῇ τὸ τοιοῦτον ὡς δυνατόν, ὀφείλει μετακινεῖν τὰ ὧν μεταξὺ τάσσεται, εἴπερ ἐστὶ συνεχῆ· ταῦτα δ' ἐστὶν ἀκίνητα. τοίνυν ἄπορος καὶ ὁ περὶ τοῦ τέμνοντός
116 ἐστι λόγος. οὐ μὴν ἀλλὰ κἂν δῶμεν αὐτοῖς τὰς ἀφαιρέσεις ποιεῖσθαι ἐπὶ τῶν αἰσθητῶν τούτων γραμμῶν, οὐ δυνήσονται οὐδ' οὕτως εὐοδῆσαι. ἢ

[1] τὸ καθ' οὗ φέρεται μ. εἶναι καὶ μ. ἀδ. τυγχάνειν scripsi: τὸ καθόλου φέρεσθαι μ. ἐὰν ᾖ μ. ἀδ. τυγχάνῃ MSS., Bekk.: τὸ καθ' οὗ φέρεται μ. εἶναι, ἵνα μ. cj. Bekk.

has the centre (which itself is a point), in the very middle, the centre will certainly be annexed either to this section or to that, or else it will itself be cut in two. But the fact of its being annexed to this section or that makes the bisection unequal; and that it should itself be bisected is in conflict with the fact that the sign is without dimensions and without parts.—Also, 113
the secant which cuts the line is either a body or incorporeal. But it cannot be a body; for, if so, it will not cut a thing without parts and incorporeal and on which it cannot strike; nor yet can it be incorporeal. For this, again, if it is a point, will not cut owing to its being without parts and striking on what is without parts; and if it is a line, again it does not cut since it must cut with its limit, and its limit is without parts.—
Besides, the limit which cuts bisects the line either by 114
falling between the two points, or by striking on the middle of the sign. But that it should strike on the middle of the sign is a thing impossible. For, as we said before, that on which it strikes will have to possess parts and be no longer without dimensions.
And that it should strike between the two points is 115
much more irrational. For, firstly, no limit can fall in the middle of what is continuous; and secondly, even if we allow that such a thing is possible, it must move apart the things between which it posts itself, if they are continuous; but these are immovable. So then, the account given of the secant is dubious.
—Moreover, even if we grant them that subtrac- 116
tions are made in the case of these sensible lines, even so they will be unable to make progress. For

γὰρ ἀφ' ὅλης τῆς γραμμῆς ἡ ἀφαίρεσις γενήσεται ἢ ἀπὸ μέρους, καὶ τὸ ἀφαιρούμενον ἢ ἴσον ἀπὸ ἴσου ἢ ἄνισον ἀπὸ ἀνίσου [ἢ ἐναλλὰξ][1] γενήσεται· οὐδὲν δὲ τούτων ἐστὶν εὔπορον, ὡς ἐν τῷ πρὸς τοὺς γραμματικοὺς καὶ ἐν τῷ πρὸς τοὺς φυσικοὺς ὑπομνήματι παρεστήσαμεν· οὐκ ἄρα δυνατόν ἐστι τοῖς γεωμέτραις ἀφαιρεῖν τι καὶ τέμνειν ἀπὸ γραμμῆς.

[1] [ἢ ἐναλλὰξ] del. Heintz.

the subtraction will either be from the whole line or from a part, and the part subtracted will be either an equal part from an equal, or an unequal from an unequal; but none of these is feasible, as we have established in our treatise *Against the Grammarians* [a] and in that *Against the Physicists*; therefore it is not possible for the Geometers to subtract or cut off anything from the line.

[a] *Cf. Adv. Gram.* 162 ff.; *Adv. Phys.* i. 280 ff.; *P.H.* iii. 85 ff.

Δ

ΠΡΟΣ ΑΡΙΘΜΗΤΙΚΟΥΣ

1 Ἐπειδὴ τοῦ ποσοῦ τὸ μέν ἐστιν ἐν τοῖς συνεχέσι σώμασιν, ὃ δὴ μέγεθος καλεῖται, περὶ ὅ ἐστι μάλιστα ἡ γεωμετρία, τὸ δὲ ἐν διεστῶσιν, ὅπερ ἀριθμὸς καθέστηκεν, περὶ ὃν ἡ ἀριθμητικὴ καταγίνεται, σκοπῶμεν ἀπὸ τῶν γεωμετρικῶν ἀρχῶν καὶ θεωρημάτων μετελθόντες καὶ τὰ περὶ ἀριθμοῦ· τούτου γὰρ ἀναιρεθέντος οὐθ' ἡ περὶ αὐτὸν συνισταμένη γενήσεται τέχνη.

2 Καθόλου μὲν οὖν οἱ ἀπὸ τῶν μαθημάτων Πυθαγορικοὶ μεγάλην ἀπονέμουσι δύναμιν τοῖς ἀριθμοῖς ὡς τῆς τῶν ὅλων φύσεως κατ' αὐτοὺς διοικουμένης. ὅθεν καὶ ἀεί ποτε ἐπεφώνουν τὸ

ἀριθμῷ δέ τε πάντ' ἐπέοικεν,

ὀμνύοντες οὐ μόνον τὸν ἀριθμὸν ἀλλὰ καὶ τὸν ὑποδείξαντα αὐτοῖς τοῦτον Πυθαγόραν ὡς θεὸν διὰ τὴν ἐν ἀριθμητικῇ δύναμιν, λέγοντες

οὐ μὰ τὸν ἁμετέρᾳ ψυχᾷ παραδόντα τετρακτύν,
πηγὴν ἀενάου φύσεως ῥιζώματ' ἔχουσαν.

3 τετρακτὺς δὲ προσηγορεύετο παρ' αὐτοῖς ὁ ἐκ τῶν πρώτων τεσσάρων ἀριθμῶν συγκείμενος δέκα ἀρι-

BOOK IV

AGAINST THE ARITHMETICIANS

SINCE one kind of quantity, which is called " magnitude," and which is the chief concern of geometry, belongs to continuous bodies, and another kind, which is number, the subject of arithmetic, belongs to discontinuous, let us pass on from the principles and theorems of geometry and examine also those which deal with number ; for if this is destroyed, the art which is constructed to handle it will not exist. 1

Now, speaking generally, the mathematical Pythagoreans [a] ascribe great power to numbers, as though the nature of all things was governed in conformity with them. Hence, they constantly kept repeating— 2

> All things, too, are like unto number.[b]

And they swear not only by number but also by Pythagoras, the man who showed it to them, as though he were a god because of the power of arithmetic, saying,[c]—

> Nay, by the man I swear who bequeathed to our soul the Tetraktys,
> Fount containing the roots of Nature ever-enduring.

And " tetraktys " was the name given by them to 3
the number ten, it being composed of the first four

[a] *e.g.* Philolaüs and Archytas.
[b] *Cf. Adv. Log.* i. 94, 109. [c] *Cf. Adv. Log.* i. 94.

θμός. ἓν γὰρ καὶ δύο καὶ τρία καὶ τέσσαρα δέκα
γίνεται· ὅς ἐστι τελειότατος ἀριθμός, ἐπείπερ ἐπ'
αὐτὸν φθάσαντες πάλιν ἀναλύομεν ἐπὶ τὴν μονάδα
καὶ ἐξ ὑπαρχῆς ποιούμεθα τὰς ἀριθμήσεις. πηγήν
τ' ἀενάου φύσεως ῥιζώματ' ἔχουσαν εἰρήκασιν αὐ-
τὸν διὰ τὸ κατ' αὐτοὺς ἐν αὐτῷ τὸν λόγον τῆς
ἁπάντων κεῖσθαι συστάσεως, οἷον εὐθέως τοῦ τε
σώματος καὶ τῆς ψυχῆς· ἀπαρκέσει γὰρ τούτων
4 ὑποδειγματικῶς μεμνῆσθαι. ἡ μὲν οὖν μονὰς
ἀρχή τις ὑπόκειται τῆς τῶν ἄλλων ἀριθμῶν ἀπερ-
γαστικὴ συστάσεως, ἡ δὲ δυὰς μήκους ἐστὶν ἀπ-
εργαστική. καθάπερ γὰρ ἐπὶ τῶν γεωμετρικῶν
ἀρχῶν ὑπεδείξαμεν πρῶτον τίς ἐστιν ἡ στιγμή,
εἶτα μετ' αὐτὴν ἡ γραμμὴ μῆκος ἀπλατὲς τυγ-
χάνουσα, τὸν αὐτὸν τρόπον καὶ ἐπὶ τοῦ παρόντος
ἡ μὲν μονὰς τὸν τῆς στιγμῆς ἐπέχει λόγον, ἡ δὲ
δυὰς τὸν τῆς γραμμῆς καὶ τοῦ μήκους· ποθὲν γάρ
ποι[1] ἐχώρησεν ἡ διάνοια ταύτην ἐννοουμένη, τοῦτο
5 δ' ἦν μῆκος. ἡ δὲ τριὰς ἐπὶ τοῦ πλάτους καὶ τῆς
ἐπιφανείας ἐτέτακτο· ποθὲν γάρ ποι[1] ⟨καὶ πάλιν
ποι⟩[2] ἐφέρετο ὁ νοῦς, καὶ προστιθεμένης τῇ κατὰ
μῆκος διαστάσει τῆς κατὰ πλάτος διαστάσεως
ἐπιφάνεια νοεῖται. ἀλλὰ κἂν ἐπιθεωρήσῃ τις τῇ
τριάδι τετάρτην μονάδα, τουτέστι τέταρτον ση-
μεῖον, γίνεται πυραμίς, στερεὸν σῶμα καὶ σχῆμα·
καὶ γὰρ μῆκος ἔχει καὶ πλάτος καὶ βάθος· ὥστε
ἐν τῷ τέσσαρα ἀριθμῷ τὸν τοῦ σώματος περιέχε-
6 σθαι λόγον. καὶ μὴν καὶ τὸν τῆς ψυχῆς· ὡς γὰρ
τὸν ὅλον κόσμον κατὰ ἁρμονίαν λέγουσι διοικεῖ-
σθαι, οὕτω καὶ τὸ ζῷον ψυχοῦσθαι. δοκεῖ δὲ ἡ

[1] ποι scripsi : ποῦ mss., Bekk.
[2] ⟨καὶ πάλιν ποι⟩ addidi : ⟨κ. π. ποῦ⟩ add. cj. Bekk.

numbers. For one and two and three and four make
up ten; and this is the most perfect number, since,
when we have reached it, we revert again to the one
and make our numerations afresh. And they have
called it the "fount containing the roots of Nature
ever-enduring" because, according to them, the
reason of the structure of all things resides in it, as for
instance that of the body and the soul[a]; for it will
suffice to mention these by way of example. Now 4
the monad (or one) is an underlying principle which
produces the structure of all the other numbers, and
the dyad (or two) is productive of length. For as in
the case of the geometrical principles we explained[b]
first what the point is, and next, after it, the line
which is length without breadth, similarly, in the
case before us, the monad corresponds to the point
and the dyad to the line and length; for thought in
conceiving this moves from some place to some place,
and this is length. And the triad (or three) is set 5
over breadth and the plane; for the mind has moved
from here to there ⟨and on again to some other
place⟩, and when the distance in breadth is added to
the distance in length the plane is conceived. But
if, in addition to the triad one imagines a fourth
monad, that is, a fourth sign, the pyramid is formed,
a solid body and figure; for it possesses length and
breadth and depth; so that the formula of the body
is comprised in the number four. And so also is that 9
of the soul; for they declare that as the whole
Universe is governed according to harmony, so too
the living creature is ensouled. And the perfect

[a] *Cf. Adv. Log.* i. 99 ff. [b] *Cf. Adv. Geom.* 19 ff.

τέλειος ἁρμονία ἐν τρισὶ συμφωνίαις λαμβάνειν τὴν
ὑπόστασιν, τῇ τε διὰ τεττάρων καὶ τῇ διὰ πέντε
καὶ τῇ διὰ πασῶν. ἡ μὲν οὖν διὰ τεσσάρων ἐν
ἐπιτρίτῳ κεῖται λόγῳ, ἡ δὲ διὰ πέντε ἐν ἡμιολίῳ,
7 ἡ δὲ διὰ πασῶν ἐν διπλασίονι. ἐπίτριτος δὲ λέγεται
ἀριθμὸς ὁ ἐξ ὅλου τινὸς ἀριθμοῦ συνεστηκὼς καὶ
ἐκ τοῦ τρίτου μέρους ἐκείνου, ὡς ἔχει ὁ ὀκτὼ
πρὸς τὸν ἕξ· καὶ γὰρ αὐτὸν τὸν ἓξ περιέσχηκε καὶ
τὸ τρίτον αὐτοῦ, τουτέστι τὴν δυάδα. ἡμιόλιος
δὲ καλεῖται, ὅταν περιέχῃ ἀριθμὸς ἀριθμὸν καὶ τὸ
ἥμισυ ἐκείνου, ὡς ἔχει ὁ ἐννέα πρὸς τὸν ἕξ· συν-
έστηκε γὰρ ἐκ τοῦ ἓξ καὶ ἐκ τοῦ ἡμίσεος αὐτοῦ,
τουτέστι τῶν τριῶν. διπλασίων δὲ προσαγορεύεται
ὁ δυσὶν ἀριθμοῖς ἴσος,[1] ὡς ὁ τέσσαρα πρὸς τὸν
8 δύο· δὶς γὰρ τὸν αὐτὸν περιέσχηκεν. ἀλλὰ γὰρ
τούτων οὕτως ἐχόντων, καὶ κατὰ τὴν ἀρχῆθεν ὑπό-
θεσιν τεσσάρων ὄντων ἀριθμῶν, τοῦ τε ἑνὸς καὶ
δύο καὶ τρία καὶ τέσσαρα, ἐν οἷς ἐλέγομεν καὶ τὴν
τῆς ψυχῆς ἰδέαν περιέχεσθαι κατὰ τὸν ἐναρμόνιον
λόγον, ὁ μὲν τέσσαρα τοῦ δύο καὶ ὁ δύο τῆς μονά-
δος ἐστὶ διπλασίων, ἐν ᾧ ἔκειτο ἡ διὰ πασῶν
συμφωνία, ὁ δὲ τρία τοῦ δύο ἡμιόλιος (καὶ γὰρ
αὐτὸν τὸν δύο περιέσχηκε καὶ τὸ ἥμισυ τούτου, ὅθεν
9 καὶ τὴν διὰ πέντε συμφωνίαν ὑπέβαλλεν), ὁ δὲ
τέσσαρα τοῦ τρία ἐπίτριτος, ὑπέκειτο δὲ καὶ ἐν
τούτῳ ἡ διὰ τεσσάρων συμφωνία. ὥστε εἰκότως
τὸν τέσσαρα ἀριθμὸν παρὰ τοῖς Πυθαγορικοῖς
εἰρῆσθαι πηγὴν ἀενάου φύσεως ῥιζώματ' ἔχουσαν.
10 Ἀλλ' ὅτι μὲν πολλὴν δύναμιν ἀπένεμον τοῖς
ἀριθμοῖς, ἐκ τούτων συμφανὲς ὑποδειγματικώ-
τερον εἰρημένων· πολὺς γὰρ ὁ περὶ ἀριθμῶν παρ'

[1] fortasse ⟨ἴσοις⟩ ἴσος.

harmony is held to consist in three symphonies [a]—that of the " By-Fours " and that of the " By-Fives " and that of the " By-Alls." Now the " By-Fours " symphony consists of the " epitrite " (4 : 3 ratio), and that of the " By-Fives " in the ratio 3 : 2, and the " By-Alls " in the ratio 2 : 1. The number called 7
" epitrite " is that composed of a certain number taken as a whole *plus* its third part—which is the ratio of eight to six ; for the eight includes the six *plus* the third part of it, that is the dyad. And a number is said to be in the ratio 3 : 2 when the number includes a number *plus* its half,—the relation of nine to six ; for the nine is composed of the six *plus* its half, that is, three. And that called " double " is that which is equal to two ⟨equal⟩ numbers,—the relation of four to two ; for it includes the same number twice. Such, then, being the facts, and 8
there being, according to the original assumption, four numbers—one, two, three and four—in which is included, as we said, the form of the soul according to the harmonical formula, the four is double the two and the two double the monad, and therein consists the " By-Alls " symphony ; and the three is to the two in the 3 : 2 ratio (for it includes the two itself *plus* its half, and thus it supplies the " By-Fives " symphony) ; and the four is to the three in the " epitrite " 9
or 4 : 3 ratio, on which is based the " By-Fours " symphony. So that naturally the number four is called by the Pythagoreans the " fount containing the roots of Nature ever-enduring."

From what has been said by way of brief illustration 10
it is clear that they ascribed much power to numbers ; for the account they give of numbers is voluminous,

[a] *Cf. P.H.* iii. 155 ; *Adv. Log.* i. 95 ff.

αὐτοῖς ἐστὶ λόγος, ὃν ἐάσαντες τὰ νῦν μηκύνειν ἁπτώμεθα τῆς ἀντιρρήσεως, τὴν ἀρχὴν τῶν λόγων ἀπὸ μονάδος ποιησάμενοι, ἥτις ἀρχὴ παντὸς ἀριθμοῦ καθέστηκε καὶ ἧς ἀναιρουμένης οὐδ' ἔστιν ἀριθμός.

11 Τὴν τοῦ ἑνὸς τοίνυν νόησιν διατυπῶν ἡμῖν πυθαγορικώτερον ὁ Πλάτων φησὶν "ἕν ἐστιν οὗ μηδὲν χωρὶς λέγεται ἕν" ἢ "οὗ μετοχῇ ἕκαστον ἕν τε καὶ πολλὰ λέγεται." τὸ γὰρ φυτόν, εἰ τύχοι, καὶ τὸ ζῷον καὶ ὁ λίθος προσαγορεύεται μὲν ἕν, οὐκ ἔστι δὲ κατὰ τὸν ἴδιον λόγον ἕν, ἀλλ' ἑν[1] μετοχῇ ἑνὸς νοεῖται, τούτου μηδενὸς τούτων καθεστῶτος.
12 οὔτε γὰρ φυτὸν οὔτε ζῷον οὔτε λίθος οὔτε ἄλλο τι τῶν ἀριθμητῶν τὸ ὄντως ἕν ἐστιν. εἰ γὰρ φυτόν ἐστιν ἢ ζῷον τὸ ἕν, πάντως ὃ μὴ φυτόν ἐστι μηδὲ ζῷον οὐ ῥηθήσεται ἕν· λέγεται δὲ καὶ φυτὸν ἓν καὶ ζῷον καὶ ἄλλα μυρία· οὐδὲν ἄρα τῶν
13 ἀριθμητῶν ἐστὶ τὸ ἕν. τὸ δὲ οὗ ἕκαστον, ἓν μὲν καθ' ἑαυτὸ ἕκαστον πολλὰ δὲ ἀθροισμῷ, μετέχον, ἕν τε καὶ πολλὰ γίνεται τῶν καθ' ἕκαστον. ὅπερ πάλιν πλῆθος οὐδέν ἐστι τῶν πολλῶν, οἷον φυτῶν ζώων λίθων· κατὰ μετοχὴν μὲν γὰρ ἐκείνου ταῦτα λέγεται πολλά, αὐτὸ δὲ οὐκ ἔστιν ἐν τούτοις.
14 πλὴν τοιαύτη μὲν ἡ τοῦ ἑνὸς ἰδέα νοεῖται τοῖς περὶ τὸν Πλάτωνα· ἐπισυνάπτοντες δὲ ἡμεῖς λέγωμεν. ἤτοι ἑτέρα τῶν κατὰ μέρος ἀριθμητῶν[2] ἐστὶν ἡ τοῦ ἑνὸς ἰδέα, ἢ σὺν ἐκείνοις τοῖς μετέχουσιν αὐτῆς νοεῖται. ἀλλὰ καθ' αὑτὴν μὲν ⟨οὐχ ὑφέστηκεν, εἴγε⟩[3] παρὰ τὰ κατὰ μέρος ἀριθμητὰ οὐδὲν νοεῖται

[1] ἓν Heintz : ἐν mss., Bekk.
[2] ἀριθμητῶν Heintz : ἀριθμῶν mss., Bekk.
[3] ⟨οὐχ ὑφέστηκεν, εἴγε⟩ add. cj. Bekk.

but forbearing for the present to dwell on it, let us take up the confutation, beginning our argument with the monad, which is the principle of all number and with the destruction of which number ceases to exist.

Now Plato, in formulating in rather Pythagorean 11
fashion the concept of the one, declares that " One is that without which nothing is termed one," or " by participation in which each thing is termed one or many." For the plant, let us say, or the animal, or the stone is called one, yet is not one according to its own proper description, but is conceived as one by participation in the One, none of them being actually
the One. For neither plant nor animal nor stone nor 12
any other numerable object is the essential One. For if a plant or an animal is the One, what is not a plant or an animal will certainly not be termed one[a]; but a plant is termed one, as is an animal and countless other things; therefore none of the numerables is
the One. But that by participation in which each 13
thing is by itself each one thing, and a plurality by aggregation, is the One and Many of the individual things. But this Plurality, again, is none of the many things, such as plants, animals, stones; for it is by participation in it that these things are termed " many," but the Plurality itself is not one of them.—
Such, then, is the Idea of the One as conceived by 14
Plato; so let us subjoin our argument. Either the Idea of the One is different from the particular numerables, or it is conceived along with those things which participate in it. But it does not subsist by itself, since no One other than the particular numer-

[a] *Cf. P.H.* iii. 156 ff.

ἓν ὑποκείμενον. λείπεται ἄρα ἐν ἐκείνοις τοῖς μετ-
15 έχουσιν αὐτοῦ νοεῖσθαι, ὃ πάλιν τῶν ἀπόρων. τὸ γὰρ ἀριθμητὸν ξύλον εἰ μετοχῇ μονάδος ἔστιν ἕν, ὃ μή ἐστι ξύλον οὐ λεχθήσεται ἕν· λέγεται δέ γε, ὡς ἀνώτερον ὑποδέδεικται· οὐκ ἄρα ἔστιν ἡ μονὰς ἧς μετοχῇ ἕκαστον τῶν κατὰ μέρος ἀριθμητῶν
16 μονὰς προσαγορεύεται. εἶτα τὸ πολλοῖς μετεχόμενον πολλά ἐστι καὶ οὐχ ἕν, τὰ δὲ ἀριθμητὰ πολλά τέ ἐστι καὶ ἄπειρα· οὐκ ἄρα μετοχῇ τῆς
17 μονάδος ἕκαστον τῶν ἀριθμητῶν ἕν ἐστιν. ὥσπερ οὖν ὁ γενικὸς ἄνθρωπος, ὅν τινες νοοῦσι ζῷον θνητόν, οὔτε Σωκράτης ἐστὶν οὔτε Πλάτων, ἐπειδὴ οὐδεὶς λεχθήσεται ἄνθρωπος,[1] οὔτε καθ' ἑαυτὸν ὑφέστηκεν οὔτε μετὰ Πλάτωνος καὶ Σωκράτους, ἐπεὶ ἐθεωρεῖτο ἂν ὡς ἄνθρωπος, οὕτω καὶ τὸ ἓν μήτε σὺν τοῖς κατὰ μέρος ἀριθμητοῖς μήτε καθ' ἑαυτὸ ὑφεστηκὸς νοούμενον εὐθέως ἐστὶν ἀνεπι-
18 νόητον. τὰ δὲ αὐτὰ λεκτέον καὶ ἐπὶ τοῦ δύο ἢ καὶ τρία καὶ καθόλου ἐπὶ παντὸς ἀριθμοῦ, ἵνα μὴ μηκύνωμεν. ἔνεστι δὲ καὶ οὕτως συνερωτᾶν. ἡ τοῦ ἑνὸς ἰδέα, ἧς μετοχῇ τι ἓν λέγεται, ἤτοι μία ἐστὶν ἰδέα ἢ τοῦ ἑνὸς πλείους ἰδέαι τυγχάνουσιν. ἀλλ' εἰ μὲν μία ἐστίν, οὐ πολλοῖς μετέχεται· τοῦ γὰρ Α, εὐσήμου χάριν διδασκαλίας, τὴν ὅλην τοῦ ἑνὸς ἰδέαν ἔχοντος, τὸ Β μὴ μετέχον ταύτης οὐκέτ'
19 ἔσται ἕν. καὶ μὴν οὐδὲ πολυμερὴς καθέστηκεν, ἵνα πολλὰ ᾖ τὰ μετέχοντα ταύτης· πρῶτον μὲν γὰρ ἔσται ἕκαστον οὐ τῆς τοῦ ἑνὸς ἰδέας μετέχον, μέρους δὲ αὐτῆς, εἶτα καὶ ἡ μονὰς ἀδιαίρετος καὶ

[1] ⟨ἄλλος⟩ add. cj. Warmington.

[a] *Cf.* the "generic Line" of *Adv. Geom.* 92.
[b] For this definition *cf.* *P.H.* ii. 26.

ables is conceived as subsisting.—It remains, then, to say that it is conceived as included in those things which partake of it, which, again, is dubious. For 15
if the numerable log is one by participation in the Monad, what is not a log will not be termed one; but, as has been shown above, it is so termed; therefore the Monad, by participation in which each of the particular numerables is called a monad, does not exist.—Further, that in which many participate 16
is Many and not One, and the numerables are both many and infinite; each of the numerables, therefore, is not one by participation in the Monad. So, 17
just as the generic Man[a]—whom some conceive as "a mortal animal"[b]—is not Socrates or Plato (for, if so, nobody else will be termed man), and does not subsist of himself nor together with Plato and Socrates (for then he would have been observed as a man), so likewise the One, not being conceived either as subsisting by itself or along with the particular numerables, is *ipso facto* inconceivable. And the same must 18
be said of the Two and the Three, and in general—not to make a long story of it—of every number.
—One may also propound the following argument: The Idea of the One, by participation in which a thing is termed one, either is one Idea, or there are several Ideas of the One.[c] But if it is one, many do not participate in it; for (to explain the point clearly) if A possesses the whole of the Idea of the One, B, which does not participate in it, will no longer be one.
—Nor yet is it multipartite, so that the things par- 19
ticipating in it might be many; for, firstly, each thing will be participating not in the Idea of the One but in a part of it; and secondly, the Monad, according

[c] *Cf. P.H.* iii. 158 ff.

ἀμερὴς ἐνοεῖτο κατʼ αὐτούς. εἰ δὲ πλείους εἰσὶν
ἰδέαι τοῦ ἑνός, ἕκαστον τῶν καθʼ ἓν τασσομένων
ἀριθμητῶν, ἤτοι τοῦ ἑνὸς ἢ τοῦ δύο, καθʼ ἓν ἑκά-
τερον, μετέχει τινὸς κοινῆς ἰδέας ἢ οὐ μετέχει.
20 καὶ εἰ μὲν οὐ μετέχει, δεήσει καὶ ἅπαντα δίχα τοῦ
μετέχειν ἰδέας κατὰ τὸ ἓν τετάχθαι, ὅπερ οὐ
θέλουσιν. εἰ δὲ μετέχει, ἡ ἐξ ἀρχῆς συναχθήσεται
ἀπορία· πῶς γὰρ μιᾶς τὰ δύο μεθέξει;

Ταῦτα μὲν οὖν περὶ μονάδος, ἧς ἀνῃρημένης πᾶς
21 ἀνῄρηται ὁ ἀριθμός· ὅμως δʼ οὖν ἐπισυνάπτωμεν
καὶ τὰ περὶ τῆς δυάδος. ἀπόρως γάρ πως καὶ αὕτη[1]
συνίσταται κατὰ τὴν τῶν μονάδων σύνοδον, ὥσπερ
καὶ Πλάτων διὰ τοῦ περὶ ψυχῆς πρότερον ἠπόρη-
κεν. παρατεθείσης γὰρ μονάδος ἑτέρᾳ μονάδι ἤτοι
προσγίνεταί τι κατὰ τὴν παράθεσιν ἢ ἀπογίνεται ἢ
22 οὔτε προσγίνεταί τι οὔτε ἀπογίνεται. ἀλλʼ εἰ μὲν
οὔτε προσγίνεταί τι οὔτε ἀπογίνεται, οὐκ ἔσται
κατὰ παράθεσιν τῆς ἑτέρας μονάδος τῇ ἑτέρᾳ ἡ
δυάς. εἰ δὲ ἀπογίνεταί τι κατὰ τὴν παράθεσιν,
ἐλάττωσις ἔσται τοῦ ἑνὸς καὶ ἑνός, καὶ δυὰς οὐ
γενήσεται. εἰ δὲ προσγίνεταί τι, τὰ δύο οὐ γενή-
σεται δύο ἀλλὰ τέσσαρα· δυὰς γὰρ ἡ ἐπιγινομένη
καὶ μονὰς καὶ ἑτέρα μονὰς τὸν τῶν τεσσάρων
ἀριθμὸν συνίστησιν. οὐδὲν ἄρα ἔσται δυάς. ἡ
δὲ αὐτὴ γένοιτʼ ἂν ἀπορία καὶ ἐπὶ παντὸς ἀριθμοῦ,
ὥστε μηδὲν εἶναι κατὰ τοῦτο ἀριθμόν.

23 Οὐ μὴν ἀλλʼ ἐπεὶ κατὰ πρόσθεσιν μονάδος καὶ
κατὰ ἀφαίρεσιν ὁ ἀριθμὸς νοεῖται, δῆλον ὡς ἐὰν
τούτων ἑκάτερον παραστήσωμεν ἀδύνατον, οἰχή-

[1] ἀπόρως . . . αὕτη scripsi : ἄπορος . . . αὐτὴ MSS., Bekk.

[a] *Cf.* §§ 18, 16.

to them, is conceived as indivisible and without parts. And if there are several Ideas of the One, each of the numerables ranked as unities (whether it be a one or a two, both taken singly) participates in a certain common Idea, or it does not participate.
But if it does not participate, all things, apart from 20
participating in an Idea, will have to be ranked as unities,—a conclusion which they reject. And if they participate, the original difficulty[a] will recur; for how will the twos participate in one Idea?

So much, then, concerning the monad, and if it is
destroyed all number is destroyed; all the same, let 21
us subjoin an attack on the dyad. For it is formed in a doubtful way by the conjunction of the monads, even as Plato formerly expressed doubts about it in his book *On the Soul*.[b] For when a monad is set beside another monad, either something is added by the juxtaposition or something is subtracted, or
nothing is either added or subtracted. But if nothing 22
is either added or subtracted, the dyad will not exist through the juxtaposition of the one monad with the other. And if something is subtracted through the juxtaposition, there will be a diminution of the one and one, and a dyad will not be formed. And if something is added, the two will become not two but four; for the additional dyad *plus* the monad and the second monad make up the number four. Therefore nothing will be a dyad. And the same difficulty will exist in the case of every number, so that owing to this number is nothing.

Since, however, number is conceived as a result of 23
the addition or subtraction of the monad, it is plain that if we shall establish that each of these two pro-

[b] *Cf. Phaedo* 96 E ff.; *Adv. Phys.* ii. 302 ff.

σεται καὶ ἡ τῶν ἀριθμῶν ὑπόστασις. λέγωμεν δὲ πρῶτον εὐθὺς περὶ ἀφαιρέσεως, ὑποδειγματικῇ

24 χρώμενοι τῇ διδασκαλίᾳ. ἡ τοίνυν ἀπὸ τῆς ὑποκειμένης δεκάδος ἀφαιρουμένη μονὰς ἤτοι ἀπὸ ὅλης τῆς δεκάδος ἢ ἀπὸ τῆς περιλειπομένης ἐννεάδος ἀφαιρεῖται· οὔτε δὲ ἀφ' ὅλης, ὡς παραστήσομεν, οὔτε ἀπὸ τῆς ἐννεάδος, ὡς διδάξομεν· οὐκ ἄρα ἀφαιρεῖταί τι ἀπὸ τῆς ὑποκειμένης δεκάδος. εἰ γὰρ ἀφ' ὅλης ταύτης ἀφαιρεῖται ἡ μονάς, ἤτοι ἡ δεκάς ἐστιν ἑτέρα τῶν κατὰ μέρος μονάδων ἢ ὁ ἀθροισμὸς τούτων δεκὰς προσαγο-

25 ρεύεται. ἀλλ' ἑτέρα μὲν τῶν κατὰ μέρος μονάδων οὐκ ἔστιν ἡ δεκάς· καὶ γὰρ ἀναιρεθεισῶν ἐκείνων οὐδ' ἔστι δεκάς, καὶ τῆς δεκάδος ἀναιρεθείσης ὁμοίως αἱ μονάδες οὐκέτι ὑπάρχουσιν. εἰ δὲ ἡ αὐτὴ ταῖς μονάσιν ἐστὶν ἡ δεκάς, τουτέστιν εἰ αἱ κατὰ μέρος μονάδες εἰσὶ δεκάς, δῆλον ὡς εἴπερ ἀπὸ τῆς δεκάδος ἡ τῆς μονάδος ἀφαίρεσις γίνεται, ἀφ' ἑκάστης μονάδος ἀφαιρεθήσεται (αἱ γὰρ κατὰ μέρος μονάδες ἦσαν ἡ δεκάς), καὶ οὕτως οὐκέτι ἔσται μονάδος ἆρσις ἀλλὰ δεκάδος. ὥστε οὐκ ἀπὸ

26 ὅλης τῆς δεκάδος αἴρεται ἡ μονάς. καὶ μὴν οὐδ' ἀπὸ τῆς ὑπολειπομένης ἐννεάδος ἡ ἆρσις αὐτῆς γίνεται· πῶς γὰρ ἔτι μετὰ τὴν ἆρσιν αὐτῆς σῶός ἐστιν ἡ ὑποκειμένη ἐννεάς; ἀλλ' εἰ μήτε ἀφ' ὅλης τῆς δεκάδος αἴρεται ἡ μονὰς μήτε ἀπὸ τῆς ὑπολειπομένης ἐννεάδος, οὐδεὶς ἀριθμὸς κατὰ ἀφαίρεσιν

27 συνίσταται. ἄλλως τε, εἰ ἀπὸ τῆς ἐννεάδος αἴρεται ἡ μονάς, ἤτοι ἀπὸ ὅλης αἴρεται ἢ ἀπὸ τῆς τελευταίας αὐτῆς μονάδος. καὶ εἰ μὲν ἀπὸ ὅλης τῆς ἐννεάδος ἀφαιρεῖται ἡ μονάς, ἔσται ἆρσις τῆς ἐννεάδος· τὸ γὰρ ἀφαιρούμενον ἀφ' ἑκάστης μονάδος,

cesses is impossible, the reality of number, too, will be abolished. Let us, for instance, deal first with subtraction, using the method of demonstration by
examples. The monad, then, which is being sub- 24
tracted from the decad assumed is subtracted either from the whole decad or from the nine left over [a]; but it is not subtracted from the whole, as we shall establish, nor from the nine, as we shall demonstrate; nothing, therefore, is subtracted from the decad assumed. For if the monad is subtracted from this as a whole, either the decad is other than the particular monads or the aggregate of these is termed a
decad. But the decad is not other than the particular 25
monads; for if these are destroyed the decad does not exist, and similarly if the decad is destroyed the monads no longer exist. And if the decad is the same as the monads,—that is to say, if the particular monads are the decad,—it is plain that if the subtraction of the monad is from the decad, it will be subtracted from each monad (for the particular monads are the decad), and thus it will no longer be a subtraction of the monad but of the decad. Consequently, the monad is not subtracted from the whole
decad. Nor, indeed, is it subtracted from the nine 26
left over; for how will the assumed nine be still preserved after the subtraction? But if the monad is not subtracted either from the decad as a whole or from the nine left over, no number subsists through
subtraction.—Besides, if the monad is subtracted 27
from the nine, it is subtracted either from the whole or from its last monad. But if the monad is subtracted from the whole nine, there will be a subtraction of the nine; for that which is subtracted from

[a] *Cf. P.H.* iii. 88 ff.; *Adv. Phys.* i. 312 ff.

τῶν κατὰ μέρος μονάδων ἐννέα οὐσῶν, τὸν τῆς
28 ἐννεάδος ἀριθμὸν συντίθησιν. εἰ δὲ ἀπὸ τῆς τελευταίας μονάδος γίνεται τὰ τῆς ἀφαιρέσεως, πρῶτον μὲν καὶ ἡ τελευταία μονάς, ἀμερὴς οὖσα, δειχθήσεται μεριστὴ τυγχάνειν, ὅπερ ἄτοπον· εἶτα εἰ ἀπὸ τῆς τελευταίας μονάδος αἴρεται ἡ μονάς, οὐ δυνή-
29 σεται ἔτι ὁλόκληρος μένειν ἡ ἐννεάς. καὶ ἄλλως, εἴπερ ἀπὸ τῆς δεκάδος γίνεται ἡ τῆς μονάδος ἆρσις, ἤτοι ἀπὸ οὔσης γίνεται τῆς δεκάδος [ἆρσις] ἢ ἀπὸ μὴ οὔσης· οὔτε δὲ ἀπὸ τῆς οὔσης γένοιτ' ἂν (ἐφ' ὅσον γὰρ μένει χρόνον δεκάς, οὐδὲν ἀπ' αὐτῆς ἀφαιρεθῆναι δύναται ὡς δεκάδος, ἐπεὶ οὐκ-
30 έτι ἔσται δεκάς) οὔτε ἀπὸ μὴ οὔσης· ἀπὸ γὰρ τοῦ μὴ ὄντος οὐδὲ ἀρθῆναί τι πέφυκεν. καὶ μὴν παρὰ τὸ εἶναι ἢ μὴ εἶναι οὐδὲν ἔστι νοῆσαι· οὐκ ἄρα αἴρεταί τι ἀπὸ τῆς δεκάδος.

'Αλλ' ὅτι μὲν ἀμήχανόν ἐστι κατ' ἀφαίρεσιν
31 ἀριθμόν τινα νοεῖν, ἐκ τούτων δέδεικται· ὅτι δὲ οὐδὲ κατὰ πρόσθεσιν, ῥᾴδιον δεῖξαι τῶν ἀναλόγων ἐχομένους ἀποριῶν. πάλιν γὰρ μονάδος προστιθεμένης δεκάδι ἤτοι τῇ ὅλῃ δεκάδι ῥητέον γίνεσθαι τὴν πρόσθεσιν ἢ τῷ τελευταίῳ μέρει τῆς δεκάδος. ἀλλ' εἰ μὲν ὅλῃ τῇ δεκάδι προστίθεται ἡ μονάς, ἐπεὶ ἡ ὅλη δεκὰς σὺν πάσαις ταῖς κατὰ μέρος μονάσι νοεῖται, δεήσει τὴν τῆς μονάδος πρόσθεσιν γινομένην πάσαις ταῖς κατὰ μέρος μονάσι δεκάδος
32 εἶναι πρόσθεσιν, ὅπερ ἄτοπον· ἀκολουθήσει γὰρ τῇ τῆς μονάδος προσθέσει τὴν δεκάδα εἰκοσάδα γίνεσθαι, ὃ τῶν ἀμηχάνων ὑπῆρχεν. οὐ τοίνυν ὅλῃ τῇ δεκάδι τὴν μονάδα προστίθεσθαι ῥητέον. καὶ μὴν οὐδὲ τῷ τελευταίῳ μέρει τῆς δεκάδος, ἐπεὶ οὐκ αὐξηθήσεται ἡ δεκὰς διὰ τὸ μὴ τὴν τοῦ ἑνὸς

each monad makes up the number of the nine, as the particular monads are nine. And if the subtraction 28
is from the last monad, then, firstly, the last monad, which is indivisible, will be shown to be divisible, which is absurd ; and secondly, if the monad is subtracted from the last monad, the nine will no longer be able to remain complete.—Further, if the sub- 29
traction of the monad is from the decad, it is from the decad either as existent or as non-existent; but it will not be from the existent (for so long as the decad remains a decad nothing can be subtracted from it as a decad, for if so it will no longer be a decad), nor from the non-existent decad ; for from what is non- 30
existent nothing can be subtracted. And of course it is impossible to conceive anything other than existence and non-existence ; therefore nothing is subtracted from the decad.

Now by these arguments it has been shown that it is not feasible to conceive any number by subtraction ; and that it is not feasible by addition either is easy to 31
show by continuing to raise difficulties of a like kind. For, again, if the monad is added to the decad, one must say that the addition is made either to the whole decad or to the last part of the decad. But if the monad is added to the whole decad, then, since the whole decad is conceived along with all the particular monads, the addition which is being made of the monad will have to be an addition to all the particular monads of the decad, which is absurd ; for it will 32
follow that by the addition of the monad the decad becomes twenty, which is a thing impossible. We must say, then, that the monad is not added to the whole decad. Nor yet to the last part of the decad, since the decad will not be increased owing to the

μέρους αὔξησιν εὐθὺς καὶ τῆς ὅλης δεκάδος αὔξησιν
33 εἶναι. [καὶ] καθόλου τε[1] ἐπὶ πᾶσιν, ἢ μενούσῃ τῇ δεκάδι προστίθεται ἡ μονὰς ἢ μὴ μενούσῃ. οὔτε δὲ μενούσῃ προστεθείη ποτ' ἄν, ἐπεὶ οὐκέτι μένει δεκάς, οὔτε μὴ μενούσῃ· τὴν γὰρ ἀρχὴν μὴ μενούσῃ οὐδὲ πρόσθεσις δύναται γενέσθαι.

34 Ἀλλ' εἴπερ ὁ ἀριθμὸς κατὰ πρόσθεσιν, ὡς ἔφην, καὶ κατ' ἀφαίρεσιν ὑφιστάμενος νοεῖται, ἐδείξαμεν δὲ ἡμεῖς ὅτι οὐθέτερον ἔστι τούτων, ῥητέον μηδὲν εἶναι ἀριθμόν. ὅθεν τοσαῦτα καὶ πρὸς γεωμέτρας καὶ ἀριθμητικοὺς ἀπορητικῶς διεξελθόντες ἀπ' ἄλλης ἀρχῆς καὶ τὴν πρὸς τοὺς μαθηματικοὺς ἀντίρρησιν ποιησώμεθα.

[1] καὶ καθόλου τε καὶ cj. Warmington.

fact that the increase of the one part is not *ipso facto* an increase of the whole decad.—Generally, too, 33
and finally, the monad is added to the decad either remaining as it is or not remaining. But it will never be added to it while it remains, since in that case it will no longer remain a decad; nor yet while it does not remain, for it is absolutely impossible for an addition to be made to it if it does not remain.

But if number is conceived as subsisting through 34
addition, as I said, and subtraction, and we have shown that neither of these exists, one must declare that number is nothing. Hence, now that we have stated at length all these sceptical arguments against the Geometers and Arithmeticians, let us start afresh and deliver our attack on the *Mathematici* (or "Astrologers").

Ε

ΠΡΟΣ ΑΣΤΡΟΛΟΓΟΥΣ

1 Περὶ ἀστρολογίας ἢ μαθηματικῆς πρόκειται
ζητῆσαι οὔτε τῆς τελείου ἐξ ἀριθμητικῆς καὶ γεω-
μετρίας συνεστώσης (ἀντειρήκαμεν γὰρ πρὸς τοὺς
ἀπὸ τούτων τῶν μαθημάτων) οὔτε τῆς παρὰ τοῖς
περὶ Εὔδοξον καὶ Ἵππαρχον καὶ τοὺς ὁμοίους
προρρητικῆς δυνάμεως, ἣν δὴ καὶ ἀστρονομίαν
2 τινὲς καλοῦσι, (τήρησις γάρ ἐστιν ἐπὶ φαινομένοις
ὡς γεωργία καὶ κυβερνητική, ἀφ' ἧς ἔστιν αὐχ-
μούς τε καὶ ἐπομβρίας λοιμούς τε καὶ σεισμοὺς
καὶ ἄλλας τοιουτώδεις τοῦ περιέχοντος μεταβο-
λὰς προθεσπίζειν), ἀλλὰ πρὸς γενεθλιαλογίαν, ἣν
σεμνοτέροις κοσμοῦντες ὀνόμασιν οἱ Χαλδαῖοι μα-
θηματικοὺς καὶ ἀστρολόγους σφᾶς αὐτοὺς ἀν-
αγορεύουσιν, ποικίλως μὲν ἐπηρεάζοντες τῷ βίῳ,
μεγάλην δ' ἡμῖν ἐπιτειχίζοντες δεισιδαιμονίαν,
μηδὲν δὲ ἐπιτρέποντες κατὰ τὸν ὀρθὸν λόγον
3 ἐνεργεῖν. καὶ τοῦτ' εἰσόμεθα μικρὸν ἄνωθεν προ-
λαβόντες περὶ τῶν συντεινόντων πρὸς τὴν ἐπι-
σκεπτικὴν αὐτῶν μέθοδον. ἔσται δὲ ἐπιδρομικώ-
τερον καὶ ὁλοσχερέστερον τὸ τῆς ὑφηγήσεως· τοῖς

BOOK V

AGAINST THE ASTROLOGERS

The task before us is to inquire concerning astrology 1
or the "Mathematical Art"—not the complete Art as composed of arithmetic and geometry (for we have confuted the professors of these subjects); nor yet that of prediction practised by Eudoxus and Hipparchus[a] and men of their kind, which some
also call "astronomy" (for this, like Agriculture and 2
Navigation, consists in the observation of phenomena, from which it is possible to forecast droughts and rainstorms and plagues and earthquakes and other changes in the surrounding vault of a similar character); it is rather the casting of nativities, which the Chaldeans adorn with more high-sounding titles, describing themselves as "mathematicians" and "astrologers," treating ordinary folk with insolence in various ways, building a great bulwark of superstition against us, and allowing us to do nothing
according to right reason. This we shall understand 3
after we have first traced back a little the things which contribute to their method of speculation; but our exposition will be somewhat cursory and sum-

[a] Eudoxus of Cnidos was a pupil of Plato and Archytas, *circ.* 370 B.C., and a famous astronomer, as was also Hipparchus, of Nicaea in Bithynia, *circ.* 150 B.C.

γὰρ προηγουμένως μετιοῦσι τὸ μάθημα τοῦτο τὰ τῆς ἀκριβείας συγκεχωρήσθω, ἡμῖν δὲ αὔταρκές ἐστι τούτων ἐπιμνησθῆναι ὧν χωρὶς οὐ δυνατὸν ἐπιβάλλειν ταῖς πρὸς τοὺς Χαλδαίους ἀντιρρήσεσιν.

4 Ἐπὶ προϋποκειμένῳ τοίνυν τῷ συμπαθεῖν τὰ ἐπίγεια τοῖς οὐρανίοις καὶ κατὰ τὰς ἐκείνων ἀπορροίας ἑκάστοτε ταῦτα νεοχμοῦσθαι

(τοῖος γὰρ νόος ἐστὶν ἐπιχθονίων ἀνθρώπων
οἷον ἐπ' ἦμαρ ἄγησι πατὴρ ἀνδρῶν τε θεῶν τε)

5 οἱ περιεργότερον ἀναβλέψαντες εἰς τὸ περιέχον Χαλδαῖοι δραστικῶν μὲν αἰτιῶν λόγον ἐπέχειν φασὶν εἰς τὸ ἕκαστον τῶν κατὰ τὸν βίον συμβαινόντων ἐκβαίνειν τοὺς ἑπτὰ ἀστέρας, συνεργεῖν δὲ τὰ τοῦ ζωδιακοῦ μέρη. τὸν μὲν οὖν ζωδιακὸν κύκλον, ὥσπερ κατηχήμεθα, διαιροῦσιν εἰς δεκαδύο ζώδια, ἕκαστον δὲ ζώδιον εἰς μοίρας τριάκοντα (ἔστω γὰρ τοῦτο ἐπὶ τοῦ παρόντος σύμφωνον αὐτοῖς), ἑκάστην δὲ μοῖραν εἰς ἑξήκοντα λεπτά· οὕτω
6 γὰρ καλοῦσι τὰ ἐλάχιστα καὶ ἀμερῆ. τῶν δὲ ζωδίων τὰ μέν τινα ἀρρενικὰ καλοῦσι τὰ δὲ θηλυκά, καὶ τὰ μὲν δίσωμα τὰ δὲ οὔ, καὶ τινὰ μὲν
7 τροπικὰ τινὰ δὲ στερεά. ἀρρενικὰ μὲν οὖν καὶ θηλυκὰ ἅπερ συνεργὸν ἔχει φύσιν πρὸς ἀρρενογονίαν ἢ θηλυγονίαν· κριὸς γὰρ ἀρρενικόν ἐστι ζώδιον, ταῦρος δέ, φασί, θηλυκόν, δίδυμοι ἀρρενικόν, καὶ ἐναλλὰξ τὰ λοιπὰ κατὰ τὴν ὁμοίαν ἀναλογίαν,

---

[a] *Cf. Adv. Phys.* i. 79 ff.

[b] *Cf.* Hom. *Od.* xviii. 136 f.; *P.H.* iii. 244.

mary ; for the exact details may be left to those who specialize in this branch of study, and it is enough for us to call attention to those points without which it is impossible to set about making our attack on the Chaldeans.

It being previously assumed, then, that things on 4
earth " sympathize " with those in the heavens,[a] and that the former are always newly affected by the effluences of the latter

> (As is the day brought on by the Sire of gods and of mortals,
> So are the thoughts of the hearts of us earth-inhabiting creatures),[b]—

on this assumption the Chaldeans, having too curi- 5
ously gazed up into the surrounding vault, declare that the seven stars [c] stand in the relation of efficient causes for the bringing about of everything which occurs in life, and that with them the parts of the zodiac co-operate. Now, as we have been informed, they divide the circle of the zodiac into twelve "*zodia*" (or "Signs"), and each sign into thirty degrees (let this be near enough to their theories, for the present), and each degree into sixty *lepta* (or "minutes"),—for so they call what is minimal and
without parts. And of the Signs some they term 6
masculine, some feminine, and some bi-corporal and some not, and some "tropical" and others fixed.
Masculine and feminine are those which possess a 7
nature which aids the birth of males or females ; thus the Ram is a masculine Sign, but the Bull, they say, is feminine, the Twins masculine, and the rest alternate in a similar proportion, some masculine, others femi-

[c] *i.e.* Sun, Moon, Mars, Jupiter, Venus, Saturn and Mercury.

8 τὰ μὲν ἀρρενικὰ τὰ δὲ θηλυκά. ἀφ' ὧν, οἶμαι,
καὶ οἱ Πυθαγορικοὶ κινηθέντες τὴν μὲν μονάδα
ἄρρεν προσαγορεύουσι, τὴν δὲ δυάδα θῆλυ, τὴν δὲ
τριάδα πάλιν ἄρρεν, καὶ ἀναλόγως [πάλιν] τοὺς
λοιποὺς τῶν τε ἀρτίων καὶ περιττῶν ἀριθμῶν.
9 ἔνιοι δὲ καὶ ἕκαστον ζώδιον εἰς δωδεκατημόρια
διελόντες τῇ αὐτῇ σχεδὸν ἐφόδῳ χρῶνται, οἷον ἐπὶ
κριοῦ τὸ μὲν πρῶτον δωδεκατημόριον αὐτοῦ κριόν
τε καλοῦσι καὶ ἄρρεν, τὸ δὲ δεύτερον ταῦρόν τε
καὶ θῆλυ, τὸ δὲ τρίτον διδύμους τε καὶ ἄρρεν· καὶ
10 ἐπὶ τῶν ἄλλων μοιρῶν ὁ αὐτὸς λόγος. δίσωμα
δὲ λέγουσιν εἶναι ζώδια διδύμους τε καὶ τὸν δια-
μετροῦντα τούτοις τοξότην, παρθένον τε καὶ ἰχ-
11 θύας, οὐ δίσωμα δὲ τὰ λοιπά. καὶ τροπικὰ μὲν
ἐν οἷς γινόμενος ὁ ἥλιος μεταλλάσσει καὶ ποιεῖ τοῦ
περιέχοντος τροπάς, οἷόν ἐστι ζώδιον ὅ τε κριὸς
καὶ τὸ τούτου διάμετρον, καθάπερ ζυγός, αἰγόκερώς
τε καὶ καρκίνος· ἐν κριῷ μὲν γὰρ ἐαρινὴ γίνεται
τροπή, ἐν αἰγοκέρῳ δὲ χειμερινή, ἐν καρκίνῳ δὲ
θερινὴ καὶ ἐν ζυγῷ φθινοπωρινή. στερεὰ δὲ ὑπ-
ειλήφασι ταῦρόν τε καὶ τὸ διαμετροῦν, τουτέστι
σκορπίον, λέοντα καὶ ὑδροχόον.

12 Οὐ μὴν ἀλλὰ καὶ πάντων τούτων τὰ ἐπὶ ἑκάστης
γενέσεως κυριεύοντα ζώδια πρὸς τὴν τῶν ἀπο-
τελεσμάτων ἔκβασιν, καὶ ἀφ' ὧν μάλιστα τὰς
προαγορεύσεις ποιοῦνται, τέσσαρά φασιν εἶναι τὸν
ἀριθμόν, ἅπερ κοινῷ μὲν ὀνόματι κέντρα καλοῦσιν,
ἰδιαίτερον δὲ τὸ μὲν ὡροσκόπον τὸ δὲ μεσουράνημα
τὸ δὲ δῦνον τὸ δὲ ὑπόγαιον καὶ ἀντιμεσουράνημα,

nine. And by them, I suppose, the Pythagoreans 8
were moved to call the monad " male," and the dyad " female," and the triad again " male," and the rest of the even and odd numbers according to the same rule. Some, too, divide each Sign into twelve 9
parts and use much the same method; as, for instance, in the case of the Ram, the first twelfth part of it they describe as the Ram and male, the second as the Bull and female, the third as the Twins and male; and the same rule holds for the other portions. And they say that the Twins, and the Archer dia- 10
metrically opposite to it, and the Virgin and the Fishes are bi-corporal Signs, but the rest are not bi-corporal. And " tropical " Signs are those which 11
when the sun enters it changes its course and produces " tropes " (or " turnings ") in the surrounding vault; such a Sign is the Ram and its diametrically opposite, that is the Scales, and Capricorn and the Crab. For the spring " trope " (or " solstice ") occurs in the Ram, and the winter one in Capricorn, the summer one in the Crab, and the autumn one in the Scales. But they have assumed that the Bull and the Sign diametrically opposite—that is, Scorpio, Leo and Aquarius—are fixed Signs.

However, of all these Signs those which are 12
dominant at each geniture for the production of effective influences and from which they principally frame their prognostications are, they say, four in number; and to these they give the generic name of " Centres," and more specifically they call them " horoscope," " mid-heaven," " setting," " subterranean," and " anti-mid-heaven," this last being itself

13 ὃ καὶ αὐτὸ μεσουράνημα ἐστίν. ὡροσκόπος μὲν οὖν ἐστὶν ὅπερ ἔτυχεν ἀνίσχειν καθ' ὃν χρόνον ἡ γένεσις συνετελεῖτο, μεσουράνημα δὲ τὸ ἀπ' ἐκείνου τέταρτον ζῴδιον σὺν αὐτῷ ἐκείνῳ, δῦνον δὲ τὸ διαμετροῦν τῷ ὡροσκόπῳ, ὑπὸ γῆν δὲ καὶ ἀντιμεσουράνημα τὸ διαμετροῦν τῷ μεσουρανήματι, οἷον (ἔσται γὰρ σαφὲς ἐπὶ παραδείγματος) καρκίνου ὡροσκοποῦντος μεσουρανεῖ μὲν κριός, δύνει
14 δὲ αἰγόκερως, ὑπὸ γῆν δέ ἐστι ζυγός. οὐ μὴν ἀλλὰ καὶ ἑκάστου τούτων τῶν κέντρων τὸ μὲν προάγον ζῴδιον ἀπόκλιμα καλοῦσι, τὸ δὲ ἑπόμενον

also "mid-heaven."[a] Now the "horoscope" is the Sign which happens to arise at the time when the birth is completed; the "mid-heaven" is the fourth Sign therefrom, it being included; the "setting" Sign is that diametrically opposite to the "horoscope"; and the "subterranean" or "anti-mid-heaven" is that diametrically opposite to the "mid-heaven": thus (for an example will make it clear), when the Crab is the "horoscope" the Ram is the "mid-heaven," and Capricorn is the "setting," and the Scales is the "subterranean." Moreover, in the case of each of these "Centres" they call the preceding Sign "declination" and the following one 13 14

[a] The diagram below, taken from Fabricius, will help to explain this passage.

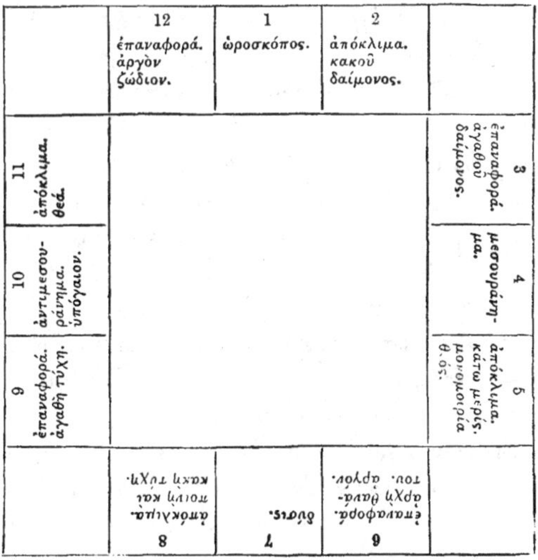

15 ἐπαναφοράν. ἤδη δὲ τὸ μὲν προαναφερόμενον τοῦ
ὡροσκοποῦντος ζωδίου, ἐν τῷ φανερῷ ὄν, κακοῦ
δαίμονός φασιν εἶναι, τὸ δὲ μετὰ τοῦτο, ἑπόμενον
δὲ τῷ μεσουρανοῦντι, ἀγαθοῦ δαίμονος, τὸ δὲ
προάγον τοῦ μεσουρανοῦντος κάτω μερίδα καὶ
μονομοιρίαν καὶ θεόν, τὸ δὲ ἐρχόμενον ἐπὶ τὴν
16 δύσιν ἀργὸν ζώδιον καὶ ἀρχὴν θανάτου, τὸ δὲ μετὰ
τὴν δύσιν ἐν τῷ ἀφανεῖ ποινὴν καὶ κακὴν τύχην,
ὅπερ καὶ διάμετρόν ἐστι τῷ κακῷ δαίμονι, τὸ δὲ
ἐρχόμενον ὑπὸ γῆν ἀγαθὴν τύχην, διαμετροῦν τῷ
17 ἀγαθῷ δαίμονι, τὸ δὲ ἀποχωροῦν ἀπὸ τοῦ ἀντιμεσ-
ουρανήματος ὡς ἐπ' ἀνατολὴν θεάν, διαμετροῦν τῷ
θεῷ, τὸ δὲ ἐπιφερόμενον τῷ ὡροσκόπῳ ἀργόν, ὃ
18 πάλιν διαμετρεῖ τῷ ἀργῷ. ἢ ἵνα συντομώτερον
φῶμεν, τοῦ ὡροσκοποῦντος ζωδίου τὸ μὲν ἀπό-
κλιμα καλεῖται κακὸς δαίμων, ἡ δ' ἐπαναφορὰ
ἀργόν· ὡσαύτως τοῦ μεσουρανήματος τὸ μὲν ἀπό-
19 κλιμα θεός, ἡ δ' ἐπαναφορὰ ἀγαθὸς δαίμων· κατὰ
τὰ αὐτὰ δὲ καὶ τοῦ ἀντιμεσουρανήματος τὸ μὲν
ἀπόκλιμα θεά, ἡ δὲ ἐπαναφορὰ ἀγαθὴ τύχη· ὁ-
μοίως τοῦ δύνοντος τὸ μὲν ἀπόκλιμα κακὴ τύχη,
20 ἡ δὲ ἐπαναφορὰ ἀργόν. ταῦτα δ' οἴονται οὐ παρ-
έργως ἐξετάζειν· οὐ γὰρ τὴν αὐτὴν δύναμιν ἔχειν
ἡγοῦνται τοὺς ἀστέρας πρὸς τὸ κακοποιεῖν ἢ μὴ
ἐπί τε τῶν κέντρων θεωρουμένους καὶ ἐπὶ ταῖς
ἀναφοραῖς καὶ τοῖς ἀποκλίμασιν, ἀλλ' ὅπου μὲν
21 ἐνεργεστέραν ὅπου δὲ ἀπρακτοτέραν. ἦσαν δέ

" ascension." Also they say that that which ascends 15
before the Sign of the horoscope, and is in view, is
that of " the evil daemon," and that after it, which
follows the " mid-heaven " Sign, is that of " the good
daemon," and that which precedes the " mid-
heaven " sign is " inferior part " and " single portion "
and " god," and that which comes to the " set-
ting " is an " ineffective " Sign and " principle of
death," and that which comes after the " setting " 16
and is out of view is " punishment " and " ill for-
tune,"—and it is diametrically opposite to the " evil
daemon,"—and that which comes to the " subter-
ranean " is " good fortune," being diametrically
opposite to the " good daemon," and that which 17
comes next after the " anti-mid-heaven," towards
the orient, is " goddess," being diametrically oppo-
site to the " god," and that which comes next to the
" horoscope " is " ineffective," and it again is dia-
metrically opposite to the " ineffective." Or, to 18
speak more concisely, the " declination " of the Sign
of the " horoscope " is called the " evil daemon,"
and its " ascension " " ineffective " ; similarly the
declination of the " mid-heaven " is " god," and its 19
ascension " good daemon " ; and in the same way
the declination of the " anti-mid-heaven " is " god-
dess," and its ascension " good fortune " ; likewise
the declination of the " setting " is " ill fortune,"
and its ascension " ineffective." And they think 20
that their searching out of these things is no mere
by-play ; for they believe that the stars have not the
same power of doing, or not doing, harm when
observed at the " centres " or at their ascensions and
declinations, but their power is more effective in one
position and less effective in another. And there 21

τινες Χαλδαίων οἱ καὶ ἕκαστον μέρος τοῦ ἀνθρω-
πείου σώματος ἑκάστῳ τῶν ζωδίων ἀνατιθέντες
ὡς συμπαθοῦν· κριὸν μὲν γὰρ κεφαλὴν ὀνομά-
ζουσι, ταῦρον δὲ τράχηλον, διδύμους δὲ ὤμους,
καρκίνον δὲ στέρνον, λέοντα δὲ πλευράς, παρθένον
22 δὲ γλουτούς, ζυγὸν δὲ λαγόνας, σκορπίον αἰδοῖον
καὶ μήτραν, τοξότην μηρούς, αἰγόκερων γόνατα,
ὑδροχόον κνήμας, ἰχθύας δὲ πόδας. καὶ ταῦτα
πάλιν οὐκ ἀσκόπως, ἀλλ' ἐπείπερ, ἐὰν ἔν τινι
τούτων τῶν ζωδίων γένηται τῶν κατὰ τὴν γένεσιν
κακοποιῶν ἀστέρων τις, πήρωσιν τοῦ ὁμωνυμοῦν-
τος ἀπεργάζεται μέρους.

Ταῦτα μὲν οὖν περὶ τῆς φύσεως τῶν ἐν τῷ ζω-
23 διακῷ κύκλῳ κεφαλαιωδέστερον ὑποδεδείχθω· οὐκ
ἄτοπον δὲ ἑξῆς διελθεῖν καὶ περὶ τῆς διαιρέσεως
αὐτῶν. ἐπιστάσεως γὰρ οὔσης ὡς τῶν ζωδίων
μὴ κατ' ἰδίαν περιγραφὴν θεωρουμένων ἀλλ' ἑπτὰ
ἀστέρων διεσπαρμένων παρατηρήσει, ἐπῆλθεν αὐ-
τοῖς εἰς δώδεκα μοίρας τὸν ὅλον καταδιελεῖν κύ-
24 κλον. ὑποδεικνύντες γὰρ τὴν ἔφοδον φασὶν ⟨ὅτι⟩[1]
ἕνα τινὰ τῶν ἐν τῷ ζωδιακῷ κύκλῳ λαμπρὸν
ἀστέρα παρατηρήσαντες ἀνατέλλοντα οἱ πάλαι,
εἶτα ἀμφορέα τετρημένον πληρώσαντες ὕδατος
εἴασαν ῥεῖν εἴς τι ἕτερον ὑποκείμενον ἀγγεῖον μέχρι
τοῦ τὸν αὐτὸν ⟨αὖτις⟩[2] ἀνασχεῖν ἀστέρα, στοχασά-
μενοί τε ἀπὸ τοῦ αὐτοῦ σημείου ἐπὶ τὸ αὐτὸ ση-
μεῖον γεγονέναι τὴν τοῦ κύκλου περιστροφὴν πάλιν
25 ἐλάμβανον τὸ δωδέκατον τοῦ ῥυέντος, καὶ ἐσκέ-
πτοντο ἐν πόσῳ τοῦτο ἔρρευσε χρόνῳ· ἐν τοσούτῳ
γὰρ ἔλεγον καὶ τὸ δωδέκατον μέρος ἀνεληλυθέναι

[1] ⟨ὅτι⟩ add. cj. Bekk. [2] ⟨αὖτις⟩ add. Nebe.

have been some Chaldeans who have referred each part of the human body to one of the Signs as " sympathizing " therewith ; thus they call the head the Ram, the neck the Bull, the shoulders the Twins, the breast the Crab, the sides the Lion, the buttocks
the Virgin, the flanks the Scales, the pudenda and 22
womb the Scorpion, the thighs the Archer, the knees Capricorn, the shins Aquarius, the feet the Fishes.[a] And this again is not done at random, but for the reason that if any one of the stars which are maleficent at the time of nativity is in any of these Signs it produces an imperfection in the part which bears the same name.

Let this then serve as a rather summary account of the nature of the things within the circle of the
zodiac ; and, next, it is not out of place to explain 23
the division of them. Being halted in their observations, because the Signs were not being viewed according to their own proper determinations but by keeping watch on seven dispersed stars, it occurred to them
to divide up the whole circle into twelve portions. For 24
in indicating the method of their approach they say that the ancients, after observing the rising of some particular bright star in the circle of the zodiac, proceeded next to fill with water a jar which had holes in it and then let the water flow into another receptacle placed underneath until the same star had arisen again ; and as they conjectured that the revolution of the circle was from the same Sign to the same Sign,
they next took the twelfth part of the water which had 25
flowed through and calculated how long a time it took in flowing ; for that, they said, was the length of time taken in covering the twelfth part of the circle,

[a] *Cf.* Manilius ii. 456 ff., iv. 702 ff.

*τοῦ κύκλου, καὶ τοῦτον ἔχειν τὸν λόγον τὸ ἀνενεχ-
θὲν μέρος τοῦ κύκλου πρὸς τὸν ὅλον κύκλον, ὃν
ἔχει τὸ ῥυὲν τοῦ ὕδατος μέρος πρὸς τὸ ὅλον ὕδωρ.
26 ἐκ ταύτης τῆς ἀναφορᾶς, φημὶ δὲ τοῦ δωδεκατη-
μορίου, τὸ τελευταῖον πέρας ἐσημειοῦντο ἀπὸ
ἀστέρος τινὸς ἐπιφανοῦς κατ' αὐτὸ θεωρουμένου
ἢ ἀπό τινος τῶν συνανατελλόντων βορειοτέρων ἢ
νοτιωτέρων. τὸ δὲ αὐτὸ ἐποίουν καὶ ἐπὶ τῶν
ἄλλων δωδεκατημορίων.*

*Ἀλλ' ἡ μὲν ἔφοδος καθ' ἣν εἰς τοσαύτας μοίρας
τὸν ζῳδιακὸν καταδιαιροῦσι κύκλον, ἐστὶ τοιαύτη·
27 ἀνάλογος δ' ἔοικεν εἶναι καὶ καθ' ἣν τὸν ἐφ' ἑκά-
στης γενέσεως ὡροσκόπον ἀρχικῶς παρατετηρη-
κέναι λέγουσιν. νύκτωρ μὲν γὰρ ὁ Χαλδαῖος,
φασίν, ἐφ' ὑψηλῆς τινὸς ἀκρωρείας ἐκαθέζετο
ἀστεροσκοπῶν, ἕτερος δὲ παρήδρευε τῇ ὠδινούσῃ
28 μέχρις ἀποτέξοιτο, ἀποτεκούσης δὲ εὐθὺς δίσκῳ
διεσήμαινε τῷ ἐπὶ τῆς ἀκρωρείας· ὁ δὲ ἀκούσας
καὶ αὐτὸς παρεσημειοῦτο τὸ ἀνίσχον ζῴδιον ὡς
ὡροσκοποῦν. μεθ' ἡμέραν δὲ τοῖς ὡροσκοπίοις
προσεῖχε καὶ ταῖς τοῦ ἡλίου κινήσεσιν.*

*29 Ἀλλὰ ταῦτα μὲν περὶ ζῳδίων· τῶν δὲ ἀστέρων
ἐνίους μὲν ἀγαθοποιοὺς εἶναι λέγουσιν ἐνίους δὲ
κακοποιοὺς τινὰς δὲ καὶ κοινούς, οἷον ἀγαθοποιοὺς
μὲν τὸν τοῦ Διὸς καὶ τὸν τῆς Ἀφροδίτης, κακο-
ποιοὺς δὲ τὸν τοῦ Ἄρεως καὶ Κρόνου, ἐπίκοινον
δὲ τὸν τοῦ Ἑρμοῦ, ἐπείπερ μετὰ μὲν ἀγαθοποιῶν
30 ἀγαθοποιὸς μετὰ δὲ κακοποιῶν κακοποιός. ἄλλοι
δὲ τοὺς αὐτοὺς ἀστέρας κατ' ἄλλην καὶ ἄλλην σχέ-
σιν ὁτὲ μὲν ἀγαθοποιοὺς ὁτὲ δὲ κακοποιοὺς ὑπάρ-
χειν νομίζουσιν· ἢ γὰρ παρὰ τὸ ζῴδιον ἢ παρὰ
τοὺς τῶν ἄλλων ἀστέρων συσχηματισμοὺς οὔτε ὁ*

and the part of the circle covered bears to the whole circle the same ratio as the part of the water which
has flowed bears to the whole of the water. From 26
this proportion—that of the twelfth part, I mean—they marked off the final limit from some one conspicuous star observed at the time or from one of the more northerly or southerly stars which rise simultaneously. And they did the same in the case of the other twelfth portions.

Such, then, is the method of approach which led them to divide the circle of the zodiac into this num-
ber of portions ; and analogous to it seems to be the 27
method by which, as they say, they originally came to observe the horoscope at each nativity. For by night, they say, the Chaldean sat on a high peak watching the stars, while another man sat beside the woman in
labour till she should be delivered, and when she 28
had been delivered he signified the fact immediately to the man on the peak by means of a gong; and he, when he heard it, noted the rising Sign as that of the horoscope. But during the day he studied the horologes (or sun-dials) and the motions of the sun.

So much, then, for the Signs : as to the stars,[a] they 29
say that some of them are "beneficent," some "maleficent," and some "common" ; thus Jupiter and Venus are beneficent, but Mars and Saturn maleficent, while Mercury is "common" since it is beneficent when with beneficent stars, but maleficent
when with maleficent. But others believe that the 30
same stars are at one time beneficent and at another maleficent according to their varying positions ; for either by reason of the Sign, or by reason of the configurations of the other stars, the maleficent star

[a] *i.e.* the "seven stars": see p. 325 n.

κακοποιὸς ἀστὴρ πάντως κακοποιός ἐστιν οὔτε
31 ὁ ἀγαθοποιὸς πάντως ἀγαθοποιός ἐστιν. πλὴν τῶν
ἑπτὰ ἡγεῖσθαι μὲν τὸν ἥλιον καὶ τὴν σελήνην οἴονται, ἐλάττονα δὲ τούτων δύναμιν ἔχειν πρὸς τὰς τῶν ἀποτελεσμάτων ἐκβάσεις τοὺς λοιποὺς πέντε· παρ' ἣν αἰτίαν οἱ Αἰγύπτιοι βασιλεῖ μὲν καὶ δεξιῷ ὀφθαλμῷ ἀπεικάζουσι τὸν ἥλιον, βασιλείᾳ δὲ καὶ ἀριστερῷ ὀφθαλμῷ τὴν σελήνην, ῥαβδοφόροις δὲ τοὺς πέντε ἀστέρας, τῷ δὲ λοιπῷ λαῷ τοὺς ἄλλους
32 ἀπλανεῖς. καὶ τῶν πέντε ἡλίῳ μὲν συμφωνεῖν καὶ
συνεπικουρεῖν φασὶ Κρόνον τε καὶ Δία καὶ Ἑρμῆν, οὓς καὶ ἡμερινοὺς καλεῖσθαι διὰ τὸ τὸν ἥλιον, ᾧ συνεργοῦσι, τῶν μεθ' ἡμέραν γεννωμένων ἐπικρατεῖν, ⟨σελήνῃ δὲ Ἄρην τε καὶ Ἀφροδίτην⟩.[1]
33 τοὺς δὲ αὐτοὺς ἀστέρας μείζονα μᾶλλον ἴσχειν
δύναμιν ἢ παρὰ τὸ ἐν ἰδίοις οἴκοις ὑπάρχειν ἢ ὑψώμασιν ἢ ὁρίοις, ἢ παρὰ τὸ δορυφορεῖσθαί τινας ὑπό τινων, ἢ παρὰ τὸ ἐπιβλέπειν ἀλλήλους καὶ συσχηματίζεσθαι ἀλλήλοις, ἢ παρὰ τὸ ἐπὶ κέντροις
34 εἶναι. οἶκος δέ ἐστι κατ' αὐτοὺς ἡλίου μὲν λέων,
σελήνης δὲ καρκίνος, Κρόνου δὲ αἰγόκερως καὶ ὑδροχόος, Διὸς τοξότης καὶ ἰχθύες, Ἄρεως κριὸς καὶ σκορπίος, Ἀφροδίτης ταῦρος καὶ ζυγός, Ἑρ-
35 μοῦ δίδυμοι καὶ παρθένος. ὑψώματα δὲ καλοῦσιν
ἀστέρων, καὶ ταπεινώματα ὡσαύτως, τὰ ἐν οἷς χαίρουσιν ἢ ὀλίγην δύναμιν ἔχουσιν· χαίρουσι μὲν γὰρ ἐν τοῖς ὑψώμασιν, ὀλίγην δὲ δύναμιν ἔχουσιν
36 ἐν τοῖς ταπεινώμασιν. οἷον ἡλίου μὲν ὕψωμα
κριός, καὶ πρὸς ἀκρίβειαν ἡ ἐννεακαιδεκάτη τούτου μοῖρα, ταπείνωμα δὲ τὸ διαμετροῦν ζώδιον,

[1] ⟨σελήνῃ . . . Ἀφροδίτην⟩ add. cj. Bekk.

is not entirely maleficent, nor is the beneficent entirely beneficent. They suppose, however, that the 31
Sun and the Moon are the principal stars of the seven, and that the other five have less power than these for the issues of the "effects"; and for this reason the Egyptians liken the Sun to the king and the right eye, and the Moon to the queen and the left eye, and the five stars to lictors, and the other fixed stars to the rest of the people. And they say that, of the five, Saturn 32
and Jupiter and Mercury are in accord with and join in aiding the Sun, and that these stars are called "diurnal" because the Sun, with which they co-operate, governs those born by day, ⟨and that Mars and Venus aid the Moon⟩. They say also that the same stars 33
have increased power owing to their being in their proper "houses" or "elevations" or "boundaries," or owing to the fact that some are "guarded" by others, or because they "look towards" one another or are in a certain "configuration" one with another, or because they are at the "centres."[a]
And, according to them, the Lion is the house of 34
the Sun, the Crab of the Moon, Capricorn and Aquarius of Saturn, the Archer and the Fishes of Jupiter, the Ram and the Scorpion of Mars, the Bull and the Scales of Venus, and the Twins and the Virgin of Mercury.—And they term the positions in which 35
the stars rejoice or in which they have little power "elevations" and "depressions" respectively; for they rejoice in their "elevations" but in their "depressions" they possess little power. Thus the Ram is 36
the "elevation" of the Sun (or, to be exact, the nineteenth part of it), and its "depression" is the Sign diametrically opposite; and of the Moon, again, the

[a] See §§ 12 ff. (pp. 327 ff.).

σελήνης δὲ πάλιν ὕψωμα μὲν ταῦρος ταπείνωμα δὲ τὸ διαμετροῦν, Κρόνου ζυγός, Διὸς καρκίνος, Ἄρεως αἰγόκερως, Ἀφροδίτης ἰχθύες, Ἑρμοῦ παρθένος. καὶ ταπεινώματα τούτων, ὡς ἔφην, τὰ
37 διαμετροῦντα τῶν ὑψωμάτων. ὅρια δὲ ἀστέρων προσαγορεύουσιν ἐν ἑκάστῳ ζωδίῳ ἐν οἷς ἕκαστος τῶν ἀστέρων ἀπὸ ποστῆς μοίρας ἐπὶ ποστὴν μοίραν πλεῖστον δύναται· περὶ ὧν οὐχ ἡ τυχοῦσα παρ' αὐτοῖς ἐστὶ καὶ κατὰ τοὺς πίνακας διαφωνία.
38 δορυφορεῖσθαι δὲ ἀστέρας λέγουσιν, ὅταν μέσοι ὦσιν ἄλλων ἀστέρων ἐν συνεχείᾳ ζωδίων· οἷον ἐὰν τοῦ αὐτοῦ ζωδίου ὃς μέν τις ἀστὴρ τὰς πρώτας ἐπέχῃ μοίρας ὃς δὲ τὰς τελευταίας ὃς δὲ τὰς ἐν μέσῳ, δορυφορεῖσθαι λέγεται ὁ ἐν μέσῳ ὑπὸ τῶν
39 τὰς ἐπ' ἄκροις ἐπεχόντων μοίρας. ἐπιβλέπειν δὲ λέγονται ἀλλήλους καὶ συμφωνεῖν ἀλλήλοις ὡς οἱ κατὰ τρίγωνον ἢ τετράγωνον φαινόμενοι. κατὰ τρίγωνον μὲν οὖν σχηματίζονται καὶ ἐπιθεωροῦσιν ἀλλήλους ἀστέρες οἱ [ἐπὶ] τριῶν ζωδίων ἔχοντες τὸ μεταξὺ διάστημα, κατὰ τετράγωνον δὲ οἱ δυοῖν.
40 καὶ δοκεῖ κατὰ μὲν τρίγωνον ἀγαθοποιῷ κακοποιὸς συσχηματιζόμενος εὐεργετικὸς εἶναι καὶ πολὺ μᾶλλον ἀγαθοποιός, ἀγαθοποιῷ δὲ ἤπιος αὐτὸ μόνον, καὶ κακοποιὸς κακοποιῷ, κατὰ δὲ τετράγωνον ἀνάπαλιν. ἐπίκεντροι δὲ λέγονται οἱ ἐπί τινος τῶν κέντρων θεωρούμενοι, ἤτοι ἐπὶ τοῦ ὡροσκόπου ἢ τοῦ μεσουρανήματος ἢ δύσεως ἢ ἀντιμεσουρανήματος.

"elevation" is the Bull and the "depression" the Sign diametrically opposite; and the elevation of Saturn is the Scales, of Jupiter the Crab, of Mars Capricorn, of Venus the Fishes, of Mercury the Virgin. And the depressions of these, as I said, are the Signs
diametrically opposite the elevations. And the 37
"boundaries" of the stars, as they call them, in each Sign are those within which, from a certain portion to a certain portion, they possess most power; and about these there is no little disagreement amongst
them and in their tablets too. And they say that the 38
stars are "guarded" when they are in the middle of other stars and in continuity with the Signs; thus if in the same Sign one star occupies the first portions and another the last and another the midmost, then the star in the middle is said to be "guarded" by
those occupying the extreme portions. And they 39
are said to "look towards" one another and to "agree with" one another, as in the case of those which appear in a triangular or quadrangular figure. Now the stars which occupy the middle interval of three Signs form a triangular figure and look towards one another, and those between two Signs form a
quadrangular figure. And it is thought that when 40
a maleficent star is in opposition to a beneficent in a triangular figure it is "well-disposed" and much more beneficent, and that a kindly in opposition to a beneficent is just "kindly," and so with a maleficent in opposition to a maleficent; but the reverse is the case when the figure is quadrangular. And they are called "epicentric" when they are observed at one of the "centres," either at the "horoscope," or at the "mid-heaven," or at the "setting," or at the "anti-mid-heaven."

41 Ἀλλὰ γὰρ τούτων οὕτως ἡμῖν ὡς ἐν τύπῳ καὶ
ὁλοσχερῶς ἐκκειμένων προληπτέον ὡς ἀπ' αὐτῶν
ὁρμηθέντες οἱ Χαλδαῖοι τὰς προαγορεύσεις ποιοῦν-
ται τῶν ἀποτελεσμάτων. διαφορὰ δὲ ἔστιν αὐ-
τῶν, ἐπεὶ τὰ μὲν ἁπλούστερα καθειστήκει τὰ δὲ
ἀκριβέστερα, καὶ ἁπλούστερα μὲν τὰ κατὰ ζώδιον
ἢ ἁπλῆν ἀστέρος δύναμιν γινόμενα, οἷον ὅτι ὅδε ὁ
ἀστὴρ ἐν τῷδε τῷ ζωδίῳ γενόμενος τοιούτους
42 ποιεῖ, ἀκριβέστερα δὲ τὰ κατὰ συνδρομὴν καὶ ὡς
αὐτοὶ λέγουσι τὰ κατὰ σύγκρασιν πλειόνων, οἷον
" ἐὰν ὅδε μὲν ὡροσκοπῇ ὅδε δὲ μεσουρανῇ ὅδε
δὲ ἀντιμεσουρανῇ οἱ δὲ ἄλλοι οὕτως ἔχωσι, συμ-
βήσεται τάδε."
43 Ὁ μὲν οὖν χαρακτὴρ τῆς Χαλδαϊκῆς μεθόδου
τοιοῦτος ἔοικεν εἶναι· ῥᾴδιον δ' ἔστι λοιπὸν ἐπὶ
παραδοθέντι τούτῳ συμπεριφέρεσθαι ταῖς κομιζο-
μέναις ἀντιρρήσεσιν. καὶ δὴ ἔνιοι μὲν ἀγροικό-
τερον πειρῶνται διδάσκειν ὡς οὐ πάντως συμ-
44 πάσχει τοῖς οὐρανίοις τὰ ἐπίγεια· οὐδὲ γὰρ οὕτως
ἥνωται τὸ περιέχον ὡς τὸ ἀνθρώπινον σῶμα, ἵνα
ὃν τρόπον τῇ κεφαλῇ τὰ ὑποκείμενα μέρη συμ-
πάσχει καὶ τοῖς ὑποκειμένοις ἡ κεφαλή, οὕτω καὶ
τοῖς ἐπουρανίοις τὰ ἐπίγεια, ἀλλά τις ἔστι τούτων
διαφορὰ καὶ ἀσυμπάθεια ὡς ἂν μὴ μίαν καὶ τὴν
45 αὐτὴν ἐχόντων ἕνωσιν. ἄλλοι δὲ καὶ τὸν περὶ
εἱμαρμένης κινοῦσι λόγον· εἰ γὰρ μὴ πάντα γίνεται
κατὰ εἱμαρμένην, οὐκ ἔστι Χαλδαϊκὴ ἡ τοῦτο
ἀξιοῦσα [κατὰ εἱμαρμένην εἶναι]. οὐκ ὀλίγοι δὲ
46 ἦσαν οἱ κἀκεῖνο συνερωτῶντες. ἐπεὶ τῶν γινο-

But now that these matters have thus been ex- 41
pounded by us in outline and summarily, it must first be grasped that the Chaldeans start from them in making their forecasts of the "effects." And in these there is a difference, since some are more simple, others more exact; more simple are those which occur by reason of the Sign or the simple power of a star,—as, for instance, that "this particular star when it is in this particular Sign produces men of such
and such a kind"; and more exact are those which 42
occur through concurrence and, as they say, through the blending of several factors,—as, for instance, "if this star is in the horoscope, and that in mid-heaven, and that other in anti-mid-heaven, and the rest are in certain other positions, then the effects will be as follows."

Such then, it seems, is the main outline of the 43
Chaldean doctrine; and now that this has been expounded it is easy to follow intelligently the counter-arguments which are brought forward. Some people, indeed, try to argue quite bluntly that terrestrial things do not "sympathize" altogether
with things celestial; for the surrounding vault is 44
not unified in the same way as the human body, so that things on earth should "sympathize" with things in the heavens in the same way as the lower parts of the body sympathize with the head, and the head with the lower parts, but in respect of the former there exists a difference and want of sympathy, as they have not one and the same unification.
—And others raise the argument concerning destiny; 45
for unless all things happen according to destiny, astrology, which maintains this, does not exist.—And there have been not a few who propound the following

μένων τὰ μὲν κατ' ἀνάγκην γίνεται τὰ δὲ κατὰ
τύχην τὰ δὲ παρ' ἡμᾶς, πάντως οἱ Χαλδαῖοι, εἰ
δυνατῆς ἐφίενται προρρήσεως, ἤτοι ἐν τοῖς κατ'
ἀνάγκην ποιήσονται τὰς προαγορεύσεις ἢ ἐν τοῖς
47 κατὰ τύχην ἐκβαίνουσιν ἢ ἐν τοῖς παρ' ἡμᾶς. καὶ
εἰ μὲν ἐν τοῖς κατ' ἀνάγκην, ἀνωφελεῖς εἰσὶν ἐν
τῷ βίῳ· τὸ γὰρ κατ' ἀνάγκην συμβαῖνον οὐκ
ἔστιν ἐκκλῖναι, ἀλλ' ἐάν τε θέλωμεν ἐάν τε μὴ
θέλωμεν, ἐκβῆναι δεῖ τὸ τοιοῦτο. τότε δ' ἂν
χρειώδης ἐτύγχανεν ἡ πρόρρησις, εἰ πρὸς τὴν
ἔκκλισιν αὐτοῦ τὴν ἀναφορὰν ἐλάμβανεν. εἰ δ' ἐν
τοῖς τυχηροῖς, ἀδύνατόν τι ἐπαγγέλλονται· ἄστατα
γὰρ τὰ τυχηρῶς γινόμενα, τῶν δὲ ἀστάτων καὶ
ἄλλοτε ἄλλως ἐκβαινόντων οὐκ ἔνεστιν ἑστῶσαν
48 ποιεῖσθαι τὴν προαγόρευσιν. λείπεται οὖν ἐν τοῖς
παρ' ἡμᾶς γιγνομένοις αὐτοὺς ποιεῖσθαι τὰς προρ-
ρήσεις. ὃ πάλιν ἀμήχανον· τὸ γὰρ ἐπ' ἐμοὶ κεί-
μενον ἐκβῆναι ἢ μή, καὶ τὸ μὴ ἔχον ἀρχῆθεν
προκαταβεβλημένην αἰτίαν, οὐκ ἂν δύναιτό τις
προλέγειν. οὐκ ἄρα δυνατῆς ἐφίενται προρρήσεως
οἱ Χαλδαῖοι.

49 Οἱ μὲν οὖν πλείους διὰ τοιούτων τινῶν ἀκροβο-
λισμῶν πειρῶνται τὴν Χαλδαϊκὴν μέθοδον ἀναι-
ρεῖν· ἡμεῖς δὲ κατὰ τὸν ὁμόθεν[1] τῆς ἐπιχειρήσεως
τρόπον τὰς ἀρχὰς καὶ ὥσπερ στοιχεῖα ταύτης
κινήσαντες ἕξομεν ⟨σὺν⟩ αὐτοῖς[2] καὶ τὴν τῶν
λοιπῶν θεωρημάτων σύστασιν ἠθετημένην.

50 Ἀρχὴ τοίνυν καὶ ὥσπερ θεμέλιος τῆς Χαλδαϊκῆς
ἐστὶ τὸ στῆσαι τὸν ὡροσκόπον· ἀπὸ τούτου γὰρ
τὰ λοιπὰ τῶν κέντρων λαμβάνεται, τά τε ἀποκλί-

[1] ὁμόθεν scripsi (cf. Xen. *Cyr.* 8. 8. 22) : ὅμοιον Bekk.
[2] ⟨σὺν⟩ αὐτοῖς scripsi : αὐταῖς mss., Bekk. : αὐτῆς Herv.

argument: Since some events occur by necessity, 46
some by chance, and some by our action, if the Chaldeans aim at a possible prophecy, they will certainly make their forecasts about events which result either from necessity or from chance or from
our action. But if they do so about necessary events, 47
their forecasts are useless in practice; for it is impossible to avert what happens by necessity, for that must take effect whether we like it or dislike it. And the prophecy would have been useful only if it had had reference to the means of averting it. And if it is about chance events, they profess what is impossible; for chance events are irregular, and of things which are irregular and turn out differently at different times it is not feasible to form a reliable
forecast. It remains, then, to say that they make 48
their prophecies about things which occur through our own action. But this again is impossible; for that which depends on me as to whether it occurs or not, and which has no original predetermined cause, no one can possibly predict. Therefore the Chaldeans do not aim at a prophecy that is possible.

The majority, then, try to abolish the Astrologers' 49
doctrine by this sort of long range fire; but we shall adopt a method of attack at close quarters, and when we have overthrown its principles and elements, so to call them, along with them we shall also find the structure of the rest of their theories demolished.

The principle and foundation, as it were, of astrology 50
is the setting up of the "horoscope"; for the rest of the "centres" are taken from this, also the declina-

ματα καὶ αἱ ἐπαναφοραὶ τά τε τρίγωνα καὶ τὰ
τετράγωνα καὶ οἱ κατ' αὐτὰ σχηματισμοὶ τῶν
ἀστέρων, ἀπὸ δὲ πάντων τούτων αἱ προαγορεύσεις.
51 ὅθεν ἀναιρεθέντος τοῦ ὡροσκόπου κατ' ἀνάγκην
οὐδὲ τὸ μεσουρανοῦν ἐστὶν ἢ δῦνον ἢ ἀντιμεσου-
ρανοῦν γνώριμον· τούτων δὲ ἀκαταληπτουμένων
52 συναφανίζεται πᾶσα ἡ Χαλδαϊκὴ μέθοδος. ὅτι δὲ
ἀνεύρετον αὐτοῖς ἐστὶ τὸ ὡροσκοποῦν ζώδιον
ποικίλως ἔνεστι διδάσκειν. ἵνα γὰρ τοῦτο κατα-
ληφθῇ, δεῖ πρῶτον μὲν τὴν γένεσιν τοῦ πίπτοντος
ὑπὸ τὴν ἐπίσκεψιν βεβαίως κατειλῆφθαι, δεύτερον
δὲ τὸ διασημαῖνον ταύτην ὡροσκόπιον ἀπλανὲς
ὑπάρχειν, τρίτον δὲ τὴν ἀναφορὰν τοῦ ζωδίου πρὸς
53 ἀκρίβειαν συνῶφθαι. ἐπὶ μὲν γὰρ τῆς ἀποτέξεως
ἡ ἀναφορὰ τοῦ κατ' οὐρανὸν ἀνίσχοντος ζωδίου
τετήρηται, καθάπερ διακόνῳ πρὸς τὴν τήρησιν τοῦ
ὡροσκόπου χρησαμένων τῶν Χαλδαίων· ἐπὶ δὲ τῇ
ἀναφορᾷ ὁ συσχηματισμὸς τῶν ἄλλων ἀστέρων,
ὅπερ διάθεμα καλοῦσι, καὶ ἐπὶ τῷ διαθέματι αἱ
54 προαγορεύσεις. οὔτε δὲ τὴν γένεσιν τῶν ὑπὸ τὴν
ἐπίσκεψιν πιπτόντων λαμβάνειν δυνατόν ἐστιν, ὡς
παραστήσομεν, οὔτε τὸ ὡροσκόπιον ἀπλανὲς καθ-
έστηκεν, οὔτε τὸ ἀνίσχον ζώδιον πρὸς ἀκρίβειαν
καταλαμβάνεται. τοίνυν ἀσύστατός ἐστιν ἡ τῶν
55 Χαλδαίων μέθοδος. λέγωμεν δὲ περὶ τοῦ πρώτου
πρῶτον.

Τὴν δὴ γένεσιν τῶν ὑπὸ τὴν ἐπίσκεψιν πεσουμένων ἀρχαϊκώτερον ἤτοι ἀπὸ τῆς τοῦ σπέρματος καταβολῆς καὶ συλλήψεως λαμβάνουσιν ἢ ἀπὸ τῆς ἐκτέξεως. ἀλλ' ἀπὸ μὲν τῆς τοῦ σπέρματος καταβολῆς καὶ συλλήψεως οὐκ ἂν εἴποιεν· ἀκατάληπτος

tions and the ascensions, and the triangular and quadrangular figures, and the resultant configurations of the stars, and the prognostications derived from all
these. Hence, if the "horoscope" is abolished, in- 51
evitably the "mid-heaven" is not known either, nor the "setting," nor the "anti-mid-heaven"; and these being non-apprehensible, therewith the whole astro-
logical doctrine disappears. That the Sign of the horo- 52
scope is indiscoverable by them one may show in many ways. For in order that it may be apprehended, in the first place the time of birth of the subject of investigation must be firmly apprehended, and secondly the "horologe" which signifies this must be unerring, and thirdly the "ascension"[a] of the Sign
must have been observed accurately. For at the 53
time of birth the ascension of the Sign which is rising in the heavens is observed, the Chaldeans using it as a minister for the observation of the horoscope; and after the ascension, the configuration of the rest of the stars, which they call the "disposition"; and
after the disposition, the predictions. But, as we 54
shall establish, it is not possible to perceive the time of birth of the subjects of the investigation, nor is the horologe unerring, nor is the rising Sign apprehended accurately. The doctrine, then, of the Chal-
deans is without substance. Let us deal with the 55
first point first.

They take the time of birth of those who are to be the subjects of the investigation, in a rather primitive way, either from that of the depositing of the seed and conception, or from that of the parturition. But they will not say that it is from the depositing of the seed and conception, for the exact time of this is not

[a] Cf. §§ 14 ff., 73 ff.

56 γάρ ἐστιν ὁ ἀκριβὴς ταύτης χρόνος. καὶ εἰκότως·
οὐ γὰρ ἔχομεν λέγειν εἴτε ἅμα τῇ θέσει τοῦ σπέρ-
ματος γέγονεν ἡ σύλληψις εἴτε καὶ μή. δύναται
μὲν γὰρ καὶ ἅμα νοήματι τοῦτο συμβαίνειν, ὥσπερ
καὶ τὸ προσαχθὲν τοῖς διαπύροις τῶν κλιβάνων
57 στέαρ (τοῦτο γὰρ εὐθὺς κολλᾶται), δύναται δὲ καὶ
μετὰ χρόνον, ἐπείπερ καὶ τὰ εἰς τὴν γῆν καταβαλ-
λόμενα τῶν σπερμάτων οὐκ εὐθὺς ῥιζοβολοῦντα
συμπλέκεται ταῖς ὑποκειμέναις βώλοις. καὶ δια-
στήματος δὲ ὄντος ἀπὸ τοῦ στόματος τῆς μήτρας
μέχρι τοῦ πυθμένος, ἔνθα καὶ τὰς συλλήψεις λέγου-
σι γίνεσθαι ἰατρῶν παῖδες, πάντως ἐν χρόνῳ τινὶ τὸ
διάστημα τοῦτο ποιεῖν πέφυκεν ἡ καταβαλλομένη
58 τοῦ σπέρματος φύσις. οἱ δὲ τούτου ἀγνοοῦντες τὴν
ποσότητα τοῦ χρόνου κατὰ τὸ ἀκριβὲς Χαλδαῖοι
τὴν σύλληψιν οὐδέποτε καταλήψονται. τοῦ ⟨γὰρ⟩[1]
σπέρματος ὁτὲ μὲν εὐθυβολουμένου καὶ αὐτοῖς
προσπίπτοντος ὑφ' ἓν τοῖς εὐφυῶς ἔχουσι πρὸς
σύλληψιν τῆς μήτρας τόποις, ὁτὲ δὲ πολυσπό-
ρως ἐμπίπτοντος, ὑπ' αὐτῆς δὲ τῆς ἐν τῇ μήτρᾳ
δυνάμεως εἰς ἕνα τόπον συνάγεσθαι δυναμένου,
τῶν ἀγνώστων τὸ πότε γίγνεται τὸ πρῶτον καὶ
πότε τὸ δεύτερον, πόσος τε ὁ εἰς ἐκείνην τὴν
σύλληψιν ἀναλισκόμενος χρόνος καὶ πόσος ὁ εἰς
59 ταύτην. ἀγνοουμένων δὲ τούτων οἴχεται καὶ ἡ
πρὸς ἀκρίβειαν τῆς συλλήψεως κατάληψις. εἴπερ
τε, ὥς τινες τῶν φυσικῶν εἰρήκασιν, ἑψόμενον
πρῶτον καὶ προμεταβάλλον ἐν μήτρᾳ τὸ σπέρμα
τότε προσέρχεται τοῖς ἀναστομωθεῖσιν αὐτῆς ἀγ-
γείοις, αὐτόθεν οὐκ εἰδότες τὴν ποσότητα τοῦ τῆς
μεταβολῆς χρόνου οὐκ εἴσονται οὐδὲ τὸν τῆς συλ-
60 λήψεως καιρόν. καὶ μὴν ὥσπερ κατὰ τὰ λοιπὰ

determinable. And naturally so ; for we have no 56
means of saying whether the conception takes place at the same time as the depositing of the seed or not. For this can occur quick as thought—like the dough that is put in very hot ovens, for this coalesces at
once,—but it can also occur after an interval, seeing 57
that the seeds deposited in the earth do not at once strike root and become entangled with the underlying soil. And as there is some distance between the mouth of the womb and its base (where, as the medical fraternity say, conceptions take place), the substance of the seed deposited will certainly occupy
some time in covering this distance. And as they 58
are ignorant of the length of this time the Chaldeans will never apprehend precisely the moment of conception. For since at one time the seed is shot straight and falls all together on the very parts of the womb which are naturally suited for conception, while at another time it is dispersed in its fall, yet can be collected into a single place by the womb's own power, there is no knowing when the first process takes place and when the second, or how much time is spent over the former conception or how much over
the latter. And when these things are unknown, 59
accurate apprehension of the conception vanishes. —Also if, as some of the physiologists affirm, the seed is first cooked and altered beforehand in the womb, and then enters those vessels of it which are gaping to receive it, as they do not know the amount of time consumed in the process of alteration, neither, of course, will they know the moment of conception.
—Moreover, as women differ from one another in 60

---

[1] ⟨γὰρ⟩ add. Herv.

μέρη τοῦ σώματος ἐν ταῖς τῶν μερῶν ἐνεργείαις διαφέρουσιν ἀλλήλων αἱ γυναῖκες, οὕτως εἰκὸς αὐτὰς καὶ κατὰ τὴν τῆς μήτρας ἐνέργειαν διαφέρειν, τὰς μὲν θᾶττον συλλαμβανούσας τὰς δὲ βράδιον. καὶ οὐ παράδοξον, ὅτε καὶ ἑαυταῖς συγκρινόμεναι νυνὶ μὲν εὐσύλληπτοι θεωροῦνται νυνὶ δὲ οὐδαμῶς.
61 τούτου δὲ οὕτως ἔχοντος τῶν ἀδυνάτων ἐστὶ λέγειν πρὸς ἀκρίβειαν τὸ πότε συνέσχηται τὸ καταβληθὲν σπέρμα, ἵνα καὶ ἀπὸ τούτου τοῦ χρόνου στήσωσιν οἱ Χαλδαῖοι τὸν τῆς γενέσεως
62 ὡροσκόπον. καὶ μὴν οὐδὲ ἔνεστι λέγειν ὡς διὰ σημείων τινῶν καταλαμβάνεσθαι πέφυκεν ὁ τῆς συλλήψεως χρόνος, καθάπερ ἐκ τοῦ κατεξηράνθαι μὲν μετὰ τὴν μῖξιν τοὺς γυναικείους κόλπους, μεμυκέναι δέ, εἰ οὕτω τύχοι, τὸ τῆς μήτρας στόμιον, ἐπεσχῆσθαι δὲ τὴν ἔμμηνον κάθαρσιν, κίσσαν
63 δὲ ἐπιγίγνεσθαι. πρῶτον μὲν γὰρ καὶ ταῦτα κοινοποιεῖται τὰ σημεῖα πρὸς τὰς μὴ συνειληφυίας· εἶτα καὶ εἰ μὴ κοινοποιοῖτο, γενομένην ἤδη κατὰ πλάτος πλειόνων ἡμερῶν διελθουσῶν σύλληψιν δηλοῖ, καὶ οὐ πρὸς ἀκρίβειαν καὶ ὑπόγυιον καὶ ἐν
64 ὡριαίοις κειμένην διαστήμασιν. χρείαν δ' ἔχουσιν οἱ Χαλδαῖοι πρὸς διάγνωσιν τῶν διαφερόντων βίων οὐ τοῦ ὁλοσχεροῦς καὶ ἐν πλάτει χρόνου τῆς συλλήψεως, τοῦ δὲ πρὸς ἀκρίβειαν.

Ἀλλὰ γὰρ ἐκ τούτων πρόδηλον ὅτι οὐχ οἷόν τέ ἐστιν ἀπὸ συλλήψεως τὸν ὡροσκόπον ἑστάναι.
65 καὶ μὴν οὐδ' ἐξ ἀποτέξεως.[1] πρῶτον μὲν γὰρ ἄπορόν ἐστι τὸ πότε ῥητέον ἀπότεξιν εἶναι, ἆρά γε ὁπόταν ἄρχηται προκύπτειν εἰς τὸν ψυχρὸν ἀέρα τὸ ἀποτικτόμενον, ἢ ὅταν ὀλίγον ἐξίσχῃ, ἢ ὅταν

[1] οὐδ' ἐξ ἀποτέξεως Nebe : οὐδὲ ἀπὸ τέξεως MSS., Bekk.

all the other parts of the body in respect of the activities of those parts, so it is likely that they differ in respect of the activity of the womb, some conceiving more quickly, others more slowly. Nor is this hard to believe, when, compared with themselves, women are seen to conceive easily at one time and by no means easily at another. And this being 61
so, it is impossible to state precisely when the deposited seed is conceived, so that (by calculating) from this time the Chaldeans might set up the horoscope of the nativity.—Nor yet can one say that 62
the time of conception may naturally be apprehended by means of certain signs,—for example, from the drying up of the uterine folds after the intercourse, and, if it should so happen, the closing of the mouth of the womb, and the cessation of the menses, and the occurrence of longings peculiar to pregnancy. For, in the first place, these signs are shared by 63
those who have not conceived; and, secondly, even if not thus shared, they indicate that conception has taken place when already, roughly speaking, several days have elapsed, and the time of it is not fixed precisely and closely and within the space of hours. But for their diagnosis of the different lives what the 64
Chaldeans need is not a rough and loose estimate of the time of conception, but an exact one.

Well then, from this it is quite plain that it is not possible for a horoscope to be set up from the time of conception. Nor yet from that of birth. For, 65
firstly, the moment when birth should be said to take place is a matter of doubt,—is it when the child begins to emerge into the cold air, or when it has emerged a little, or when it is deposited on the

66 εἰς τὴν γῆν κατενεχθῇ. εἶτα οὐδὲ ἐφ᾽ ἑκάστου
τούτων δυνατόν ἐστι τὸν ἀκριβῆ τῆς ἀποτέξεως
χρόνον ὁρίζειν· καὶ γὰρ διὰ παράστημα ψυχῆς καὶ
δι᾽ ἐπιτηδειότητα σώματος καὶ προδιάθεσιν[1] τῶν
τόπων καὶ δι᾽ ἐμπειρίαν μαίας καὶ ἄλλας ἀπείρους
προφάσεις οὐχ ὁ αὐτός ἐστι χρόνος καθ᾽ ὃν προ-
κύπτει τὸ τικτόμενον ῥαγέντων τῶν ὑμένων ἢ
ἐκτὸς ὀλίγον γίνεται ἢ εἰς τὴν γῆν καταφέρεται,
67 ἀλλ᾽ ἄλλος ἐπ᾽ ἄλλων. ὃν πάλιν μὴ δυνάμενοι
ὡρισμένως καὶ ἀκριβῶς σταθμήσασθαι οἱ Χαλ-
δαῖοι ἐκπεσοῦνται τοῦ δεόντως τὴν τῆς ἀποτέξεως
ὥραν ὁρίζειν.

Ὅτι μὲν οὖν τὸ ὅσον ἐπὶ τοῖς τῆς ἀποτέξεως
χρόνοις ἐπαγγέλλονται μὲν τὸν ὡροσκόπον γινώ-
σκειν Χαλδαῖοι, οὐκ ἴσασι δέ, ἐκ τούτων συμ-
68 φανές· ὅτι δὲ οὐδὲ τὸ ὡροσκόπιον ἀπλανές ἐστιν
αὐτοῖς, πάρεστι κατὰ τὸν ὅμοιον ἐπιλογίζεσθαι
τρόπον. ὅταν γὰρ λέγωσιν ὅτι ὁ παρεδρεύων τῇ
ὠδινούσῃ τὴν ἀπότεξιν δίσκῳ σημαίνει τῷ ἐπὶ τῆς
ἀκρωρείας ἀστεροσκοποῦντι Χαλδαίῳ, κἀκεῖνος εἰς
οὐρανὸν ἀποβλέπων ἐπισημειοῦται τὸ ἀνίσχον ζώ-
διον, τὸ μὲν πρῶτον ὑποδείξομεν αὐτοῖς ὅτι τῆς
ἀποτέξεως ἀορίστου τυγχανούσης, καθὼς μικρῷ
πρόσθεν παρεστήσαμεν, οὐδὲ τὸ δίσκῳ διασημαί-
69 νειν ταύτην εὔκολον. εἶτα ἔστω καὶ καταληπτὴν
τυγχάνειν τὴν ἀπότεξιν, ἀλλ᾽ οὔ γε πρὸς ἀκριβῆ
χρόνον ταύτην παρασημειοῦσθαι δυνατόν ἐστιν.
τὸν γὰρ τοῦ δίσκου ψόφον ἐν πλείονι χρόνῳ [καὶ
ἐν συχνῷ],[2] πρὸς αἴσθησιν δυναμένῳ[3] μερίζεσθαι, κι-
νεῖσθαι συμβέβηκεν ἐπὶ τὴν ἀκρώρειαν. τεκμή-
ριον δὲ τὸ ἐπὶ τῶν ἐν τῇ ὀρεινῇ δενδροτομούντων

[1] προδιάθεσιν scripsi : πρὸς διάθεσιν MSS., Bekk.

ground? Secondly, not even in each of these cases 66
is it possible to determine the exact time of the birth; for owing to the present state of the soul and the fitness of the body and the predisposition of the parts and the skill of the midwife and countless other causes, the time at which, after the bursting of the caul, the child is emerging, or has emerged a little, or is deposited on the ground, is not the same but different in different cases. And as the Chaldeans are again 67
unable to measure this time definitely and precisely they will fail to determine correctly the hour of birth.

From this it is evident that in so far as it depends on the times of birth, though the Chaldeans profess that they know the horoscope, they do not know it. And 68
one may argue in like manner that their "horologe" is not unerring. For when they say that the man who is sitting beside the woman in labour signifies the time of birth by means of a gong to the Chaldean who is on a peak watching the stars, and that he gazing at the heaven notes the rising Sign, we shall point out to them, firstly, that the time of birth being undetermined—as we proved a moment ago,[a]—it is not easy to signify it by means of a gong.—Next, let 69
it be granted that the time of birth is discoverable, still it is not possible to transmit it by sign at the exact time. For the fact is that in moving up to the peak the sound of the gong takes a considerable amount of time which perceptibly admits of division into parts. And what is observed in the case of those who fell trees on a mountain-side is a proof of this;

[a] *Cf.* §§ 65 ff.

[1] [καὶ ἐν συχνῷ] om. Herv. (an ὡς ἐν σ. ?).
[2] δυναμένῳ cj. Hackforth: δυνάμενον MSS., Bekk.

θεωρούμενον· μετὰ γὰρ ἱκανὴν ὥραν τοῦ κατενεχ-
θῆναι τὸν πέλεκυν ἐξακούεται ἡ τῆς πληγῆς φωνὴ
ὡς ἂν ἐν πλείονι χρόνῳ φθάνουσα ἐπὶ τὸν ἀκού-
70 οντα. καὶ διὰ τοῦτο τοίνυν οὐκ ἔστιν ἀκριβῶς
τοῖς Χαλδαίοις τὸν χρόνον τοῦ ἀνίσχοντος ζῳδίου
καὶ κατ' ἀκρίβειαν ὡροσκοποῦντος λαμβάνειν. καὶ
μὴν οὐ μόνον φθάνει πλείων διελθεῖν χρόνος μετὰ
τὴν ἀπότεξιν, ἐν ᾧ γίνεται ὁ ἦχος ἀπὸ τοῦ ταῖς
τῆς τικτούσης ὠδῖσι παρεδρεύοντος ὡς ἐπὶ τὸν
ἀστεροσκοποῦντα· ἀλλὰ καὶ ἐν ᾧ οὗτος ἀναβλέπει
καὶ περισκοπῶν ἐξετάζει τὸ ἐν τίνι τῶν ζωδίων
ἐστὶν ἡ σελήνη καὶ τῶν λοιπῶν ἀστέρων ἕκαστος,
φθάνει ἀλλοῖον γενέσθαι τὸ περὶ τοὺς ἀστέρας
διάθεμα, τῆς τοῦ κόσμου κινήσεως ἀλήπτῳ[1] τάχει
περιφερομένης, πρὶν τηρητικῶς παραπλάσασθαι τῇ
τοῦ γεννηθέντος ὥρᾳ τὰ κατ' οὐρανὸν βλεπόμενα.
71 ἄλλως τε ἡ τοιαύτη παρατήρησις νύκτωρ ἴσως
δύναται προκόπτειν τοῖς Χαλδαίοις, ὅτε τά τε ἐν
τῷ ζωδιακῷ βλέπεται κύκλῳ καὶ οἱ σχηματισμοὶ
τῶν ἀστέρων εἰσὶν ἐμφανεῖς. ἐπεὶ οὖν τινὲς καὶ
μεθ' ἡμέραν γεννῶνται, ὅτε οὐδὲν τῶν προειρη-
μένων δυνατόν ἐστι παρασημειοῦσθαι, μόνας δέ,
εἰ καὶ ἄρα, τὰς τοῦ ἡλίου κινήσεις, ῥητέον ἐπὶ
τινῶν μὲν δυνατὴν εἶναι τὴν τῶν Χαλδαίων μέθοδον
72 ἐπὶ τινῶν δὲ ἀδύνατον. ὅρα δὲ μή ποτε καὶ νύ-
κτωρ οὐκ ἰσχύουσιν ἀπλανεῖς διὰ παντὸς ποιεισθαῖ
τὰς τῶν οὐρανίων παρατηρήσεις· πολλάκις γὰρ
συννεφεῖς εἰσὶν αἱ νύκτες καὶ ἀχλυώδεις, ἀγαπη-
τὸν δὲ ἦν πάσης ἀναιρουμένης τοιαύτης προφάσεως
τὸ βέβαιον εὑρεῖν ἐν τῷ[2] μαθήματι, μήτοι γε

[1] ἀλήπτῳ Nebe : ἀλέκτῳ Bekk. (*perpetua* Herv.).
[2] ⟨τούτῳ⟩ τῷ cj. Warmington.

for the sound of the blow is heard a considerable time after the fall of the axe, so that it takes some time to reach the listener. So because of this it is 70
not possible for the Chaldeans to take the time of the rising Sign which is the exact horoscope with accuracy.—Moreover, not only does a considerable time elapse after the birth, during which the sound passes from the man seated beside the woman in travail to the man who is watching the stars; but also, while the latter is gazing upward and looking round to discover in which of the Signs the Moon lies and each of the other stars, the "disposition" of the stars changes, as the Universe in its motion revolves at an incredible speed, before he has described after observation the things seen in the heavens at the child's natal hour.—Furthermore, the Chaldeans can, 71
perhaps, have some success with this sort of observation by night, when the objects within the circle of the zodiac are seen and the configurations the of stars are plain to view. Since, however, some are born in the daytime, when none of the objects mentioned above can be noted, but only, if anything, the motions of the sun, one must declare that the Chaldeans' method is possible in some cases, impossible in others. But beware lest even at night some- 72
times they are unable to make observations of celestial objects that are entirely correct; for the nights are often clouded over and misty, and one might be well satisfied if, with all such occasions eliminated, one could find in this science substantial truth; but when there exists some obstacle to the

καὶ κωλύματός τινος ὄντος πρὸς τὴν ἀκριβῆ τῶν οὐρανίων κατάληψιν.

73 Ἀθετήσαντες δὴ καὶ τὸ κατὰ τοὺς Χαλδαίους ὡροσκόπιον, συντόμως τε παραστήσαντες ὅτι μετὰ τῶν τῆς γενέσεως χρόνων ἄληπτόν ἐστιν αὐτοῖς, ἐπὶ τὸ λειπόμενον τῆς ὑποσχέσεως μέρος χωρῶμεν. ἐλείπετο δὲ καὶ περὶ τῆς ἀναφορᾶς τῆς ἐν τῷ ζωδιακῷ κύκλῳ διελθεῖν, ἀποστάντας τῶν ἔμπροσθεν
74 ἐκκειμένων ἡμῖν ἐλέγχων. φαμὲν τοίνυν ὅτι δυσδιόριστοί εἰσιν ἀπ' ἀλλήλων, μᾶλλον δὲ ἀδυνάτως ἔχουσι κατὰ τὸ ἀκριβὲς ὁρισθῆναι αἱ τῶν ζωδίων μοῖραι, ἀλλ' εἰκός ἐστιν ἤδη ἀνεσχηκὸς ζώδιον δοκεῖν μήπω ἀνατεταλκέναι, καὶ ἀνάπαλιν μήπω
75 ἀνατεταλκὸς δοκεῖν ἤδη ἀνεσχηκέναι. οὐδὲ γὰρ ἡ προειρημένη τῶν ὑδριῶν ἔφοδος δύναταί τι τοῖς Χαλδαίοις παρεπικουρεῖν, ἐπείπερ καὶ παρὰ τὸ ῥέον ὕδωρ καὶ παρὰ τὴν τοῦ ἀέρος κρᾶσιν ἀνώμαλα τὰ τῆς ῥύσεως καὶ τῶν ἀντιπαρηκόντων τῇ ῥύσει χρόνων. τὴν μὲν γὰρ τοῦ ὕδατος φορὰν εἰκός ἐστιν ἀνόμοιον γίνεσθαι κατ' ἀρχάς, ὅτε καθαρόν ἐστι τὸ ῥέον, καὶ ἐξ ὑστέρου, ὅτε ἰλυῶδες καὶ
76 δυσρευστότερον· τὴν δὲ τοῦ ἀέρος κρᾶσιν πιθανὸν ἀχλυώδους μὲν καὶ παχυτέρου ὄντος ἀντιπίπτειν τῇ ἐκρύσει, τρόπον τινὰ ἐμφράττουσαν αὐτήν, διαυγοῦς δὲ καὶ λεπτομοροῦς καθεστῶτος συνεργεῖν
77 μᾶλλον. καὶ αὐτὸς δὲ ὁ ἀμφορεὺς οὐχ ὡσαύτως ῥυήσεται πλήρης καθεστώς, ὡσαύτως δὲ ἡμίκενος ἢ πρὸς τῷ κενοῦσθαι τυγχάνων, ἀλλ' ὁτὲ μὲν ὀξύτερον ὁτὲ δὲ βραδύτερον ὁτὲ δὲ μέσως, τῆς οὐρανίου φορᾶς ἰσοταχῶς διὰ παντὸς ἐλαυνομένης.
78 τὸ δὲ πάντων κυριώτατον, ἕκαστον τῶν ζωδίων οὐ

accurate perception of celestial objects, it is far otherwise.

And now that we have demolished the "horologe" 73
of the Chaldeans and have shown concisely that it, together with the times of nativity, is beyond their grasp, let us proceed to the remaining part of our undertaking. What remains was to discuss the rising in the circle of the Zodiac without touching on the criticisms we have stated above. We assert, then, 74
that the portions of the Signs are hard to mark off from one another, or rather cannot possibly be defined with accuracy; indeed it is likely that a Sign which has already ascended should appear not to have risen as yet, and conversely that a Sign which has not yet risen should appear to have already ascended. For 75
the scheme of the waterpots, mentioned above,[a] is of no avail to rescue the Chaldeans, since owing to the flow of the water, and owing to the mixture of the air, the flow itself and the times parallel to the flow do not correspond. For as regards the motion of the water, it is likely that it is not the same at the beginning, when the flowing water is clear, and later on, when it is turbid and flows less easily; and as to the 76
mixture of the air, it probably opposes the outflow, acting as a kind of block, when it is misty and rather dense, and gives it more aid when it is pellucid and of fine texture. The jar itself, too, will not leak 77
equally when it is full and when it is half-empty or nearly emptied, but more rapidly at one time and more slowly at another, and at yet another time at a medium pace, whereas the celestial motion continues constantly at an even speed. And most 78
important of all, each of the Signs is not a continu-

[a] *Cf.* § 24.

συνεχές ἐστι σῶμα, οὐδ᾽ ὥσπερ ἡρμολογημένον
τῷ πρὸ ἑαυτοῦ καὶ μεθ᾽ αὑτὸ συνῆπται μηδεμιᾶς
μεταξὺ πιπτούσης διαστάσεως, ἀλλ᾽ ἐκ διεσπαρ-
μένων ἀστέρων συνέστηκε καὶ μεταξύτητάς τινας
ἐχόντων καὶ διαλείμματα, τοῦτο μὲν κατὰ τὴν
79 μεσότητα τοῦτο δὲ πρὸς τοῖς πέρασιν. ὅθεν πάν-
τως, ἀριθμητοῖς μορίοις τῶν ἐν τῷ κύκλῳ ζωδίων
περιγραφομένων, πλάνην ἀναγκαῖόν ἐστι γίγνε-
σθαι τοῖς ἀπὸ τῆς γῆς παρατηροῦσι, λανθάνοντος
αὐτοὺς τοῦ προσπίπτοντος διαλείμματος, εἴτε τοῦ
προηγουμένου ζωδίου πέρας ἐστὶν εἴτε τοῦ ἐπ-
80 ανιόντος ἀρχή. οἱ δὲ λόφοι ἀφ᾽ ὧν αἱ ἀστεροσκο-
πίαι γίνονται, οὐχ οἱ αὐτοὶ πάντοτε διαμένουσιν,
ἀλλὰ κατὰ μοῖραν ἑτεροιουμένου καὶ μεταβάλ-
λοντος τοῦ κόσμου ἤτοι κατακλυσμοῖς ἐξ ὄμβρων
ἢ σεισμοῖς γῆς ἢ ἄλλοις τισὶ τοιούτοις παθήμασιν
ἐνοχλοῦνται, ὥστε καὶ παρὰ τὴν τούτων ἐξαλλαγὴν
μὴ τὰς αὐτὰς γίνεσθαι τῶν ἀστέρων παρατηρήσεις,
ἀλλ᾽ ἑτέραν μὲν ⟨τοῖς⟩[1] ἀφ᾽ ὕψους συμβαίνειν παρα-
τήρησιν διαφέρουσαν δὲ τοῖς ἀπὸ χθαμαλοῦ βλέ-
πουσι, καὶ τὸ ἐκείνοις ὀφθὲν μὴ πάντως καὶ τοῖς
81 ἄλλοις τεθεωρῆσθαι. συμπαραλάβοι δ᾽ ἄν τις ἐνταῦ-
θα καὶ τὴν τῶν αἰσθήσεων παραλλαγήν· ἄλλοι γὰρ
ἄλλων εἰσὶν ὀξυωπέστεροι, καὶ ὃν τρόπον τὸ μηδέ-
πω βλεπόμενον ἡμῖν διὰ ποσὴν ἀπόστασιν, τοῦτο
ὡς μέγιστον κατειλήφασιν ἀετοί τε καὶ ἱέρακες
δι᾽ ὑπερβολὴν ὀξυωπίας, οὕτω τὸ ἀνίσχον ἤδη καὶ
ὡροσκοποῦν ζώδιον ἐκ μακροῦ διαστήματος τῷ
μὴ ὀξυωποῦντι Χαλδαίῳ πιθανόν ἐστιν ὡς μηδέπω
ἀνατεταλκὸς δοξάζεσθαι, καὶ κατὰ σύγκρισιν ἀμ-
82 βλυωποῦντι. προσθετέον δὲ τούτοις ὡς ἐναργέ-
στατον τῆς Χαλδαϊκῆς ἔλεγχον καὶ τὴν περὶ τῷ

ous body, nor is it connected, as though by joints, with the one which precedes it and the one which follows it, with no space intervening, but it is composed of scattered stars which have certain interspaces and intervals, this one at the centre and that
one at the limits. Hence, as the Signs in the circle 79
are circumscribed by numerable portions, it is certainly necessary that those who are observing from the earth should fall into error, since they cannot tell whether the interval under observation is the limit of the preceding Sign or the commencement of
that which comes next. And the peaks upon which 80
the watchings of the stars take place do not remain always the same, but, as the Universe alters and changes as fate decrees, either by floods caused by rain, or by earthquakes, or by some other accidents of a similar kind, they are disturbed, so that owing to their alteration the observations of the stars do not remain the same, but the observation taken by those on the height is of one kind and that taken by those who are gazing from the plain is of a different kind, and what is seen by the former is not in all cases visible to the
others. And here one may also take account of the 81
differences in the senses ; for some are more keen of sight than others, and just as an object which is not as yet seen by us owing to its great distance is perceived as a very large object by eagles and hawks owing to their excessively keen sight, so it is probable that the Sign which has already ascended and is the horoscope should appear as not yet risen to the Chaldean, who is not keen of sight but by comparison short-sighted,
because of its vast distance. And to these we must 82
add, as the clearest disproof of astrology, the differ-

[1] ⟨τοῖς⟩ addidi.

ὁρίζοντι τοῦ ἀέρος διαφοράν· εἰκὸς γὰρ ὅτι παχυ-
μερεστάτου[1] αὐτοῦ καθεστῶτος κατὰ ἀνάκλασιν
τῆς ὄψεως τὸ ὑπὸ γῆν ἔτι καθεστὼς ζώδιον δοκεῖν
ἤδη ὑπὲρ γῆς τυγχάνειν, ὁποῖόν τι καὶ ἐπὶ τῆς ἐφ᾽
ὕδατος ἀντανακλωμένης ἡλιακῆς ἀκτῖνος γίνεται·
μὴ βλέποντες γὰρ τὸν ἥλιον αὐτὸν ⟨ταύτην⟩[2] πολ-
83 λάκις ὡς ἥλιον δοξάζομεν. τὸ δὲ πάντων συνεκ-
τικώτατον, εἰ μὲν πᾶσι τοῖς κατὰ τὴν οἰκουμένην
τὰ οὐράνια παρατηροῦσιν ἕκαστον τοῦ ζωδιακοῦ
δωδεκατημόριον ἰσοχρόνως ἐφαίνετο καὶ κατὰ τὴν
αὐτὴν εὐθεῖαν ἐθεωρεῖτο, τάχ᾽ ἴσως ἂν ἐδύναντο
Χαλδαίων παῖδες παγίως λαβεῖν τὸ περὶ τὸν ὁρί-
84 ζοντα ἀνίσχον ζώδιον. νυνὶ δὲ ἐπεὶ οὐ παρὰ πᾶσιν
ἰσοχρόνως ἀναφαίνεται ἀλλὰ τοῖς μὲν θᾶττον τοῖς
δὲ βράδιον καὶ τισὶ μὲν πλάγιον τισὶ δὲ ὀρθόν,
ἐπακολουθεῖ τὸ μὴ πᾶσι τὸ αὐτὸ δοκεῖν ὡροσκο-
πεῖν ζώδιον, ἀλλὰ τὸ τούτοις ἤδη δοκοῦν ἀνατε-
ταλκέναι, τοῦτ᾽ ἄλλοις ἀκμὴν ὑπόγειον ὑπάρχειν,
καὶ τὸ ἑτέροις φαινόμενον ἐν ἀποκλίματι τοῦ ὡρο-
σκοποῦντος ζωδίου, τοῦτο ἑτέροις θεωρεῖσθαι
85 ὡροσκοποῦν. καὶ ὅτι ταῦθ᾽ οὕτως ἔχει, πρόδηλον
ἐκ τοῦ καὶ τοὺς ἀπλανεῖς ἀστέρας, καθάπερ ἀρ-
κτοῦρον καὶ κύνα, μὴ κατὰ τὸν αὐτὸν χρόνον τοῖς
ἐν παντὶ κλίματι κατοικοῦσι φαίνεσθαι ἀλλ᾽ ἄλλοις
κατ᾽ ἄλλον.

Ὅτι μὲν οὖν οὐκ ἐνδέχεται κατ᾽ ἀκρίβειαν τὸ
ὡροσκοποῦν ζώδιον λαβεῖν, διὰ δὲ τοῦτο οὐδὲ τῶν
ἄλλων τι κέντρων, ἀφ᾽ ὧν αἱ προαγορεύσεις γίνον-
86 ται τοῖς Χαλδαίοις, αὐτάρκως παρεστήσαμεν. ἐκ

[1] ὅτι παχυμερεστάτου scripsi : ὅτι παχυμεροῦς mss., Bekk. (ἐνίοτε π. cj. Bekk.).

[2] ⟨ταύτην⟩ add. cj. Bekk.

ence of the air at the horizon, for as it is of the greatest possible density, it is likely that, owing to the reflexion of the visual stream, the Sign which is still below the earth will appear to be already above the earth,—the sort of phenomenon which also occurs in the case of the solar ray which is reflected on water; for, though we do not behold the sun itself, we
often imagine it to be the sun.—But the most con- 83
clusive argument of all is this: If each twelfth portion of the zodiac appeared at the same time to all those in the world who are observing the celestial objects, and was seen in the same straight line, then perhaps the Chadean fraternity might have been able to perceive with certainty the Sign ascending at the horizon.
But as it is, since it does not appear to all at the same 84
time, but more quickly to some and more slowly to others, and to some obliquely but vertically to others, it follows that the same Sign does not seem to all to form the horoscope, but that which to one party seems to have risen already appears to others to be quite beneath the earth, and that which to some appears in the declination[a] of the Sign of the horoscope is by others viewed as the Sign of the horoscope.
And that this is the case is quite evident from the fact 85
that the fixed stars, such as Arcturus and Canis, do not appear to the dwellers in every region at the same time, but at different times to different people.

So now we have established by proofs sufficient in themselves that it is not possible to determine accurately the Sign of the horoscope, nor, consequently, any one of the other "centres" from which the
Chaldeans derive their predictions. But over and 86

[a] *Cf.* §§ 14, 18.

περιουσίας δὲ λεκτέον ὅτι κἂν καταληπτὸς ᾖ ὁ ἀκριβὴς τῆς τούτων ἐπαναφορᾶς χρόνος, ἐκεῖνο μὲν συμφανὲς ὅτι οὐθεὶς τῶν παραγινομένων πρὸς τοὺς Χαλδαίους ἰδιωτῶν τετηρηκὼς ἐφ' ἑαυτοῦ τὸν ἀκριβῆ χρόνον παραγίνεται· πολλῆς γὰρ ἦν τεχνιτείας τὸ πρᾶγμα, ὡς πρότερον ἐδείκνυμεν, καὶ
87 πλέον ἢ κατ' ἰδιώτην ὑπέφαινεν. ἐπεὶ οὖν ὁ Χαλδαῖος οὐκ ἐτήρησε τὸν ἀκριβῆ τῆς γενέσεως χρόνον ἐπὶ τοῦδε τοῦ ἰδιώτου ἀλλὰ παρ' αὐτοῦ τοῦτον ἀκούει, οὑτοσὶ δὲ ὁ ἰδιώτης τὰ μὲν δι' ἀπειρίαν τὰ δὲ καὶ διὰ τὸ μὴ πάνυ τι ἐσπουδακέναι περὶ τὸ πρᾶγμα πάλιν οὐκ οἶδε τὸν ἀκριβῆ χρόνον, καταλείπεται ἄρα πρόρρησιν μὲν μηδ' ἡντινοῦν βεβαίαν, πλάνην δὲ καὶ φενακισμὸν ἀπὸ Χαλδαϊκῆς τοῖς ἀνθρώποις περιγίνεσθαι.

88 Εἰ δὲ ἀναστρέψαντες λέγοιεν μὴ τὸν ἀκριβῆ χρόνον λαμβάνεσθαι ἀλλὰ τὸν ὁλοσχερῆ καὶ ἐν πλάτει, ὑπ' αὐτῶν σχεδὸν ἐλεγχθήσονται τῶν ἀποτελεσμάτων· οἱ γὰρ ἐν τῷ αὐτῷ καθ' ὁλοσχέρειαν χρόνῳ γεννηθέντες οὐ τὸν αὐτὸν ἔζησαν βίον, ἀλλ' οἱ μὲν λόγου χάριν ἐβασίλευσαν οἱ δὲ ἐν πέδαις
89 κατεγήρασαν. οὐθεὶς γοῦν Ἀλεξάνδρῳ τῷ Μακεδόνι γέγονεν ἴσος, πολλῶν κατὰ τὴν οἰκουμένην συναποτεχθέντων αὐτῷ, οὐδὲ Πλάτωνι τῷ φιλοσόφῳ. ὥστε εἰ τὸν ἐν πλάτει τῆς γενέσεως χρόνον ὁ Χαλδαῖος ἐπισκέπτεται, οὐ δυνήσεται[1] παγίως λέγειν ὅτι ὁ κατὰ τοῦτον τὸν[2] χρόνον γεννηθεὶς εὐτυχήσει, πολλοὶ γὰρ κατὰ τὸν αὐτὸν χρόνον τούτῳ γεννηθέντες ἐδυστύχησαν, καὶ ἀνάπαλιν ὅτι ὅδε τις ἀπορήσει· οὐκ ὀλίγοι γὰρ τῶν τὸ αὐτὸ διάθεμα

[1] δυνήσεται Nebe : δυνηθήσεται MSS., Bekk.

above the foregoing we should add the argument that even if the exact time of the ascent of these Signs is apprehensible, yet it is plain that none of the ordinary persons who apply to the Chaldeans has observed for himself the exact time before applying; for the task calls for much expertness, as we have shown above,[a] and seems beyond the capacity of the ordinary man. Since, then, the Chaldean did not 87
observe the exact time of the birth in the case of a particular ordinary person, but hears it from the person himself, and this ordinary person again, partly through want of skill and partly through not taking very much trouble about the matter, does not know the exact time, the result is that men gain from astrology no valid prediction whatsoever but error and deception.

And if they turn round and say that the time is 88
determined not exactly but roughly and approximately, the results themselves will be enough to refute them; for those who were born at what is roughly the same time have not lived the same life, but some, for example, have been kings while others have grown old in chains. Thus, though many through- 89
out the world were born at the same time as he, none was equal to Alexander of Macedon, nor to the philosopher Plato. So that if the Chaldean considers what is roughly the time of birth, he will not be able to state definitely that he who was born at this time will have good fortune, for many who were born at the same time as he have had ill fortune; or conversely, that this particular man will be ill off; for not a few of those who shared in the same "disposi-

[a] *Cf.* §§ 27 ff.

[1] τοῦτον τὸν scripsi: τὸν αὐτὸν MSS., Bekk.

90 ἐσχηκότων εὐπορώτατοι κατεγήρασαν. καὶ μὴν
οὐδὲ μετρίως ἐλέγχειν φαίνεται τοὺς Χαλδαίους
καὶ ὁ ἀναστρέφων πρὸς τὴν εἰρημένην ἐπιχείρησιν
λόγος. εἰ γὰρ οἱ τὸ αὐτὸ διάθεμα τῆς γενέσεως
ἔχοντες τοῖς αὐτοῖς ἀποτελέσμασιν ἐν τῷ βίῳ
περιπίπτουσι, πάντως καὶ οἱ διαφόρους ἔχοντες
γενέσεις διάφοροι γίνονται. ὅπερ ἐστὶ ψεῦδος·
91 ὁρῶμεν γὰρ πολλοὺς κατά τε ἡλικίας διαφέροντας
καὶ κατὰ μορφὰς σωμάτων καὶ κατὰ ἄλλας παμ-
πληθεῖς ἰδιότητας παθῶν τῷ ὁμοίῳ τέλει περι-
πεπτωκότας καὶ ἤτοι ἐν πολέμῳ ἀπολομένους ἢ
ἐν συμπτώσεσιν οἰκιῶν ἀποληφθέντας ἢ ναυαγίαις
καταποντισθέντας· οἷς, εἴπερ ἔζων, πῶς ἂν ὁ
Χαλδαῖος προειρήκει τὴν ἐσομένην τοῦ βίου κατα-
92 στροφήν, ἄξιον διαπορεῖν. εἰ γὰρ ὁ μὲν ἐν τῇ
ἀκίδι τοῦ τοξότου γεννηθεὶς κατὰ τὸν μαθηματικὸν
σφαγήσεται λόγον, πῶς αἱ τοσαῦται μυριάδες τῶν
βαρβάρων ἀνταγωνιζόμεναι πρὸς τοὺς Ἕλληνας ἐν
Μαραθῶνι ὑφ᾽ ἓν κατεσφάγησαν; οὐ γὰρ δή γε
ἐπὶ πάντων ὁ αὐτὸς ἦν ὡροσκόπος. καὶ πάλιν εἰ
ὁ ἐν τῇ κάλπιδι τοῦ ὑδροχόου γεννηθεὶς ναυαγήσει,
πῶς οἱ ἀπὸ Τροίας ἀνακομιζόμενοι τῶν Ἑλλήνων
περὶ τὰ κοῖλα τῆς Εὐβοίας συγκατεποντίσθησαν;[1]
93 ἀμήχανον γὰρ πάντας μακρῷ διαφέροντας ἀλλήλων
ἐν τῇ κάλπιδι τοῦ ὑδροχόου γεγεννῆσθαι. καὶ μὴν
οὐδὲ ἔνεστι λέγειν ὅτι δι᾽ ἕνα πολλάκις, ᾧ εἵμαρται
κατὰ πέλαγος φθαρῆναι, πάντες οἱ ἐν τῇ νηὶ συν-
απόλλυνται· διὰ τί γὰρ ἡ τούτου εἱμαρμένη τὰς
πάντων νικᾷ, ἀλλ᾽ οὐχὶ διὰ τὸν ἕνα ᾧ εἵμαρται ἐπὶ
94 γῆς θανεῖν πάντες περισώζονται; ἄλλος δέ τις

[1] συγκατεποντίσθησαν Nebe: συγκατεποντώθησαν mss., Bekk.

tion "[a] have been exceedingly well off in their old age.—Moreover, the argument thus turned round to 90
meet the attack mentioned appears to confute the Chaldeans very completely. For if those who have the same "disposition" at birth meet with the same results during life, then certainly those whose births are different become different. But this is false; for 91
we see many, who differ as to age and bodily shape and countless other peculiar affections, yet meeting with a similar end,—either perishing in war or being crushed by the collapse of houses or drowned in shipwrecks; and if these had lived, one may well be puzzled to know how the Chaldean would have foretold what the conclusion of their life was to be. For if the man who 92
was born in the arrow's point of the Archer is doomed —according to the astrological theory—to be slain, how is it that all those myriads of barbarians who fought against the Greeks at Marathon were all slain at one time? For the horoscope was not the same for them all. And again, if he who was born in the pitcher of Aquarius is doomed to suffer shipwreck, how is it that the Greeks who were being brought back from Troy[b] were all drowned together round the "Hollows" of Euboea? For that all these men, who 93
differed greatly from one another, could have been born in the pitcher of Aquarius is impossible. Nor yet is it possible to say that because of one man who was, for instance, destined to perish at sea all those in the ship perish along with him; for what reason is there why this man's destiny overmasters the destinies of them all, rather than that they should all be saved because of one man whose destiny it is to die on dry land?—And some other person will raise a doubt re- 94

[a] *Cf.* § 53.

[b] *Cf.* Eurip. *Hel.* 1126 ff.

ἀπορήσει καὶ περὶ τῶν ἀλόγων ζώων. εἰ γὰρ
παρὰ τοὺς συσχηματισμοὺς τῶν ἀστέρων τὰ κατὰ
τὸν βίον ἀποτελέσματα πέφυκεν ἐκβαίνειν, ἐχρῆν
ἐν τῷ αὐτῷ τούτῳ τοῦ ζωδίου μορίῳ κάνθωνος
ἅμα καὶ ἀνθρώπου γεννηθέντων τὴν αὐτὴν ἐν
ἀμφοτέροις ἀκολουθεῖν τῶν βίων ἔκβασιν, καὶ μὴ
τὸν μὲν ἄνθρωπον πολλάκις ἐπιφανῶς πολιτευσά-
μενον περισπούδαστον εἶναι τοῖς δήμοις, τὸν δὲ
κάνθωνα διὰ παντὸς ἀχθοφορεῖν ἢ εἰς μύλωνας
95 ἀπάγεσθαι. τοίνυν οὐκ ἔστιν εὔλογον πρὸς τὰς
τῶν ἀστέρων κινήσεις διοικεῖσθαι τὸν βίον· ἢ εἴπερ
ἔστιν εὔλογον, ἡμῖν πάντως ἀκατάληπτον.

Ἀπὸ δὲ τῆς αὐτῆς ὁρμώμενοι δυνάμεως δυσω-
πήσομεν αὐτοὺς καὶ ἐν οἷς συνοικειοῦν θέλουσι τοῖς
τῶν ζωδίων τύποις τάς τε μορφὰς καὶ τὰ ἤθη τῶν
ἀνθρώπων, οἷον ὅταν λέγωσιν, ὁ ἐν λέοντι γεννη-
θεὶς ἀνδρεῖος ἔσται, ὁ δὲ ἐν παρθένῳ τετανόθριξ
96 χαροπὸς λευκόχρως ἄπαις αἰδήμων. ταῦτα γὰρ
καὶ τὰ τούτοις ὅμοια γέλωτος μᾶλλον ἢ σπουδῆς
ἐστὶν ἄξια. πρῶτον μὲν γάρ, εἰ ὅτι ἄλκιμον καὶ
ἀρρενωπόν ἐστιν ὁ λέων, φασὶ τὸν ἐν αὐτῷ γεννώ-
μενον ἀνδρεῖον τυγχάνειν, πῶς τὸν ταύρον ἀναλο-
97 γοῦντα τούτῳ θῆλυ νομίζουσι ζῷον; εἶτα μάταιον
τὸ[1] οἴεσθαι ζώδιον κάλλιστον τὸν λέοντα τὸν ἐν
οὐρανῷ ἀναλογίαν ἔχειν τῷ ἐπὶ γῆς· εἰκὸς γὰρ
τοὺς παλαιοὺς τὰ τοιαῦτα τῶν ὀνομάτων τίθεσθαι
κατὰ ψιλὴν τὴν τοῦ χαρακτῆρος ἐμφέρειαν, τάχα
δὲ οὐδὲ κατ' αὐτὴν ἀλλ' εὐσήμου χάριν διδα-
98 σκαλίας. τί γὰρ ὅμοιον ἔχουσιν ἄρκτῳ οἱ ἑπτὰ

[1] μάταιον τὸ Nebe (et ego): μετὰ τοῦτο MSS., Bekk. (*ineptum est* Herv.).

[a] But Manilius vi. 202 says "fecundus erit, quod mirum

garding the irrational animals. For if the effects in life naturally result from the configurations of the stars, then, when a pack-ass and a man are both born in this same portion of the Sign, the same kind of life ought to have followed as a result in both cases, instead of the man being, for example, conspicuous as a statesman and admired by the citizens, while the pack-ass is continually laden with burdens or led away to the mill-houses. So it is not reasonable that life is 95
ordered according to the motions of the stars; or if it is reasonable, certainly it is beyond our comprehension.

Starting from the same standpoint we shall also put them to shame when they propose to associate the shapes and characters of men with the figures of the Signs, as, for instance, when they say that the man born in Leo will be brave, and the man born in the Virgin will be straight-haired, bright-eyed, white-skinned, childless,[a] and modest. For this and such- 96
like notions are deserving of ridicule rather than serious attention. For, in the first place, if they assert that the man born in Leo is brave because the lion is a valiant and manly beast, how is it that they reckon the Bull, which is on a par with the Lion, to be a womanish beast? And, secondly, it is nonsense to 97
suppose that the Lion in the heavens, that most beautiful Sign, bears any analogy to the earthly lion; for it is probable that the ancients gave them names of this sort merely because of the similarity of their figures, and perhaps not even for this reason, but just for the sake of clearness in exposition. For what 98
resemblance have the seven stars to a bear, separated

in virgine, partus" (hence Bekker conjectured εὔπαις for ἄπαις).

ἀστέρες, διεστῶτες ἀπ' ἀλλήλων; ἢ δράκοντος κεφαλῇ οἱ πέντε, ἐφ' ὧν φησὶν ὁ Ἄρατος

ἀλλὰ δύο κροτάφους, δύο δ' ὄμματα, εἷς δ' ὑπ-
ένερθεν
ἐσχατιὴν ἐπέχει γένυος δεινοῖο πελώρου.

99 οὐ μὴν ἀλλὰ καὶ ὡς ἀνώτερον ἐλέγομεν, τῶν ἐν τῷ αὐτῷ ζῳδίῳ γεννωμένων οὔθ' αἱ μορφαί εἰσιν αἱ αὐταὶ οὔτε τὰ ἤθη ἐστὶν ὅμοια, ἐκτὸς εἰ μὴ τὰς μοίρας εἰς ἃς ἕκαστον διαιρεῖται ζῴδιον καὶ τὰ λεπτὰ φήσουσι τῆς τοιαύτης διαφορᾶς εἶναι ποιητικά. ὃ πάλιν ἐστὶν ἀδύνατον· ἐδείξαμεν γὰρ τὴν ἐν τοῖς αὐτοῖς χρόνοις τῆς ἀποτέξεως καὶ ὡροσκο-
100 πήσεως ἀκρίβειαν ἀσύστατον. δυοῖν τε θάτερον· ἢ γὰρ ὅτι λέων λέγεται τὸ ζῴδιον, καὶ ὁ γεννηθεὶς ἀνδρεῖος γίνεται, ἢ ὅτι τραπέντος τοῦ ἀέρος ὑπὸ τοῦ κατ' οὐρανὸν λέοντος τοιαῦται συμβαίνουσι καὶ περὶ τὸν ἀποτικτόμενον ἄνθρωπον διαθέσεις. ἀλλὰ διὰ μὲν τὸ λέοντα καλεῖσθαι τὸ ὡροσκοποῦν ζῴδιον οὐ πιθανὸν ἀνδρεῖον γίνεσθαι· τούτῳ γὰρ τῷ λόγῳ ἐχρῆν καὶ τοὺς τῷ ἐπιγείῳ λέοντι συναποτεχθέντας ἢ συντραφέντας ἀνδρείους ὑπάρχειν παρ-
101 όσον λέων λέγεται τὸ ᾧ συνετράφησαν ζῷῳ. εἰ δὲ διὰ τὴν τοῦ ἀέρος τροπήν, τί τοῦτο πρὸς τὴν τοῦ βίου διαφοράν; εἰς μὲν γὰρ τὸ ἰσχυρὸν τοῖς σώμασι γίνεσθαι τὸ γεννώμενον καὶ θηριῶδες τοῖς ἤθεσιν τάχα συμβάλλεται ἡ ποιὰ τοῦ ἀέρος κρᾶσις, εἰς δὲ τὸ δανείοις κατάχρεων γενέσθαι τὸ γεννώμενον ἢ βασιλεύειν ἢ δεθῆναι ἢ σπανότεκνον ἢ σπανάδελφον ὑπάρχειν οὐδ' ὁτιοῦν φαίνεται συνερ-
102 γεῖν ὁ ἀήρ. καὶ πάλιν εἰ ὁ παρθένου ὡροσκοπού-

as they are from one another? Or the five to a dragon's head,—the five stars whereof Aratus says [a]

> But two sit on the temples and two on the eyes, and below them
> One has its seat on the base of the jaw of the terrible monster.

Nor, in fact, as we said above,[b] are those born in the 99
same Sign of the same shape or of similar character, unless they shall say that the sections and subsections into which each Sign is divided [c] are capable of producing differences of the sort. But this again is impossible; for we have proved [d] that accuracy as regards the identical times of the birth and of the
horoscope-taking is not to be had.—Also, of two 100
things one: either the man born is brave because the Sign is called the Lion, or because, when the air under the celestial Lion is changed, dispositions of that sort are brought about in the man who is being born. But it is not credible that he should be brave because the Sign of his horoscope is called the Lion; for by this reasoning those who were born or reared along with the earthly lion ought also to have been brave, inasmuch as the animal along with which they
were reared is called a lion. And if it is because of 101
the change of the air, what has this to do with a difference in the life? For though a certain blend of the air possibly contributes to the bodily strength and beast-like character of the creature born, yet the air does not seem to co-operate at all in causing the creature to be involved in debt or to be a king or to be put in gaol or to be lacking in children or brethren.
—And again, if he who has the Virgin for horoscope 102

[a] *Cf.* Aratus, *Phaenom.* 56 f. Aratus, *circ.* 270 B.C., wrote astronomical poems.

[b] *Cf.* §§ 89 ff.

[c] *Cf.* § 5.

[d] *Cf.* §§ 74 ff.

σης τετανόθριξ χαροπὸς λευκόχρως, δεήσει μηδένα τῶν Αἰθιόπων παρθένον ἔχειν ὡροσκοποῦσαν, ἐπεὶ δώσουσιν Αἰθίοπα λευκὸν εἶναι καὶ χαροπὸν καὶ
103 τετανότριχα, ὃ πάντων ἐστὶν ἀτοπώτατον. καθόλου δέ, ἐπεὶ οὐδ᾽ ἐνδείκνυσθαι λέγουσιν αὐτοῖς τοὺς ἀστέρας τὰς τῶν ἀνθρωπίνων βίων διαφοράς, ἀλλ᾽ αὐτοὶ ταύτας συμπαρατετηρηκέναι ταῖς τῶν ἀστέρων σχέσεσι, φημὶ ὅτι εἰ μελλήσει βέβαιος πρόρρησις γίνεσθαι, δεῖ τὴν αὐτὴν τῶν ἀστέρων σχέσιν μὴ ἅπαξ συμπαρατετηρηκέναι τῷ τινὸς βίῳ ἀλλὰ καὶ δεύτερον δευτέρου καὶ τρίτον τρίτου, ἵνα ἐκ τοῦ διομαλίζειν ἐπὶ πάντων τὰς τῶν ἀποτελεσμάτων ἐκβάσεις μάθωμεν ὅτι τῶν ἀστέρων τοιοῦτον ἀναδεξαμένων τὸν σχηματισμὸν τόδε πάντως
104 ἔσται τὸ ἀποβησόμενον· καὶ ὃν τρόπον ἐν τῇ ἰατρικῇ ἐτηρήσαμεν ὅτι ἡ τῆς καρδίας τρῶσις αἴτιόν ἐστι θανάτου, οὐ τὴν Δίωνος μόνον τελευτὴν αὐτῇ συμπαρατηρήσαντες ἀλλὰ καὶ Θέωνος καὶ Σωκράτους καὶ ἄλλων πολλῶν, οὕτω καὶ ἐν μαθηματικῇ εἰ πιστόν ἐστιν ὅτι ὅδε ὁ συσχηματισμὸς τῶν ἀστέρων τοιούτου βίου μηνυτικὸς καθέστηκεν, πάντως οὐχ ἅπαξ ἐφ᾽ ἑνὸς ἀλλὰ πολλάκις ἐπὶ
105 πολλῶν παρετηρήθη. ἐπεὶ οὖν ὁ αὐτὸς τῶν ἀστέρων συσχηματισμὸς διὰ μακρῶν, ὥς φασί, χρόνων θεωρεῖται, ἀποκαταστάσεως γινομένης τοῦ μεγάλου ἐνιαυτοῦ δι᾽ ἐννεακισχιλίων ἐνακοσίων καὶ ἑβδομήκοντα καὶ ἑπτὰ ἐτῶν, οὐ φθάσει ἀνθρωπίνη τήρησις τοῖς τοσούτοις αἰῶσι συνδραμεῖν ἐπὶ μιᾶς γενέσεως, καὶ ταῦτα οὐχ ἅπαξ ἀλλὰ πολλάκις ἤτοι

---

[a] *Cf.* § 95.

[b] These are used as "stock names" (as we might say

is straight-haired, bright-eyed, and white-skinned,[a] it must follow that none of the Ethiopians has the Virgin for horoscope, else they will be granting that an Ethiopian is white, bright-eyed and straight-haired, which is of all things the most absurd.—And 103
in general, since they declare that it is not the stars that inform them of the differences in men's lives but they themselves observe them together with the positions of the stars, I affirm that if the prediction is to be reliable, the same position of the stars ought not to be observed once only in connexion with the life of some one person, but a second time with a second life, and a third time with a third, so that from the equality of the resultant effects in all the cases we might learn that when the stars have assumed a certain configuration the result will certainly be of one particular kind; and just as in medicine we have observed that 104
a puncture of the heart is the cause of death, after having observed together with it not only the death of Dion but also of Theon and Socrates[b] and many others, so also in astrology, if it is credible that this particular configuration of the stars is indicative of that particular kind of life, then it certainly has been observed not once only in one single case but many times in many cases. Since, then, the same con- 105
figuration of the stars is seen, as they say, at long intervals—the recurrence of "The Great Year"[c] taking place after 9977 years,—human observation will not succeed in traversing so many centuries even in the case of one nativity, and that, too, when it is interrupted not once but oftentimes, either by the

"Smith and Jones and Robinson"); they do not refer to the historical persons named: *cf. Adv. Phys.* i. 269, ii. 289.

[c] For various theories of "The Great Year" see Adam's *Republic of Plato*, vol. ii. p. 304.

⟨τῆς⟩[1] τοῦ κόσμου φθορᾶς, εἰρήκασιν ὥς τινες, μεσολαβούσης αὐτήν, ἢ πάντως γε τῆς κατὰ μέρος μεταβολῆς ἐξαφανιζούσης τὸ συνεχὲς τῆς ἱστορικῆς παραδόσεως.

106 Τοσαῦτα μὲν οὖν ἐστὶ καὶ τὰ πραγματικῶς δυνάμενα λέγεσθαι πρὸς τοὺς Χαλδαίους. μεθ' ἃ πάλιν ἀπ' ἄλλης ἀρχῆς σύντομον οὖσαν καὶ τὴν πρὸς τοὺς μουσικοὺς ζήτησιν ἀποδώσομεν.

destruction of the Universe, as some have declared, or certainly by a partial upheaval which wholly does away with the continuity of historical tradition.

Such, then, are the many valid objections which can be brought against the Chaldeans. After which, making a fresh start, we shall set out our criticism—which is brief—of the Musicians. 106

---

[1] ⟨τῆς⟩ add. cj. Bekk.

Ζ

ΠΡΟΣ ΜΟΥΣΙΚΟΥΣ

1 Ἡ μουσικὴ λέγεται τριχῶς, καθ' ἕνα μὲν τρόπον ἐπιστήμη τις περὶ μελῳδίας καὶ φθόγγους καὶ ῥυθμοποιίας καὶ τὰ παραπλήσια καταγιγνομένη πράγματα, καθὸ καὶ Ἀριστόξενον τὸν Σπινθάρου λέγομεν εἶναι μουσικόν, καθ' ἕτερον δὲ ἡ περὶ ὀργανικὴν ἐμπειρία,[1] ὡς ὅταν τοὺς μὲν αὐλοῖς καὶ ψαλτηρίοις χρωμένους μουσικοὺς ὀνομάζωμεν, τὰς δὲ ψαλτρίας μουσικάς. ἀλλὰ κυρίως κατὰ ταῦτα[2] τὰ σημαινόμενα καὶ παρὰ πολλοῖς λέγεται μουσική·
2 καταχρηστικώτερον δὲ ἐνίοτε προσαγορεύειν εἰώθαμεν τῷ αὐτῷ ὀνόματι καὶ τὴν ἔν τινι πράγματι κατόρθωσιν. οὕτω γοῦν μεμουσωμένον τι ἔργον φαμέν, κἂν ζωγραφίας μέρος ὑπάρχῃ, καὶ μεμουσῶσθαι τὸν ἐν τούτῳ κατορθώσαντα ζωγράφον.
3 ἀλλὰ δὴ κατὰ τοσούτους τρόπους νοουμένης τῆς μουσικῆς, πρόκειται νῦν ποιεῖσθαι τὴν ἀντίρρησιν οὐ μὰ Δία πρὸς ἄλλην τινὰ ἢ πρὸς τὴν κατὰ τὸ πρῶτον νοουμένην σημαινόμενον· αὕτη γὰρ καὶ ἐντελεστάτη παρὰ τὰς ἄλλας μουσικὰς δοκεῖ καθ-
4 εστηκέναι. τῆς δὲ ἀντιρρήσεως, καθάπερ καὶ ἐπὶ

[1] ἐμπειρία cj. Bekk. : ἐμπειρίαν MSS., Bekk.
[2] κατὰ ταῦτα cj. Bekk. : κατ' αὐτὰ MSS., Bekk.

BOOK VI

AGAINST THE MUSICIANS

The term "Music" is used in three senses; in one 1
as a science dealing with melodies and notes and rhythm-making and similar things, in which sense we say that Aristoxenus,[a] son of Spintharus, was a musician; in another sense it connotes instrumental skill, as when we describe those who use flutes and harps as musicians and female harp-players as
musicians. It is with these significations that the 2
term "Music" is properly and generally used. But it is sometimes our habit to apply the same name in a loose sense to correctness in some performance. Thus we speak of a work as "musical," even though it be a piece of painting, and of the painter who has
achieved therein correctness as "musical." While 3
music, then, is conceived in all these ways, it is certainly not our present purpose to frame our refutation of it if conceived in any other way than that first signified; for Music in this sense, as compared with the other kinds, seems to be the most
perfect. And, as in the case of Grammar, the 4

[a] Aristoxenus of Tarentum, *circ.* 320 B.C., was a musician and disciple of Aristotle.

γραμματικῆς, διττόν ἐστι τὸ εἶδος. οἱ μὲν οὖν
δογματικώτερον ἐπεχείρησαν διδάσκειν ὅτι οὐκ ἀνα-
γκαῖόν ἐστι μάθημα πρὸς εὐδαιμονίαν μουσική,
ἀλλὰ βλαπτικὸν μᾶλλον, καὶ τοῦτο δείκνυσθαι ἔκ
τε τοῦ διαβάλλεσθαι τὰ πρὸς τῶν μουσικῶν λεγό-
μενα καὶ ἐκ τοῦ τοὺς προηγουμένους λόγους ἀνα-
5 σκευῆς ἀξιοῦσθαι· οἱ δὲ ἀπορητικώτερον πάσης
ἀποστάντες τῆς τοιαύτης ἀντιρρήσεως ἐν τῷ σα-
λεύειν τὰς ἀρχικὰς ὑποθέσεις τῶν μουσικῶν
6 ᾠήθησαν καὶ τὴν ὅλην ἀνῃρῆσθαι μουσικήν. ὅθεν
καὶ ἡμεῖς ὑπὲρ τοῦ μὴ δοκεῖν τι τῆς διδασκαλίας
χρεωκοπεῖν, τὸν ἑκατέρου δόγματος ἢ πράγματος
χαρακτῆρα κεφαλαιωδέστερον ἐφοδεύσομεν, μήτε
ἐν τοῖς παρέλκουσιν ὑπερεκπίπτοντες εἰς μακρὰς
διεξόδους μήτε ἐν τοῖς ἀναγκαιοτέροις ὑστεροῦντες
πρὸς τὴν τῶν ἐπειγόντων ἔκθεσιν, ἀλλὰ μέσην
καὶ μεμετρημένην κατὰ τὸ δυνατὸν ποιούμενοι τὴν
διδασκαλίαν.

7 Τάξει δὲ ἀρχέτω πρῶτον τὰ ὑπὲρ μουσικῆς εἰω-
θότα παρὰ τοῖς πολλοῖς θρυλεῖσθαι. εἴπερ τοίνυν,
φασί, φιλοσοφίαν ἀποδεχόμεθα σωφρονίζουσαν τὸν
ἀνθρώπινον βίον καὶ τὰ ψυχικὰ πάθη καταστέλ-
λουσαν, πολλῷ μᾶλλον ἀποδεξόμεθα[1] τὴν μουσι-
κήν, ὅτι οὐ βιαστικώτερον ἐπιτάττουσα ἡμῖν ἀλλὰ
μετὰ θελγούσης τινὸς πειθοῦς τῶν αὐτῶν ἀπο-
τελεσμάτων περιγίνεται ὧνπερ καὶ ἡ φιλοσοφία.
8 ὁ γοῦν Πυθαγόρας μειράκια ὑπὸ μέθης ἐκβεβακ-
χευμένα ποτὲ θεασάμενος ὡς μηδὲν τῶν μεμηνότων
διαφέρειν, παρῄνεσε τῷ συνεπικωμάζοντι τούτοις
αὐλητῇ τὸ σπονδεῖον αὐτοῖς ἐπαυλῆσαι μέλος· τοῦ
δὲ τὸ προσταχθὲν ποιήσαντος οὕτως αἰφνίδιον

[1] ἀποδεξόμεθα cj. Bekk.: ἀποδεχόμεθα MSS., Bekk.

refutation is of two kinds. Some have tried in a dogmatical way to show that Music is a subject not necessary for happiness, but harmful rather, and to prove this both by finding fault with the statements made by the Musicians and by claiming to refute their leading arguments; whereas others, avoiding 5
all that sort of refutation, have held, in more sceptical fashion, that through the wrecking of the principal assumptions of the Musicians the whole of Music is likewise destroyed. Accordingly, in order to avoid 6
the appearance of cutting down at all our debt of exposition, we shall discuss in summary fashion the character of each of these dogmas or operations, neither divagating into long disquisitions regarding irrelevant matters, nor, as regards such as are relevant, failing to give a full account of the pressing points, but making our exposition as moderate and measured as possible.

First in order, let us state the views commonly ex- 7
pressed concerning Music by the majority of people. If, they say, we welcome Philosophy as regulating human life and repressing the passions of the soul, much more shall we welcome Music because it produces the same results as Philosophy not by commanding us in a violent manner but by means of a seductive persuasiveness. Thus Pythagoras, having noticed on 8
one occasion that the youths who were in a state of Bacchic frenzy from drunkenness differed not at all from madmen, advised the flute-player who was with then in their revels to play them the "spondean" tune[a]; and when he had done as instructed, they

[a] The sort of slow, solemn melodies used at *spondai* ("libations").

μεταβαλεῖν σωφρονισθέντας ὡς εἰ καὶ τὴν ἀρχὴν
9 ἔνηφον. οἵ τε τῆς Ἑλλάδος ἡγούμενοι καὶ ἐπ'
ἀνδρίᾳ διαβόητοι Σπαρτιᾶται μουσικῆς ἀεί ποτε
στρατηγούσης αὐτῶν ἐπολέμουν. καὶ οἱ ταῖς
Σόλωνος χρώμενοι παραινέσεσι πρὸς αὐλὸν καὶ
λύραν παρετάσσοντο, ἔνρυθμον ποιούμενοι τὴν
10 ἐνόπλιον κίνησιν. καὶ μὴν ὥσπερ σωφρονίζει μὲν
τοὺς ἄφρονας ἡ μουσική, εἰς ἀνδρίαν δὲ προτρέπει
τοὺς δειλοτέρους, οὕτω καὶ παρηγορεῖ τοὺς ὑπ'
ὀργῆς ἐκκαιομένους. ὁρῶμεν γοῦν ὡς καὶ ὁ παρὰ
τῷ ποιητῇ μηνίων Ἀχιλλεὺς καταλαμβάνεται ὑπὸ
τῶν ἐξαποσταλέντων πρεσβευτῶν

φρένα τερπόμενος φόρμιγγι λιγείῃ
καλῇ δαιδαλέῃ· ἐπὶ δ' ἀργύρεον ζυγὸν ἦεν.
τὴν ἕλετ' ἐξ ἐνάρων, πόλιν Ἠετίωνος ὀλέσσας.
τῇ ὅ γε θυμὸν ἔτερπεν

ὡς ἂν σαφῶς γινώσκων τὴν μουσικὴν πραγματείαν
μάλιστα δυναμένην περιγίνεσθαι τῆς περὶ αὐτὸν
11 διαθέσεως. καὶ μὴν δι' ἔθους ἦν καὶ τοῖς ἄλλοις
ἥρωσιν, εἴ ποτε ἀποδημοῖεν καὶ μακρὸν πλοῦν
στέλλοιντο, ὡς πιστοτάτους φύλακας καὶ σω-
φρονιστῆρας τῶν γυναικῶν αὐτῶν ἀπολείπειν τοὺς
μουσικούς. Κλυταιμνήστρᾳ γέ τοι παρῆν ἀοιδός,
ᾧ πολλὰ ἐπέτελλεν Ἀγαμέμνων περὶ τῆς κατὰ
12 ταύτην σωφροσύνης. ἀλλ' ὁ Αἴγισθος πανοῦργος
ὢν αὐτίκα τὸν ἀοιδὸν τοῦτον

ἄγων εἰς νῆσον ἐρήμην
κάλλιπεν οἰωνοῖσιν ἕλωρ καὶ κῦρμα γενέσθαι·

---

[a] i.e. the Athenians.

suddenly changed and became sober just as if they had been sober from the beginning. The Spartans, 9
too, the leaders of Hellas and renowned for their bravery, always had music to lead their army when they went to war. And those [a] who followed the counsels of Solon formed up their ranks to the sound of the flute and the lyre and made their movements under arms rhythmical.[b] Moreover, as Music gives 10
sober sense to those lacking in sense and incites the cowards to courage, so also it calms down those who are burning with anger. Thus we see how Achilles in his rage (as the poet describes him) is found by the messengers who had been sent to him [c]—

> Giving delight to his soul with the tuneful notes of his zither;
> Goodly and cunningly wrought it was, with its cross-bar of silver,
> Which he did choose from the spoils when he sack'd Eëtion's city.
> With this now he was soothing his soul,—

—as if he knew full well that the practice of music was the thing best able to overcome his state of mind. Furthermore, it was customary for the other 11
heroes, whenever they left home and set out on a long voyage, to leave the musicians behind as being the most trusty guardians and controllers of their wives. Thus Clytaemnestra was accompanied by a minstrel, to whom Agamemnon gave strict orders regarding the chastity of his wife.[d] But Aegisthus, who was a 12
crafty rascal, straightway took this minstrel

> Unto a desolate island,
> There did he leave him a prey to become and a spoil to the vultures;

[a] The rhythm of war songs and dances was that of the "Cretic" (–⏑–) or "Paean" (⏑⏑⏑–).

[c] *Cf.* Hom. *Il.* ix. 186 ff. [d] *Cf.* Hom. *Od.* iii. 267 ff.

εἶθ' οὕτως ἀφύλακτον λαβὼν τὴν Κλυταιμνήστραν διέφθειρεν, προτρεψάμενος αὐτὴν ἐπιθέσθαι τῇ
13 ἀρχῇ τοῦ Ἀγαμέμνονος. οἵ τε μέγα δυνηθέντες ἐν φιλοσοφίᾳ, καθάπερ καὶ Πλάτων, τὸν σοφὸν ὅμοιόν φασιν εἶναι τῷ μουσικῷ, τὴν ψυχὴν ἡρμοσμένην ἔχοντα. καθὸ καὶ Σωκράτης καίπερ βαθυγήρως ἤδη γεγονὼς οὐκ ᾐδεῖτο πρὸς Λάμπωνα τὸν κιθαριστὴν φοιτῶν, καὶ πρὸς τὸν ἐπὶ τούτῳ ὀνειδίσαντα λέγειν ὅτι κρεῖττόν ἐστιν ὀψιμαθῆ μᾶλλον
14 ἢ ἀμαθῆ διαβάλλεσθαι. οὐ χρὴ μέντοι, φασίν, ἀπὸ τῆς νῦν ἐπιτρίπτου καὶ κατεαγυίας μουσικῆς τὴν παλαιὰν διασύρειν, ὅτε καὶ Ἀθηναῖοι πολλὴν πρόνοιαν σωφροσύνης ποιούμενοι καὶ τὴν σεμνότητα τῆς γε μουσικῆς κατειληφότες ὡς ἀναγκαιότατον
15 αὐτὴν μάθημα τοῖς ἐκγόνοις παρεδίδοσαν. καὶ τούτου μάρτυς ὁ τῆς ἀρχαίας κωμῳδίας ποιητής, λέγων

> λέξω τοίνυν βίον ἐξ ἀρχῆς ὃν ἐγὼ θνητοῖσι παρεῖχον.
> πρότερον γὰρ ἔδει παιδὸς φωνὴν γρύσαντος μηδέν' ἀκοῦσαι,
> εἶτα βαδίζειν ἐν ταῖσιν ὁδοῖς εὐτάκτως ἐς κιθαριστοῦ.

ὅθεν εἰ καὶ κεκλασμένοις τισὶ μέλεσι νῦν καὶ γυναικώδεσι ῥυθμοῖς θηλύνει τὸν νοῦν ἡ μουσική, οὐδὲν τοῦτο πρὸς τὴν ἀρχαίαν καὶ ἔπανδρον μου-
16 σικήν. εἴπερ τε ἡ ποιητικὴ βιωφελής ἐστι, ταύτην δὲ φαίνεται κοσμεῖν ἡ μουσικὴ μελίζουσα καὶ ἐπῳδὸν παρέχουσα, χρειώδης γενήσεται ἡ μουσική. ἀμέλει γέ τοι καὶ οἱ ποιηταὶ μελοποιοὶ

---

[a] *Cf.* Plato, *Rep.* 410 E, 443 E, 554 E.

and then he took Clytaemnestra, who was now without a protector, and seduced her, having urged her to seize Agamemnon's dominion. Those also who have been 13
men of great ability in philosophy, such as Plato, say that the sage resembles the musician as he has his soul "harmonized."[a] Accordingly, Socrates, though he was already far gone in years, was not ashamed of going to get lessons from Lampon the lyre-player, and in reply to one who jeered at him for this he said that it was better to be accused of being late-learned than unlearned.—Nor indeed, they say, ought we 14
to run down the ancient music because the present-day music is hackneyed and effeminate, when the Athenians, who devote great care to temperance, appreciating the dignity of music have handed it down to their descendants as a most necessary branch of learning. A witness to this is the poet of the Old 15
Comedy,[b] who says—

> I will now relate from the start the life which I have provided for mortals.
> The first rule was that none should hear from an urchin the sound of a mutter,
> Next, they must walk in order good on their way to their master of music.

Hence, even if music now weakens the mind by its effeminate tunes and womanish rhythms, this is no argument against the ancient and virile music.—Also, if poetry is useful for life,[c] and music appears to 16
adorn it by its melodies and and by making it fit for singing, then music will be beneficial. And, of course, the poets[d] are called "tune-makers," and of

[b] V. 1 comes from Telecleides (*ap.* Athen. vi. p. 268 b), vv. 2, 3, from Aristoph. *Nub.* 963 f.

[c] *Cf. Adv. Gram.* 272 ff.

[d] *i.e.* lyric (or "melic") poets, like Pindar.

λέγονται, καὶ τὰ Ὁμήρου ἔπη τὸ πάλαι πρὸς λύραν
17 ᾔδετο· ὡσαύτως δὲ καὶ τὰ παρὰ τοῖς τραγικοῖς
μέλη καὶ στάσιμα, φυσικόν τινα ἐπέχοντα λόγον,
ὁποῖά ἐστι τὰ οὕτω λεγόμενα,

> γαῖα μεγίστη καὶ Διὸς αἰθήρ,
> ὁ μὲν ἀνθρώπων καὶ θεῶν γενέτωρ,
> ἡ δ' ὑγροβόλους σταγόνας νοτίας
> παραδεξαμένη τίκτει θνατούς,
> τίκτει δὲ βορὰν φῦλά τε θηρῶν,
> ὅθεν οὐκ ἀδίκως
> μήτηρ πάντων νενόμισται.

18 καθόλου γὰρ οὐ μόνον χαιρόντων ἐστὶν ἄκουσμα,
ἀλλ' ἐν ὕμνοις καὶ εὐωχίαις καὶ θεῶν θυσίαις ἡ
μουσική· διὰ δὲ τοῦτο καὶ ἐπὶ τὸν τῶν ἀγαθῶν
ζῆλον τὴν διάνοιαν προτρέπεται. ἀλλὰ καὶ λυπουμένων παρηγόρημα· ὅθεν καὶ τοῖς πενθοῦσιν αὐλῷ[1]
μελῳδοῦσιν οἱ τὴν λύπην αὐτῶν ἐπικουφίζοντες.

19 Τοιαῦτα μὲν ὑπὲρ μουσικῆς· λέγεται δὲ πρὸς
ταῦτα τὸ μὲν πρῶτον ὅτι οὐκ ἔστιν ἐκ προχείρου
διδόμενον τὸ φύσει τῶν μελῶν τὰ μὲν εἶναι διεγερτικὰ τῆς ψυχῆς τὰ δὲ κατασταλτικά. παρὰ γὰρ
τὴν ἡμετέραν δόξαν τὸ τοιοῦτο γίνεται. ὥσπερ
γὰρ ὁ τῆς βροντῆς κτύπος, καθά φασιν Ἐπικουρείων παῖδες, οὐ θεοῦ τινὸς ἐπιφάνειαν σημαίνει
ἀλλὰ τοῖς ἰδιώταις καὶ δεισιδαίμοσι τοιοῦτος εἶναι
20 δοξάζεται, ἐπεὶ καὶ ἄλλων σωμάτων ἐπ' ἴσης ἀλλήλοις προσκρουσάντων ὅμοιος ἀποτελεῖται κτύπος, ὥσπερ καὶ μύλου περιαγομένου ἢ χειρῶν συμπαταγουσῶν, τὸν αὐτὸν τρόπον καὶ τῶν κατὰ μουσικὴν μελῶν οὐ φύσει τὰ μὲν τοιά ἐστι τὰ δὲ τοῖα, ἀλλ' ὑφ' ἡμῶν προσδοξάζεται. τὸ αὐτὸ γοῦν

old the verses of Homer were sung to the lyre. So 17
likewise were the songs and choral odes of the tragic poets, preserving a natural relation, such as are those in verses like these [a]—

> Mightiest Earth and Aether of Heaven,—He of mortal men and of gods is the Sire,
> She takes to her bosom the drippings of rain, And she bears as her offspring mortal men,
> And food does she bear, herds too of wild-beasts;
> Wherefore not without cause Mother of all is she deemèd.

For, in sum, music is not only a sound of rejoicing, 18
but is heard also in sacred hymns and feasts and sacrifices to the gods; and because of this it incites the mind to emulate the good. It is, too, a consolation to those in grief; and for this reason those who are trying to lighten the grief of mourners sing for them to the flute.

Such are the arguments in defence of music; but 19
in reply to these it is argued, firstly, that it is not conceded off-hand that some tunes are in their nature stimulating, others repressive. For such a thing is contrary to our belief. For just as a clap of thunder—as the Epicurean fraternity declare [b]—does not betoken the epiphany of a god, though supposed to do so by ignorant and superstitious folk,
since a similar clap is produced by other bodies clash- 20
ing together in the same way, as by a mill-stone revolving or the clapping of hands, so likewise in the case of musical tunes it is not by nature that some are of this kind and others of that kind, but it is we ourselves who suppose them to be such. Thus the same

[a] *Cf.* Eurip. *Frag.* 836 (Nauck).
[b] *Cf.* Lucret. vi. 96 ff.

[1] αὐλῷ scripsi: αὐλοὶ MSS., Bekk. (αὐλοῖς Heintz).

μέλος τῶν μὲν ἵππων διεγερτικόν ἐστι, τῶν δὲ ἀνθρώπων ἐν θεάτροις ἀκουόντων οὐδαμῶς. καὶ τῶν ἵππων δὲ τάχα οὐ διεγερτικόν ἐστιν ἀλλὰ
21 ταρακτικόν. εἶτα κἂν τοιαῦτα ᾖ τὰ τῆς μουσικῆς μέλη, οὐ διὰ τοῦτο καὶ ἡ μουσικὴ βιωφελὴς καθέστηκεν. οὐ γὰρ ὅτι δύναμιν ἔχει σωφρονιστικήν, καταστέλλει τὴν διάνοιαν, ἀλλ' ὅτι[1] περισπαστικήν· παρὸ καὶ ἡσυχασθέντων πως τῶν τοιούτων μελῶν πάλιν ὁ νοῦς, ὡς ἂν μὴ θεραπευθεὶς ὑπ' αὐτῶν, ἐπὶ τὴν ἀρχῆθεν ἀνακάμπτει
22 διάνοιαν. ὅνπερ οὖν τρόπον ὁ ὕπνος ἢ ὁ οἶνος οὐ λύει τὴν λύπην ἀλλ' ὑπερτίθεται, κάρον ἐμποιῶν καὶ ἔκλυσιν καὶ λήθην, οὕτω τὸ ποιὸν μέλος οὐ καταστέλλει λυπουμένην ψυχὴν ἢ περὶ ὀργὴν σεσοβημένην τὴν διάνοιαν, ἀλλ' εἴπερ, περισπᾷ.
23 ὅ τε Πυθαγόρας τὸ μὲν πρῶτον μάταιος ἦν, τοὺς μεθύοντας ἀκαίρως σωφρονίζειν βουλόμενος ἀλλὰ μὴ ἐκκλίνων· εἶτα καὶ τούτῳ τῷ τρόπῳ ἐπανορθούμενος αὐτοὺς ὁμολογεῖ πλεῖόν τι δύνασθαι τῶν φιλοσόφων πρὸς ἐπανόρθωσιν ἠθῶν τοὺς αὐλητάς.
24 τό τε τοὺς Σπαρτιάτας πρὸς αὐλὸν καὶ λύραν πολεμεῖν τοῦ μικρῷ πρότερον εἰρημένου τεκμήριόν ἐστιν, ἀλλ' οὐχὶ τοῦ βιωφελῆ τυγχάνειν τὴν μουσικήν. καθάπερ δ' οἱ ἀχθοφοροῦντες ἢ ἐρέσσοντες ἢ ἄλλο τι τῶν ἐπιπόνων δρῶντες ἔργων κελεύουσιν εἰς τὸ ἀνθέλκειν τὸν νοῦν ἀπὸ τῆς κατὰ τὸ ἔργον βασάνου, οὕτω καὶ ⟨οἱ⟩[2] αὐλοῖς ἢ σάλπιγξιν ἐν πολέμοις χρώμενοι οὐ διὰ τὸ ἔχειν τι τῆς διανοίας ἐπεγερτικὸν τὸ μέλος καὶ ἀνδρικοῦ λήματος αἴτιον ὑπάρχειν τοῦτο ἐμηχανήσαντο, ἀλλ' ἀπὸ τῆς ἀγω-

[1] ἀλλ' ὅτι scripsi : ἀλλὰ ἢ mss., Bekk. (ἀλλὰ ᾖ Shorey).

tune serves to excite horses, but not at all to excite men who hear it in a theatre. And even to horses it may prove not exciting but disturbing.—Secondly, 21
even if the tunes of music are of the kind stated, music is not on that account useful for life. For it does not repress the mental state because it possesses a moderating influence, but because its influence is distracting; consequently, when tunes of that sort have ceased to sound, the mind, as though not cured by them, reverts to its original state. Thus, just as 22
sleep or wine does not banish grief but puts it off by inducing torpor and languor and oblivion, so a certain kind of tune does not repress a soul in grief or a mind that is shaken by anger but, if anything, distracts it. And, as to Pythagoras,[a] in the first 23
place he was foolish in desiring to render drunkards sober at the wrong moment, instead of quitting the place; and secondly, by trying to reform them in this way he confesses that flute-players have more influence than philosophers for the reforming of morals. Also, the fact that the Spartans[b] make war 24
to the sound of flute and lyre is a proof of what was said a little while ago, but not of the usefulness for life of music. And just as men carrying loads or rowing a boat or doing any other toilsome work make rhythmical calls[c] in order to divert their minds from the distress caused by their work, so also when men make use of flutes or trumpets in war, it is not because the tune has any power to excite the mind, or because it produces a courageous spirit, that they have employed this device, but through anxiety to divert

[a] *Cf.* § 8. [b] *Cf.* § 9.
[c] *e.g.* ῥυππαπαί Aristoph. *Ran.* 1073.

[2] ⟨οἱ⟩ add. cj. Bekk.

νίας καὶ ταραχῆς ἀνθέλκειν ἑαυτοὺς σπουδάσαντες,
εἴγε καὶ στρόμβοις τινὲς τῶν βαρβάρων βουκινί-
ζοντες[1] καὶ τυμπάνοις κτυποῦντες πολεμοῦσιν· ἀλλ'
25 οὐδὲν τούτων ἐπ' ἀνδρίαν προτρέπεται. τὰ δὲ
αὐτὰ λεκτέον καὶ ἐπὶ τοῦ μηνίοντος Ἀχιλλέως·
καίτοι ἐρωτικοῦ ὄντος καὶ ἀκρατοῦς οὐ παράδοξον
26 τὴν μουσικὴν σπουδάζεσθαι. νὴ Δί', ἀλλὰ καὶ
οἱ ἥρωες τὰς ἑαυτῶν γυναῖκας ᾠδοῖς τισὶν ὡς
σώφροσι φύλαξι παρακατετίθεντο, καθάπερ ὁ
Ἀγαμέμνων τὴν Κλυταιμνήστραν. ταῦτα δὲ ἤδη
μυθολογούντων ἐστὶν ἀνδρῶν, εἶτα καὶ παρὰ πόδας
αὑτοὺς διελεγχόντων· πῶς γάρ, εἴπερ μουσικὴ περὶ
τῆς τῶν παθῶν ἐπανορθώσεως ἐπιστεύετο, τὸν μὲν
Ἀγαμέμνονα ἡ Κλυταιμνήστρα ἐπὶ τῆς ἰδίας ἑστίας
κατέκτανεν ὥσπερ βοῦν ἐπὶ φάτνῃ, εἰς δὲ τοὺς
Ὀδυσσέως οἴκους ἡ Πηνελόπη ὄχλον ἄσωτον ἐπι-
δέχεται μειρακίων, ἀεὶ δὲ τὰς ἐπιθυμίας αὐτῶν
ἐλπιδοκοποῦσα καὶ παραύξουσα μοχθηρότερον καὶ
χαλεπώτερον τῆς ἐπὶ Ἴλιον στρατείας τὸν ἐν
27 Ἰθάκῃ πόλεμον ἤγειρε τῷ γήμαντι; καὶ μὴν οὐδ',
εἰ[2] οἱ περὶ τὸν Πλάτωνα μουσικὴν ἀπεδέξαντο,
ῥητέον [οὐ] πρὸς εὐδαιμονίαν αὐτὴν συντείνειν,
ἐπεὶ καὶ μὴ λειπόμενοι τῆς τούτων ἀξιοπιστίας,
καθάπερ οἱ περὶ τὸν Ἐπίκουρον, ἠρνήσαντο ταύτην
τὴν ἀντιποίησιν, λέγοντες τοὐναντίον αὐτὴν ἀσύμ-
φορον εἶναι καὶ

ἀργήν, φίλοινον, χρημάτων ἀτημελῆ.

28 εὐήθεις δέ εἰσι καὶ οἱ τὴν ἀπὸ ποιητικῆς χρείαν
συμπλέκοντες αὐτῇ πρὸς εὐχρηστίαν, ἐπείπερ δύ-
ναται μέν τις, ὡς καὶ ἐν τῷ πρὸς τοὺς γραμματι-

themselves from the struggle and turmoil, seeing that some barbarians blow the trumpet with conches and beat drums as they make war; but none of these practices impels men to bravery. And of Achil- 25
les [a] in his anger the same may be said; although for one who was amorous and incontinent it is not surprising that he was devoted to music. Yet (say 26
they) the heroes entrusted their wives, as did Agamemnon Clytaemnestra, to minstrels, thinking them to be sober-minded guardians. But these, of course, are the fictions of story-tellers who at the next moment confute themselves; for, if music was trusted as a means of rectifying the passions, how was it that Clytaemnestra slew Agamemnon on his own hearth-stone "like an ox at the stall," [b] and that Penelope admits into the halls of Odysseus a crowd of dissolute youths [c] and by contantly luring on and strengthening their desires by false hopes raised up for her husband the war in Ithaca which was more grievous and bitter than the campaign against Troy? Moreover, if Plato [d] admitted music, we should not 27
therefore assert that music contributes to happiness, since others who are not inferior to him in trustworthiness,—such as Epicurus,—have denied this contention, and declared on the contrary that music is unprofitable and

> Wine-loving, idle, having no regard for wealth.

Simple-minded, too, are those who link with it 28
the profit derived from poetry to prove its utility, since—as we stated in our book *Against the Gram*

[a] *Cf.* § 10. [b] *Cf.* Hom. *Od.* xi. 411.
[c] *i.e.* Antinoüs and the other "Suitors." [d] *Cf.* § 13.

[1] *βουκινίζοντες* cj. Bekk.: *βουκινίζουσι* MSS., Bekk.
[2] *οὐδ' εἰ . . .* [*οὐ*] Herv.: *εἰ οὔτε . . . οὐ* MSS., Bekk.

κοὺς ἐλέγομεν, ἀνωφελῆ διδάσκειν τὴν ποιητικήν, οὐδὲν δὲ ἔλαττον κἀκεῖνο δεικνύναι ὅτι ἡ μὲν μουσικὴ περὶ μέλος καταγινομένη μόνον τέρπειν πέφυκεν, ἡ δὲ ποιητικὴ καὶ περὶ διάνοιαν καταγινομένη δύναται συνωφελεῖν τε καὶ σωφρονίζειν.

29 Ἀλλ' ὁ μὲν πρὸς τὰ ἐγκεχειρημένα λόγος ἐστὶ τοιοῦτος· προηγουμένως δὲ λέγεται [καὶ] κατὰ μουσικῆς ὡς εἴπερ ἔστι χρειώδης, ἤτοι κατὰ τοῦτο λέγεται χρειοῦν παρόσον ὁ μουσικευσάμενος πλεῖον παρὰ τοὺς ἰδιώτας τέρπεται πρὸς μουσικῶν ἀκροαμάτων, ἢ παρόσον οὐκ ἔστιν ἀγαθοὺς γενέσθαι
30 μὴ προπαιδευθέντας ὑπ' αὐτῶν, ἢ τῷ τὰ αὐτὰ στοιχεῖα τυγχάνειν τῆς μουσικῆς καὶ ⟨τῆς⟩[1] τῶν κατὰ φιλοσοφίαν πραγμάτων εἰδήσεως, ὁποῖόν τι καὶ περὶ γραμματικῆς ἀνώτερον ἐλέγομεν, ἢ τῷ κατὰ ἁρμονίαν διοικεῖσθαι τὸν κόσμον, καθὼς φάσκουσι Πυθαγορικῶν παῖδες, δέεσθαί τε ἡμᾶς τῶν μουσικῶν θεωρημάτων πρὸς τὴν τῶν ὅλων εἴδησιν, ἢ τῷ τὰ ποιὰ μέλη ἠθοποιεῖν τὴν ψυχήν.
31 οὔτε δὲ τῷ τοὺς μουσικοὺς πλέον τέρπεσθαι παρὰ τοὺς ἰδιώτας ἀπὸ τῶν ἀκροαμάτων λέγοιτ' ἂν χρειοῦν ἡ μουσική. πρῶτον μὲν γὰρ οὐκ ἀναγκαία ἰδιώταις ἡ τέρψις καθάπερ αἱ ἐπὶ λιμῷ ἢ δίψει ἢ κρύει γινόμεναι ὑπὸ ⟨βρώματος ἢ⟩[2] πόματος ἢ
32 ἀλέας· εἶτα κἂν τῶν ἀναγκαίων ὑπάρχωσι, δυνάμεθα χωρὶς μουσικῆς ἐμπειρίας αὐτῶν ἀπολαύειν. νήπια γοῦν ἐμμελοῦς μινυρίσματος κατακούοντα κοιμίζεται, καὶ τὰ ἄλογα τῶν ζώων ὑπὸ αὐλοῦ καὶ σύριγγος κηλεῖται, εἴ γε[3] δελφῖνες, ὡς λόγος, αὐ-

[1] ⟨τῆς⟩ add. cj. Bekk.
[2] ⟨βρώματος ἢ⟩ add. Herv., Fabr.
[3] εἴ γε scripsi: οἵ τε mss., Bekk.

marians [a]—one can argue that poetry is useless, and prove equally well that while music, being concerned with melody only, naturally serves to give pleasure, poetry which is concerned with thoughts as well, is able to be of benefit and teach prudence.

Such, then, is the line of reasoning against the 29
attempts at defence; but the principal argument against music is that if it is useful it is alleged to be useful on the ground that he who has practised music compared with ordinary people gets more pleasure from hearing musical performances; or on the ground that it is not possible for men to be good unless they have already been trained by musicians; or because 30
the elements of music are the same as those of the science of the subject-matter of philosophy, (which is much like what we previously said about grammar); or because the Universe is ordered according to harmony,[b] as the Pythagorean fraternity declare, and we need the theorems of Music in order to understand the Whole of things; or because tunes of a certain kind affect the character of the soul. But it 31
will not be stated that music is useful because musicians as compared with ordinary people get more pleasure from listening to performances. For, firstly, the pleasure felt by ordinary people is not inevitable as are those caused by food, drink and warmth after hunger, thirst and cold; and secondly, even if they 32
are inevitable we can enjoy them without musical skill; infants, certainly, are lulled to sleep by listening to a tuneful cradle-song, and irrational animals are charmed by the sound of the flute and pipe, seeing that dolphins, as we are told, swim up to ships as they

[a] *Cf. Adv. Gram.* 277 ff., 297 ff.
[b] *Cf. Adv. Arithm.* 3.

λῶν μελῳδίαις τερπόμενοι προσνήχονται τοῖς ἐρεσσομένοις σκάφεσιν· ὧν οὐδὲ ὁπότερον ἔοικε μου-
33 σικῆς ἔχειν ἐμπειρίαν ἢ ἔννοιαν. καὶ διὰ τοῦτο μή ποτε, ὃν τρόπον χωρὶς ὀψαρτυτικῆς καὶ οἰνογευστικῆς ἡδόμεθα ὄψου ἢ οἴνου γευσάμενοι, ὧδε καὶ χωρὶς μουσικῆς ἡσθείημεν ἂν τερπνοῦ μέλους ἀκούσαντες, τοῦ μὲν ὅτι τεχνικῶς γίνεται ⟨τοῦ τεχνίτου⟩[1] μᾶλλον παρὰ τὸν ἰδιώτην ἀντιλαμβανομένου, τοῦ δὲ πλεῖον[2] ἡστικοῦ πάθους μηδὲν κερ-
34 δαίνοντος. ὥστε οὐχ αἱρετὸν μουσικὴ παρόσον τοὺς εἰδήμονας αὐτῆς ἐπὶ πλεῖον τέρπεσθαι συμβέβηκεν. καὶ μὴν οὐδὲ τῷ προοδοποιεῖν τὴν ψυχὴν εἰς σοφίαν· ἀνάπαλιν γὰρ ἀντικόπτει καὶ ἀντιβαίνει πρὸς τὸ τῆς ἀρετῆς ἐφίεσθαι, εὐαγώγους εἰς ἀκολασίαν καὶ λαγνείαν παρασκευάζουσα τοὺς νέους, ἐπείπερ ὁ
35 μουσικευσάμενος

μολπαῖσιν ἡσθεὶς τοῦτ' ἀεὶ θηρεύεται.
ἀργὸς μὲν οἴκοις καὶ πόλει γενήσεται,
φίλοισι δ' οὐθείς, ἀλλ' ἄφαντος οἴχεται,
ὅταν γλυκείας ἡδονῆς ἥσσων τις ᾖ.

36 κατὰ ταὐτὰ δὲ οὐδὲ ⟨ἐκ τοῦ⟩[3] ἀπὸ τῶν αὐτῶν στοιχείων ὁρμᾶσθαι ταύτην τε καὶ φιλοσοφίαν εἰσακτέον τὸ κατ' αὐτὴν χρειῶδες, ὡς αὐτόθεν ἐστὶ συμφανές. λείπεται ἄρα τῷ καθ' ἁρμονίαν τὸν κόσμον διοικεῖσθαι ἢ τῷ ἠθοποιοῖς μέλεσι κεχρῆσθαι χρειώδη πρὸς εὐδαιμονίαν λέγειν αὐτὴν τυγχάνειν. ὧν τὸ μὲν τελευταῖον ἤδη διαβέβληται ὡς οὐχ
37 ὑπάρχον ἀληθές, τὸ δὲ κατὰ ἁρμονίαν διοικεῖσθαι

[1] ⟨τοῦ τεχνίτου⟩ . . . ἀντιλαμβανομένου . . . κερδαίνοντος Heintz : ἀντιλαμβανόμενοι . . . κερδαίνοντες MSS., Bekk.

[2] πλεῖον cj. Bekk., Heintz : πλείω MSS., Bekk.

[3] ⟨ἐκ τοῦ⟩ add. cj. Bekk.

are being rowed along because of the pleasure they take in the tunes played by flutes [a]; yet neither the infants nor the animals are likely to have skill in music or understanding of it. And for this reason it 33
may be that, just as we enjoy tasting food or wine though without the art of cooking food or that of wine-tasting, so also, though without the art of music, we take pleasure in hearing a delightful melody; for though the expert musician understands that it is artistically performed better than the ordinary man, he gets from it no greater feeling of pleasure. So 34
that music is not choiceworthy on the ground that those who understand it receive in consequence more pleasure.—Nor yet because it prepares and leads on the soul to wisdom; for, on the contrary, it resists and opposes the striving after virtue, making young people easily led into incontinence and debauchery, seeing that he who cultivates music [b]—

In songs delighting makes this always his pursuit. 35
Whene'er a man is master'd by sweet pleasure's charm
At home he will be idle, idle too in town,
To friends no friend, he speeds away and disappears.—

In the same way, the usefulness of music must not be 36
inferred (as is obvious at once) from the fact that it and philosophy are derived from the same elements. [c]—It only remains, then, to say that it is useful for happiness because the Universe is ordered according to harmony, or because it employs tunes which form character. But the last of these reasons has already been denounced [d] as not being true, and that the 37

[a] So dolphins are called φίλαυλοι, "flute-loving," in Eurip. *Electra* 435.
[b] *Cf*. Eurip. *Antiopê* (*Frag*. 187, vv. 3 ff. Nauck).
[c] *Cf*. § 30. [d] *Cf*. § 28.

τὸν κόσμον ποικίλως δείκνυται ψεῦδος, εἶτα καὶ
ἂν ἀληθὲς ὑπάρχῃ, οὐδὲν τοιοῦτον δύναται πρὸς
μακαριότητα, καθάπερ οὐδὲ ἡ ἐν τοῖς ὀργάνοις
ἁρμονία.

Ἀλλὰ τὸ μὲν πρῶτον εἶδος τῆς πρὸς τοὺς μου-
38 σικοὺς ἀντιρρήσεως τοιουτότροπόν ἐστιν, τὸ δὲ
δεύτερον καὶ τῶν τῆς μουσικῆς ἀρχῶν καθαπτό-
μενον πραγματικωτέρας μᾶλλον ἔχεται ζητήσεως.
οἷον ἐπεὶ ἡ μουσικὴ ἐπιστήμη τίς ἐστιν ἐμμελῶν
τε καὶ ἐκμελῶν ἐνρύθμων τε καὶ ἐκρύθμων, πάντως
ἐὰν δείξωμεν ὅτι οὔτε τὰ μέλη ὑποστατά ἐστιν
οὔτε οἱ ῥυθμοὶ τῶν ὑπαρκτῶν πραγμάτων τυγχά-
νουσιν, ἐσόμεθα παρεστακότες καὶ τὴν μουσικὴν
ἀνυπόστατον. λέγωμεν δὲ πρῶτον περὶ μελῶν καὶ
τῆς τούτων ὑποστάσεως, μικρὸν ἄνωθεν καταρξά-
μενοι.

39 Φωνὴ τοίνυν ἐστίν, ὡς ἄν τις ἀναμφισβητήτως
ἀποδοίη, τὸ ἴδιον αἰσθητὸν ἀκοῆς· καθάπερ γὰρ
μόνης ὁράσεως ἔργον ἐστὶ τὸ χρωμάτων ἀντιλαμ-
βάνεσθαι καὶ μόνης ὀσφρήσεως τὸ εὐωδῶν καὶ
δυσωδῶν ἀντιποιεῖσθαι καὶ ἤδη γεύσεως τὸ γλυ-
κέων ἢ πικρῶν αἰσθάνεσθαι, οὕτω γένοιτ' ἂν ἴδιον
40 αἰσθητὸν ἀκοῆς ἡ φωνή. τῆς δὲ φωνῆς ἡ μέν τίς
ἐστιν ὀξεῖα ἡ δὲ βαρεῖα, μεταφορικώτερον ἀπὸ τῶν
περὶ τὴν ἁφὴν αἰσθητῶν ἑκατέρου τούτων λαμβά-
νοντος τὴν προσηγορίαν· καθάπερ γὰρ τὸ κεντοῦν
καὶ τέμνον τὴν ἁφὴν ὀξὺ προσηγόρευσεν ὁ βίος καὶ
τὸ θλάσιν ἐμποιοῦν καὶ πιέζον βαρύ, τὸν αὐτὸν
τρόπον καὶ τῆς φωνῆς τὴν μὲν οἱονεὶ τέμνουσαν
τὴν ἀκοὴν ὀξεῖαν, τὴν δὲ ὥσπερ θλῶσαν βαρεῖαν.
41 καὶ οὐ ξένον εἰ ὥσπερ φαιάν τινα καὶ μέλαιναν καὶ
λευκὴν φωνὴν ἀπὸ τῶν πρὸς τὴν ὅρασιν αἰσθητῶν

Universe is ordered according to harmony is shown to be false by a variety of proofs ; and further, even if it be true, a thing of this kind can be of no help towards felicity, just as the harmony in instruments is of no help. 38

Such, then, is the first kind[a] of argumentation against the Musicians : the second kind, which deals with the principles of music, involves an inquiry of a more practical nature. Thus, since Music is a science of what is " in tune " and " out-of-tune," " in rhythm " and " out-of-rhythm," if we shall prove that neither are tunes existent nor rhythms realities, then we shall certainly have established that Music is non-existent. And first let us discuss tunes and their existence, starting with some preliminary matters.

Now sound, as one will grant without dispute, is the 39
sense-object peculiar to hearing ; for as it is the task of vision alone to perceive colours, and of smell alone to apprehend odours good and bad, and of taste to sense sweet things and bitter, so sound will be the proper sense-object of hearing. And of sound one 40
kind is " sharp," another " grave," each of these deriving its name by transference from the sense-objects of touch ; for just as custom has given the name of " sharp " to what pricks or cuts the sense of touch, and " grave " to that which causes crushing and compresses, in the same way it calls the sound which cuts, as it were, the hearing " sharp," and that which crushes it, as it were, " grave." Nor is it 41
strange if, just as we describe sounds as grey or black or white, calling them after the sense-objects of

[a] For the two kinds *cf.* §§ 4, 5.

κεκλήκαμεν, ὧδε καὶ ἀπὸ τῶν πρὸς τὴν ἁφὴν ἐχρησάμεθά τισι μεταφοραῖς. ὅταν μὲν οὖν ἐπ' ἴσης ἐκφέρηται ἡ φωνὴ καὶ ὑπὸ μίαν τάσιν, ὡς μηδένα περισπασμὸν γίνεσθαι τῆς αἰσθήσεως ἤτοι ἐπὶ τὸ βαρύτερον ἢ τὸ ὀξύτερον, τότε ὁ τοιοῦτος
42 ἦχος φθόγγος καλεῖται, παρὸ καὶ οἱ μουσικοὶ ὑπογράφοντές φασι "φθόγγος ἐστὶν ἐμμελοῦς φωνῆς πτῶσις ὑπὸ μίαν τάσιν." τῶν δὲ φθόγγων οἱ μέν εἰσιν ὁμόφωνοι οἱ δὲ οὐχ ὁμόφωνοι, καὶ ὁμόφωνοι μὲν οἱ μὴ διαφέροντες ἀλλήλων κατ' ὀξύτητα καὶ βαρύτητα, οὐχ ὁμόφωνοι δὲ οἱ μὴ οὕτως ἔχοντες.
43 τῶν δὲ ὁμοφώνων, ὡς καὶ τῶν οὐχ ὁμοφώνων, τινὲς μὲν ὀξεῖς τινὲς δὲ βαρεῖς καλοῦνται, καὶ πάλιν τῶν οὐχ ὁμοφώνων οἱ μὲν διάφωνοι προσαγορεύονται οἱ δὲ σύμφωνοι, καὶ διάφωνοι μὲν οἱ ἀνωμάλως καὶ διεσπασμένως τὴν ἀκοὴν κινοῦντες,
44 σύμφωνοι δὲ οἱ ὁμαλώτερον καὶ ἀμερίστως. σαφέστερον δὲ μᾶλλον ἔσται τὸ ἑκατέρου γένους ἰδίωμα τῇ ἀπὸ τῶν πρὸς γεῦσιν ποιοτήτων μεταβάσει χρησαμένων ἡμῶν. ὥσπερ τοίνυν τῶν γευστῶν τὰ μὲν τοιαύτην ἔχει κρᾶσιν ὥστε μονοειδῶς καὶ λείως κινεῖν τὴν αἴσθησιν, ὁποῖον τὸ οἰνόμελι καὶ ὑδρόμελι, τὰ δὲ οὐχ ὡσαύτως οὐδὲ ὁμοίως, καθάπερ τὸ ὀξύμελι (ἑκάτερον γὰρ τούτων τῶν μιγμάτων τὴν ἴδιον ἐντυποῖ ποιότητα τῇ γεύσει), οὕτω τῶν φθόγγων διάφωνοι μέν εἰσιν οἱ ἀνωμάλως τὴν ἀκοὴν καὶ διεσπασμένως κινοῦντες, σύμφωνοι δὲ οἱ ὁμαλώτεροι. ἀλλὰ γὰρ ἡ μὲν διαφορὰ τῶν
45 φθόγγων τοιαύτη τίς ἐστι παρὰ μουσικοῖς· περιγράφεται δέ τινα πρὸς τούτων διαστήματα, καθ' ἃ καὶ ἡ φωνὴ κινεῖται ἤτοι ἐπὶ τὸ ὀξύτερον ἀνα-

vision, so also we employ names derived by transference from those of touch. Whenever, then, the sound is expelled evenly and at one " tension " (or " pitch "), so that there is no diversion of the sense towards the " grave " or the " sharp," then such a sound is called a " note," and so the Musicians say 42
in their definitions " A note is the fall under one tension of a musical sound." And of notes some are " homophonous " (or " in unison "), others not " homophonous "; " homophonous " being those which do not differ from one another in respect of " sharpness " or " gravity," and not " homophonous " those which are not of this character. And of the " homophonous,"—as also of the not 43
" homophonous,"—some are termed " sharp," others " grave "; and of the not " homophonous," again, some are called " dissonant," others " consonant," " dissonant " being such as affect the hearing unevenly and intermittedly, and " consonant " such as affect it evenly and continuously. But the special 44
character of each kind will be more clear if we make use of a comparison with the qualities relating to taste. Thus, just as some of the objects of taste are of such a blend as to affect the sense uniformly and smoothly, like honey-wine and honey-water, whereas others affect it in distinct and dissimilar ways, as for instance honey-vinegar (for each of these constituents stamps its own quality on the sense of taste),—so also with notes, those which affect the hearing unevenly and intermittedly are " dissonant," and those which are even are " consonant." Such, then, is the distinction as to " notes " as laid down by the Musicians. —And by these, certain " intervals " are determined 45
within which the sound moves, either ascending

βαίνουσα ἢ ἐπὶ τὸ βαρύτερον ἀνιεμένη. παρ' ἣν αἰτίαν κατὰ τὸ ἀνάλογον τῶν διαστημάτων τούτων τὰ μὲν σύμφωνα τὰ δὲ διάφωνα προσηγόρευται,
46 καὶ σύμφωνα μὲν ὁπόσα ὑπὸ συμφώνων φθόγγων περιέχεται, διάφωνα δὲ ὁπόσα ὑπὸ διαφώνων. τῶν δὲ συμφώνων διαστημάτων τὸ μὲν πρῶτον καὶ ἐλάχιστον διὰ τεσσάρων οἱ μουσικοὶ προσαγορεύουσι, τὸ δὲ μετὰ τοῦτο μεῖζον διὰ πέντε, καὶ τοῦ
47 διὰ πέντε μεῖζον τὸ διὰ πασῶν. πάλιν τε τῶν διαφώνων διαστημάτων ἐλάχιστον μέν ἐστι καὶ πρῶτον παρ' αὐτοῖς ἡ καλουμένη δίεσις, δεύτερον δὲ τὸ ἡμιτόνιον, ὅ ἐστι διπλοῦν τῆς διέσεως, τρίτον
48 ὁ τόνος, ὅς ἐστι διπλασίων τοῦ ἡμιτονίου. οὐ μὴν ἀλλ' ὃν τρόπον ἅπαν διάστημα κατὰ μουσικὴν ἐν φθόγγοις ἔχει τὴν ὑπόστασιν, οὕτω καὶ πᾶν ἦθος· τὸ δ' ἔστι τι γένος μελῳδίας. καθὰ γὰρ τῶν ἀνθρωπίνων ἠθῶν τινὰ μέν ἐστι σκυθρωπὰ καὶ στιβαρώτερα, ὁποῖα τὰ τῶν ἀρχαίων ἱστοροῦσιν, τὰ δὲ εὐένδοτα πρὸς ἔρωτας καὶ οἰνοφλυγίας καὶ ὀδυρμοὺς καὶ οἰμωγάς, οὕτω τὶς μὲν μελῳδία σεμνά τινα καὶ ἀστεῖα ἐμποιεῖ τῇ ψυχῇ κινήματα,
49 τὶς δὲ ταπεινότερα καὶ ἀγεννῆ. καλεῖται δὲ κατὰ κοινὸν ἡ τοιουτότροπος μελῳδία τοῖς μουσικοῖς ἦθος ἀπὸ τοῦ ἤθους εἶναι ποιητική, καθάπερ καὶ τὸ χλωρὸν δέος τὸ χλωροποιόν, καὶ τὸ "νότοι βαρυήκοοι ἀχλυώδεις καρηβαρικοὶ νωθροὶ διαλυτι-
50 κοί" ἀντὶ τοῦ τούτων δραστικοί. τῆς δὲ κοινῆς μελῳδίας ταύτης τὸ μέν τι χρῶμα λέγεται τὸ δὲ ἁρμονία τὸ δὲ διάτονον, ὧν ἡ μὲν ἁρμονία αὐστη-

---

[a] *Cf. P.H.* iii. 155 ; *Adv. Arithm.* 6-9.
[b] *Cf.* Hom. *Il.* vii. 479.

towards the "sharp" or descending towards the "grave." And because of this some of these intervals are, by analogy, called "consonant," others "dissonant"; and "consonant" are all those con- 46
tained by consonant notes, and "dissonant" all those contained by dissonant notes. Of consonant intervals the first and least is termed by the Musicians "By-Fours" (or "The Fourth"), the greater one which comes next to this "By-Fives" (or "The Fifth"), and the one which is greater than the By-Fives the "By-Alls" (or "The Octave").[a] Also,
of the dissonant intervals, again, the least and first 47
is what is called by them "Diesis" (or "Quarter-tone"); second is the "Semi-tone," which is double the "Diesis"; third, the "Tone," which is
double the "Semi-tone."—Moreover, just as every 48
interval in music consists of notes, so also does every "Êthos" (or "character"); and it is a certain "Genus" of melody. For just as of human characters some are gloomy and stubborn, such as those of the ancients are reported to have been, while others yield easily to lusts and debauchery and lamentations and groanings, so a certain kind of melody produces in the soul stately and refined motions, another kind
motions that are base and ignoble. Melody of this 49
sort is called, in general, by the Musicians "Êthos" from the fact that it is productive of "character," just as fear is called "pallid"[b] for "pallor-causing," and South-winds "deaf" and "foggy" and "headachey" and "dull" and "relaxing," instead of being
said to be causes of these effects. And of this general 50
type of melody one kind is called "Chroma" (or "Colour," Chromatic), another "Harmony" (or "Enharmonic"), another "Diatonic"; and of these

ῥοῦ τινὸς ἤθους καὶ σεμνότητος κατασκευαστική
πως ὑπῆρχεν, τὸ δὲ χρῶμα λιγυρόν τί ἐστι καὶ
θρηνῶδες, τὸ δὲ διάτονον ἔντραχυ καὶ ὑπάγροικον.
51 ἀλλὰ δὴ πάλιν τὸ μὲν ἁρμονικὸν μέλος τῶν μελῳ-
δουμένων ἀδιαίρετόν ἐστι, τὸ δὲ διάτονον καὶ τὸ
χρῶμα εἰδικωτέρας τινὰς εἶχε διαφοράς, δύο μὲν
τὸ διάτονον, τήν τε τοῦ μαλακοῦ διατόνου καλου-
μένην καὶ τὴν τοῦ συντόνου, τρεῖς δὲ τὸ χρῶμα·
τὸ μὲν γάρ τι αὐτοῦ τονικὸν καλεῖται τὸ δὲ ἡμιτό-
νιον τὸ δὲ μαλακόν.

52 Πλὴν ἐκ τούτων συμφανὲς ὅτι πᾶσα ἡ κατὰ μελῳ-
δίας θεωρία παρὰ τοῖς μουσικοῖς οὐκ ἐν ἄλλῳ τινὶ
τὴν ὑπόστασιν εἶχεν εἰ μὴ ἐν τοῖς φθόγγοις. καὶ
διὰ τοῦτο ἀναιρουμένων αὐτῶν τὸ μηδὲν ἔσται ἡ
μουσική. πῶς οὖν καὶ ἐρεῖ τις ὅτι οὐκ εἰσὶ φθόγ-
γοι; ἐκ τοῦ φωνὴν αὐτοὺς κατὰ γένος ὑπάρχειν,
φήσομεν, καὶ τὴν φωνὴν ἀνύπαρκτον ἡμῖν ἐν τοῖς
σκεπτικοῖς ὑπομνήμασι δεδεῖχθαι ἀπὸ τῆς τῶν δογ-
53 ματικῶν μαρτυρίας. οἵ τε γὰρ ἀπὸ τῆς Κυρήνης
φιλόσοφοι μόνα φασὶν ὑπάρχειν τὰ πάθη, ἄλλο δὲ
οὐδέν· ὅθεν καὶ τὴν φωνὴν μὴ οὖσαν πάθος, ἀλλὰ
πάθους ποιητικήν, μὴ γίγνεσθαι τῶν ὑπαρκτῶν.
οἵ γέ τοι περὶ τὸν Δημόκριτον καὶ Πλάτωνα πᾶν
αἰσθητὸν ἀναιροῦντες συναναιροῦσι καὶ τὴν φωνήν,
54 αἰσθητόν τι δοκοῦσαν πρᾶγμα ὑπάρχειν. καὶ γὰρ
ἄλλως, εἰ ἔστι φωνή, ἤτοι σῶμά ἐστιν ἢ ἀσώ-
ματον· οὔτε δὲ σῶμά ἐστιν, ὡς οἱ περιπατητικοὶ διὰ
πολλῶν διδάσκουσιν, οὔτε ἀσώματος, ὡς οἱ ἀπὸ
55 τῆς στοᾶς· οὐκ ἄρα ἔστι φωνή. ἄλλος δέ τις κἂν
ἐκείνως ἐπιχειρήσειε λέγειν, ὡς εἰ μὴ ἔστι ψυχή

the Enharmonic is of an austere Êthos and productive of solemnity, while the Chromatic is shrill and dirgelike, and the Diatonic somewhat rough and unrefined. And further, the Enharmonic tune is 51
inseparable from the songs that are being sung, but the Diatonic and the Chromatic have certain distinct sub-species, the Diatonic two,—that of the so-called "Lax Diatonic" and that of the "Tense Diatonic," —and the Chromatic three, called respectively the "Tonic," the "Semi-tonic" and the "Lax."

Now from this it is evident that the whole of 52
the Musicians' theory of melody is based on nothing else than notes. And, because of this, if these are destroyed Music will be nothing. How, then, can one assert that notes do not exist? From the fact, we shall say, that they come under the genus "sound," and that sound is non-existent has been proved by us,[a] in our *Sceptic Commentaries*, from the evidence of the Dogmatists. For the 53
Cyrenaic philosophers[b] assert that only the feelings exist, and nothing else; and hence that sound, not being a feeling but productive of feeling, is not an existent. Democritus,[c] indeed, and Plato, by destroying every sense-object, destroy therewith sound, which is held to be an object of sense.—Furthermore, 54
if sound exists it is either corporeal or incorporeal; but it is not corporeal, as the Peripatetics demonstrate by numerous arguments; nor is it incorporeal, as the Stoics prove[d]; therefore sound does not exist.—And some other person will attempt to argue 55
thus: If the soul does not exist, neither do the

[a] *Cf. Adv. Log.* ii. 131 (though this is not the book referred to).
[b] *Cf.* vol. i. p. xvii, *P.H.* . 215; *Adv. Log.* i. 191.
[c] *Cf.* vol. i. pp. xi f.
[d] *Cf. Adv. Gram.* 155.

οὐδὲ αἰσθήσεις· μέρη γὰρ ταύτης ὑπῆρχον. εἰ δὲ μὴ εἰσὶν αἱ αἰσθήσεις, οὐδὲ τὰ αἰσθητά· πρὸς αἰσθήσεις γὰρ ἡ τούτων ὑπόστασις νοεῖται. εἰ δὲ μὴ αἰσθητά, οὐδὲ φωνή· εἶδος γάρ τι τῶν αἰσθητῶν ὑπῆρχεν. ἀλλὰ μὴν οὐδέν ἐστι ψυχή, καθὼς ἐν τοῖς περὶ αὐτῆς ὑπομνήμασιν ἐδείκνυμεν· οὐκ
56 ἄρα ἔστι φωνή. καὶ μὴν εἰ μήτε βραχεῖά ἐστι φωνὴ μήτε μακρά, οὐκ ἔστι φωνή· οὔτε δὲ βραχεῖά ἐστιν οὔτε μακρὰ φωνή, ὡς ἐν τοῖς πρὸς τοὺς γραμματικοὺς ὑπεμνήσαμεν, περὶ συλλαβῆς καὶ λέξεως ζητοῦντες πρὸς τούτους· οὐκ ἄρα ἔστι
57 φωνή. πρὸς τούτοις ἡ φωνὴ οὔτε ἐν ἀποτελέσματι οὔτε ἐν ὑποστάσει νοεῖται, ἀλλ' ἐν γενέσει καὶ χρονικῇ παρεκτάσει· τὸ δὲ ἐν γενέσει νοούμενον γίνεται, οὐδέπω δ' ἔστιν, ὥσπερ οὐδὲ οἰκία γινομένη ἢ ναῦς καὶ ἄλλα παμπληθῆ εἶναι λέγεται.
58 τοίνυν οὐθέν ἐστι φωνή. καὶ ἄλλοις δὲ συχνοῖς εἰς τοῦτο ἔνεστι λόγοις χρῆσθαι, περὶ ὧν, ὡς ἔφην, ἐν τοῖς Πυρρωνείοις ὑπομνηματιζόμενοι διεξῄειμεν. νυνὶ δὲ φωνῆς μὴ οὔσης οὐδὲ φθόγγος ἔστιν, ὃς ἐλέγετο φωνῆς πτῶσις ὑπὸ μίαν τάσιν· φθόγγου δὲ μὴ ὄντος οὐδὲ διάστημα μουσικὸν καθέστηκεν, οὐ συμφωνία, οὐ μελῳδία, οὐ τὰ ἐκ τούτων γένη. διὰ τοῦτο οὐδὲ μουσική· ἐπιστήμη γὰρ ἐλέγετο ἐμμελῶν τε καὶ ἐκμελῶν.

59 Ὅθεν ἀπ' ἄλλης ἀρχῆς ὑποδεικτέον ὅτι κἂν τού-

senses ; for they are parts of the soul. And if the senses do not exist, neither do the sense-objects ; for their existence is conceived as relative to the senses. And if sense-objects do not exist, neither does sound ; for it is a species of sense-object. But, in fact, soul is nothing, as we have shown in our treatise concerning it [a] ; therefore sound does not exist.—Moreover, 56
if sound is neither short nor long, sound does not exist ; but sound is neither short nor long, as we have shown in our treatise *Against the Grammarians*,[b] when questioning their theory of the syllable and the word ; therefore sound does not exist.—In 57
addition, sound is conceived neither as a completed thing nor as a substantial thing,[c] but as a thing which comes into being and has extension in time ; but that which is conceived as coming into being is coming into being, but does not as yet exist,—just as a house which is coming into being, or a ship or a multitude of other things, is not said to exist. So, then, sound is nothing. There are also many other 58
arguments one might use to prove this point, such as those with which, as I said, we have dealt fully in our comments in *Pyrrhonea*.[d]—Now, then, as sound does not exist, neither does the note, for it was described as " the fall of a sound under one tension " [e] ; and when the note does not exist, neither does the musical interval exist, nor symphony, nor melody, nor the Genera formed from these.[f] Therefore, Music does not exist either ; for it was described as " the Science of things in tune and out of tune." [g]

Hence, making a fresh start, we must show that, 59

[a] No longer extant ; but *cf. P.H.* ii. 31 ff., iii. 186.

[b] *Cf. Adv. Gram.* 124 ff.

[c] *Cf. Adv. Log.* ii. 131.

[d] Another lost work.

[e] *Cf.* § 42.

[f] *Cf.* § 50.

[g] *Cf.* § 38.

των ἀποστῶμεν, διὰ τὴν ἐγχειρηθησομένην ἐπὶ τῆς ῥυθμοποιίας ἀπορίαν ἀνυπόστατος καθέστηκεν ἡ μουσική. εἰ γὰρ μηδέν ἐστι ῥυθμός, οὐδὲ ἐπιστήμη τις ἔσται περὶ ῥυθμοῦ· ἀλλὰ μὴν οὐδέν ἐστι ῥυθμός, ὡς παραστήσομεν· οὐκ ἄρα ἔστι τις ἐπι-
60 στήμη περὶ ῥυθμοῦ. ὡς γὰρ πολλάκις εἰρήκαμεν, ῥυθμὸς σύστημά ἐστιν ἐκ ποδῶν, ὁ δὲ ποῦς τὸ συνεστὼς ἐξ ἄρσεως καὶ θέσεως· ἡ δὲ ἆρσις καὶ ἡ θέσις ἐν ποσότητι χρόνου θεωρεῖται, ὧν τινὰς μὲν ἐπεῖχεν ἡ θέσις τινὰς δὲ ἡ ἆρσις χρόνους. καθάπερ γὰρ ἐκ μὲν στοιχείων συλλαβαὶ ἐκ δὲ συλλαβῶν λέξεις συντίθενται, οὕτως ἐκ μὲν τῶν χρόνων οἱ
61 πόδες ἐκ δὲ τῶν ποδῶν οἱ ῥυθμοὶ γίνονται. ἐὰν οὖν δείξωμεν ὅτι οὐδέν ἐστι χρόνος, ἕξομεν συναποδεδειγμένον ὅτι οὐδὲ πόδες ὑπάρχουσιν, διὰ δὲ τοῦτο οὐδὲ οἱ ῥυθμοί, ἐξ ἐκείνων τὴν σύστασιν λαμβάνοντες. ᾧ ἀκολουθήσει τὸ μηδὲ ἐπιστήμην εἶναί τινα περὶ ῥυθμούς. πῶς οὖν; ὅτι οὐδέν ἐστι χρόνος, ἤδη μὲν παρεστήσαμεν ἐν τοῖς Πυρρωνείοις, οὐδὲν δὲ ἧττον καὶ τὰ νῦν παραστήσομεν
62 ἐπὶ ποσόν. εἰ γὰρ ἔστι τι χρόνος, ἤτοι πεπέρασται ἢ ἄπειρος ἐστίν. οὔτε δὲ πεπέρασται, ἐπεὶ ἐροῦμέν ποτε γεγονέναι χρόνον ὅτε χρόνος οὐκ ἦν, καὶ ἔσεσθαί ποτε χρόνον ὅτε χρόνος οὐκ ἔσται, οὔτε ἄπειρος καθέστηκεν· ἔστι γάρ τι αὐτοῦ παρῳχηκὸς [καὶ ἐνεστὼς][1] καὶ μέλλον, ὧν ἑκάτερον εἰ μὲν οὐκ ἔστιν, πεπέρασται ὁ χρόνος, εἰ δ' ἔστιν, ἔσται ἐν τῷ παρόντι καὶ ὁ παρῳχηκὼς καὶ ὁ μέλλων, ὅπερ
63 ἄτοπον. οὐκ ἄρα ἔστι χρόνος. τό γε μὴν ἐξ

[1] [καὶ ἐνεστὼς] del. Heintz.

even if we lay aside these considerations, Music is non-existent owing to the doubt regarding rhythm-making which is now to be raised. For if rhythm is nothing, no science of rhythm will exist ; but, in fact, rhythm is nothing, as we shall establish ; therefore no science of rhythm exists. For as we have fre- 60
quently stated, rhythm is a system composed of "feet," and a "foot" is that which consists of "arsis" and "thesis"[a] ; and "arsis" and "thesis" are seen in quantity of time, for the "thesis" occupies certain periods of time and the "arsis" certain other periods. For just as syllables are composed of elements, and words of syllables, so feet are composed of times, and rhythms of feet. If, then, we 61
shall show that time is nothing, we shall have shown therewith that feet also do not exist, nor, consequently, rhythms, as they derive their consistence from feet. And from this it will follow that no science of rhythms exists. How then ? That time is nothing we have already established in our *Pyrrhonea*,[b] but none the less we shall also establish it now, up to a point.—For if time is anything, it is 62
either limited or unlimited. But it is not limited, since, if so, we shall be saying that there was once a time when time did not exist, and that there will sometime be a time when time will not exist. Nor is it unlimited ; for a part of it is past, a part future, and if each of these does not exist time is limited, and if each does exist, then both the past and the future will exist in the present, which is absurd. Therefore time does not exist.—Also, that which is 63

[a] Terms derived from the motions of dancers in raising ("arsis") and setting down ("thesis") the feet.

[b] *Cf. P.H.* iii. 136 ff. ; *Adv. Phys.* ii. 169 ff.

ἀνυπάρκτων συνεστὼς ἀνύπαρκτον ἐστίν· ὁ δὲ χρόνος ἔκ τε τοῦ παρῳχημένου καὶ μηκέτ' ὄντος καὶ ἐκ τοῦ μέλλοντος μηδέπω δὲ ὄντος συνεστὼς
64 ἀνύπαρκτος ἔσται. ἄλλως τε, εἰ μὲν ἀμερής ἐστιν ὁ χρόνος, πῶς τὸ μέν τι αὐτοῦ παρῳχημένον τὸ δὲ ἐνεστὼς τὸ δὲ μέλλον λέγομεν; εἰ δὲ μεριστός ἐστιν, ἐπεὶ πᾶν τὸ μεριστὸν ὑπό τινος αὐτοῦ μέρους καταμετρεῖται, ὡς πῆχυς μὲν ὑπὸ παλαιστοῦ, ὁ παλαιστὴς δὲ ὑπὸ δακτύλου, δεήσει καὶ αὐτὸν
65 ὑπό τινος τῶν αὐτοῦ μερῶν καταμετρεῖσθαι. οὔτε δὲ τῷ ἐνεστῶτι δυνατὸν καταμετρεῖν τοὺς ἄλλους χρόνους, ἐπείπερ ὁ γινόμενος καὶ ὁ ἐνεστὼς χρόνος ὁ αὐτὸς ἔσται κατ' αὐτοὺς παρῳχημένος καὶ μέλλων, παρῳχήμενος μὲν ὅτι τὸν παρῳχημένον καταμετρεῖ χρόνον, μέλλων δὲ ὅτι τὸν μέλλοντα· ὅπερ ἄτοπον. οὔτε[1] τοίνυν τινὶ τῶν λειπομένων δυοῖν τὸν ἐνεστῶτα καταμετρητέον. δι' ἣν αἰ-
66 τίαν οὐδὲ ταύτῃ λεκτέον εἶναί τινα χρόνον. πρὸς τούτοις ὁ χρόνος τριμερής ἐστι, καὶ τὸ μὲν ἔχει παρῳχηκὸς τὸ δὲ ἐνεστὼς τὸ δὲ μέλλον, ὧν τὸ μὲν παρῳχημένον οὐκέτι ἔστιν, τὸ δὲ μέλλον οὔπω ἔστιν, τὸ δὲ ἐνεστὼς ἤτοι ἀμερές ἐστιν ἢ μεριστόν. ἀλλ' ἀμερὲς μὲν οὐκ ἂν εἴη· ἐν ἀμερεῖ μὲν γὰρ οὐδὲν δύναται γίνεσθαι μεριστόν, ὥς φησὶ Τίμων,
67 οἷον τὸ γίνεσθαι, τὸ φθείρεσθαι. καὶ ἄλλως, εἴπερ ἀμερές ἐστι τὸ ἐνεστὼς τοῦ χρόνου, οὔτε ἀρχὴν ἔχει ἀφ' ἧς ἄρχεται, οὔτε πέρας ἐφ' ὃ καταλήγει, διὰ δὲ τοῦτο οὐδὲ μέσον· καὶ οὕτως οὐκ ἔσται ὁ ἐνεστὼς χρόνος. εἰ δὲ μεριστός ἐστιν, εἰ μὲν εἰς τοὺς μὴ ὄντας χρόνους μερίζεται, οὐκ ἔσται χρόνος, εἰ δ' εἰς τοὺς ὄντας χρόνους, οὐκ ἔσται ὅλος ὁ

[1] οὔτε scripsi; οὐ mss., Bekk.

composed of non-existents is non-existent ; and time being composed of the past, which no longer exists, and of the future, which does not yet exist, will be non-existent.—Again, if time is indivisible, how is it 64
that we call a part of it "past," a part "present," and a part "future"? And if it is divisible, since everything divisible is measured by some part of itself,—as the cubit by the palm, and the palm by the finger,—time, too, will have to be measured by one of its own parts. But it is not possible to measure the 65
other times by the present, since, according to them, the same instant and present time will be both past and future,—past because it measures the past time, and future because it measures the future ; which is absurd. Nor yet can one measure the present by one of the two remaining times. And, because of this, one must declare that thus, too, no time exists.—Furthermore, time is tripartite,[a] and one part is past, 66
one present, and one future ; and of these the past no longer exists, the future does not yet exist, and the present is either indivisible or divisible. But it will not be indivisible ; for, as Timon asserts, no divisible thing, such as becoming and perishing, can become in indivisible time.—Again, if the present part of 67
time is indivisible, it has neither a beginning from which it begins, nor an end at which it ceases, nor, in consequence, a middle, and thus present time will not exist. And if it is divisible, then, if it is divided into the non-existing times, time will not exist ; while if it is divided into the existing times, time will not exist as

[a] *Cf. P.H.* iii. 144 ; *Adv. Phys.* ii. 197 ff., where, too, Timon is quoted.

χρόνος, ἀλλὰ τῶν μερῶν αὐτοῦ τινὰ μὲν ἔσται τινὰ δὲ οὐκ ἔσται. τοίνυν οὐδέν ἐστι χρόνος, διὰ δὲ τοῦτο οὐδὲ πόδες, οὐδὲ ῥυθμοί, οὐδ' ἡ περὶ τοὺς ῥυθμοὺς ἐπιστήμη.

68 Τοσαῦτα πραγματικῶς καὶ πρὸς τὰς τῆς μουσικῆς εἰπόντες ἀρχὰς ἐν τοσούτοις τὴν πρὸς τὰ μαθήματα διέξοδον ἀπαρτίζομεν.

a whole, but some of its parts will exist and others will not exist. So, then, time is nothing; and, consequently, feet, too, and rhythms, and the science of rhythms are nothing.

Having said thus much by way of effective argument against the principles of Music, we here bring to a close our disputation against the Arts and Sciences. 68

INDEX OF NAMES

(*Abbreviations used are:* Gr. = "*Against the Grammarians.*" R. = "*Against the Rhetoricians.*" Gm. = "*Against the Geometers.*" Ar. = "*Against the Arithmeticians.*" Ast. = "*Against the Astrologers.*" M. = "*Against the Musicians.*" P. = "*Against the Professors.*")

INDEX OF NAMES

INDEX OF NAMES